STATISTICAL THEORY AND METHODOLOGY OF TRACE ANALYSIS

ELLIS HORWOOD SERIES IN ANALYTICAL CHEMISTRY

EDITORS: Dr. R. A. Chalmers and Dr. Mary Masson, University of Aberdeen

"I recommend that this Series be used as reference material. Its Authors are among the most respected in Europe". *J. Chemical Ed., New York.*

Application of Ion-selective Membrane Electrodes in Organic Analysis
F. BAIULESCU and V. V. COŞOFREŢ, Polytechnic Institute, Bucharest
Industrial Methods of Microanalysis
S. BANCE, May and Baker Research Laboratories, Dagenham
Ion-Selective Electrodes in Life Sciences
J. COMER, MSE Scientific Instruments, Crawley, and D. B. KELL, University College of Wales, Aberystwyth
Inorganic Reaction Chemistry: Systematic Chemical Separation
D. T. BURNS, Queen's University, Belfast, A. G. ČATCHPOLE, Kingston Polytechnic, A. TOWNSHEND, University of Birmingham
Quantitative Inorganic Analysis
R. BELCHER, M. CRESSER, R. A. CHALMERS
Handbook of Process Stream Analysis
K. J. CLEVETT, Crest Engineering (U.K.) Inc.
Automatic Methods in Chemical Analysis
J. K. FOREMAN and P. B. STOCKWELL, Laboratory of the Government Chemist, London
Fundamentals of Electrochemical Analysis
Z. GALUS, Warsaw University
Laboratory Handbook of Thin Layer and Paper Chromatography
J. GASPARIČ, Charles University, Hradec Kralove
J. CHURAČEK, University of Chemical Technology, Pardubice
Handbook of Analytical Control of Iron and Steel Production
T. S. HARRISON, Group Chemical Laboratories, British Steel Corporation
Handbook of Organic Reagents in Inorganic Analysis
Z. HOLZBECHER et al., Institute of Chemical Technology, Prague
Analytical Applications of Complex Equilibria
J. INCZÉDY, University of Chemical Engineering, Veszprem
Particle Size Analysis
Z. K. JELÍNEK, Organic Synthesis Research Institute, Pardubice
Operational Amplifiers in Chemical Instrumentation
R. KALVODA, J. Heyrovský Institute of Physical Chemistry and Electrochemistry, Prague
Atlas of Metal-ligand Equilibria in Aqueous Solution
J. KRAGTEN, University of Amsterdam
Gradient Liquid Chromatography
C. LITEANU and S. GOCAN, University of Cluj
Laboratory Handbook of Chromatographic and Allied Methods
O. MIKEŠ, Czechoslovak Academy of Sciences, Prague
Statistical Theory and Methodology of Trace Analysis
C. LITEANU and I. RÎCĂ, University of Cluj
Spectrophotometric Determination of Elements
Z. MARCZENKO, Warsaw Technical University
Separation and Enrichment Methods of Trace Analysis
J. MINCZEWSKI et al., Institute of Nuclear Research, Warsaw
Handbook of Analysis of Organic Solvents
V. SEDIVEČ and J. FLEK, Institute of Hygiene and Epidemiology, Prague
Foundations of Chemical Analysis
O. BUDEVSKY, Academy of Medicine, Sofia, Bulgaria
Handbook of Analysis of Synthetic Polymers and Plastics
J. URBANSKI et al., Warsaw Technical University
Analysis with Ion-selective Electrodes
J. VESELÝ and D. WEISS, Geological Survey, Prague
K. ŠTULÍK, Charles University, Prague
Electrochemical Stripping Analysis
F. VYDRA, J. Heyrovský Institute of Physical Chemistry and Electrochemistry, Prague
K. ŠTULÍK, Charles University, Prague
B. JULAKOVÁ, The State Institute for Control of Drugs, Prague
Isoelectric Focusing Methods
K. W. WILLIAMS, L. SODERBERG, T. LAAS, Pharmacia Fine Chemicals, Uppsala

STATISTICAL THEORY AND METHODOLOGY OF TRACE ANALYSIS

C. LITEANU
University of Cluj-Napoca
Rumania

and

I. RÎCĂ
Chemical Laboratory
Hunedoara Ironworks
Rumania

Translation Editor: **R. A. Chalmers**
University of Aberdeen

ELLIS HORWOOD LIMITED
Publishers Chichester

Halsted Press: a division of
JOHN WILEY & SONS
New York - Chichester - Brisbane - Toronto

First published in 1980 by

ELLIS HORWOOD LIMITED

Market Cross House, Cooper Street, Chichester, West Sussex, PO19 1EB, England

The publisher's colophon is reproduced from James Gillison's drawing of the ancient Market Cross, Chichester.

Distributors:

Australia, New Zealand, South-east Asia:
Jacaranda-Wiley Ltd., Jacaranda Press,
JOHN WILEY & SONS INC.,
G.P.O. Box 859, Brisbane, Queensland 40001, Australia.

Canada:
JOHN WILEY & SONS CANADA LIMITED
22 Worcester Road, Rexdale, Ontario, Canada.

Europe, Africa:
JOHN WILEY & SONS LIMITED
Baffins Lane, Chichester, West Sussex, England.

North and South America and the rest of the world:
Halsted Press, a division of
JOHN WILEY & SONS
605 Third Avenue, New York, N.Y. 10016, U.S.A.

British Library Cataloguing in Publication Data

Liteanu, Candin
Statistical theory and methodology of trace analysis. –
(Ellis Horwood series in analytical chemistry).
I. Title II. Rica, I
543 QD139.T7 79–41461

ISBN 0–85312–108–7 (Ellis Horwood Ltd., Publishers)
ISBN 0–470–26797–6 (Halsted Press)

Typeset in Press Roman by Ellis Horwood Ltd.
Printed in Great Britain by W. & J. Mackay Ltd., Chatham

This book is dedicated to those researchers from the past,
who contributed to the progress of science no less than those of today.

Candin Liteanu, Ion Rîcă

Table of Contents

Preface . . . 13

Chapter 1 Introduction

1.1 Current importance of trace analysis . . . 15
1.2 Analytical problems in trace analysis . . . 16
1.3 Analytical errors . . . 19
1.4 Applications of mathematical statistics . . . 22

Chapter 2 Elements of Statistics Applied to Data Processing for Analytical measurements

2.1 Distribution of random variables . . . 27
2.2 The normal distribution of random variables . . . 29
2.3 Other distributions . . . 31
2.3.1 The χ^2 distribution . . . 31
2.3.2 Student's t distribution . . . 33
2.3.3 The F distribution . . . 34
2.3.4 The binomial distribution . . . 35
2.4 Sampling and estimation . . . 36
2.5 The confidence interval . . . 36
2.5.1 Confidence interval for the mean of the distribution $N(m, \sigma)$ . . . 37
2.5.2 The confidence interval of the sample variance . . . 38
2.5.3 Confidence interval of the binomial distribution parameters . . . 40
2.6 Testing statistical hypotheses . . . 41
2.6.1 Testing extreme values . . . 43
2.6.2 Testing the sample mean . . . 45
2.6.2.1 Testing the mean with respect to a known value . . . 45
2.6.2.2 Testing two means . . . 46
2.6.3 Testing sample variances . . . 49
2.6.3.1 Testing a sample variance with respect to a known value . . . 49
2.6.3.2 Testing two sample variances (the F test) . . . 50
2.6.3.3 Testing several sample variances (Bartlett's test) . . . 51

2.6.4 Testing sampling probabilities 51
2.6.4.1 Testing sampling probability with respect to a known value . 51
2.6.4.2 Testing two sampling probabilities 52
2.6.5 Testing hypotheses about the type of distribution function . . . 54
2.6.5.1 The χ^2 test 54
2.6.5.2 Kolmogorov's test 56
2.6.6 Variance analysis 56
2.6.6.1 Single-factor variance analysis 57
2.6.6.2 Two-factor variance analysis 61
2.6.7 Order statistics 65
2.6.7.1 Smirnov's test 66
2.6.7.2 Wilcoxon's test 66
2.6.7.3 Van der Waerden's test 67
2.6.8 The sequential probability ratio test 68
2.6.8.1 The sequential probability ratio test for the normal distribution 69
2.6.8.2 The sequential probability ratio test for the binomial distribution 74
2.7 Regression analysis 76
2.7.1 Regression analysis for a single independent variable 76
2.7.1.1 Linear regression 76
2.7.1.2 Non-linear regression 82
2.7.2 Multiple regression 83
2.8 Correlation 83

Chapter 3 Information Processing of Results
3.1 Introduction 89
3.2 Analogies between information transmission and analytical systems 89
3.3 Entropy as a measure of uncertainty 92
3.4 Information 95
3.5 The amount of information obtained in detection and identification 96
3.6 The amount of information given by a determination 98
3.7 Redundancy of an analysed material 104
3.8 Informational characterization of analytical systems 106
3.8.1 Informational power of an analytical instrument 106
3.8.2 Informational capacity of an analytical system 108
3.8.3 Redundancy of an analytical system 110
3.8.4 Information flow of an analytical system 110
3.8.5 Informational power of an analytical system 111
References 111

Chapter 4 Stability of Analytical Systems

4.1 Introduction . . . 115
4.2 Stability of analytical systems . . . 117
4.2.1 Stationary stability without systematic deviation . . . 118
4.2.2 Stationary stability with constant systematic deviation . . . 120
4.2.3 Stationary stability with random systematic deviation . . . 120
4.2.4 Stationary stability with intermittent gross error . . . 121
4.2.5 Non-stationary stability with time – dependent systematic deviation . . . 122
4.2.6 Non-stationary stability with time-dependent variance . . . 122
4.3 Statistical methods for assessing stability . . . 124
4.3.1 Regression analysis . . . 124
4.3.2 Statistical hypothesis for stability . . . 124
4.3.3 Variance analysis . . . 131
4.4 Reliability of analytical systems . . . 134
4.5 Deterioration and regulation of analytical systems . . . 137
4.6 Improvement of stability . . . 138
4.7 Statistical control of the stability of a working analytical system . . 139

Chapter 5 Relation between Signal and Concentration

5.1 Introduction . . . 141
5.2 Interference . . . 144
5.3 Background . . . 147
5.4 Sensitivity . . . 149
5.5 Calibration function and analytical function . . . 150
5.6 The linear calibration function . . . 151
5.7 Statistical hypothesis and the linear response function . . . 156
5.8 The linear calibration function $\overline{y} = b'c$ . . . 161
5.9 The method of multiple standard additions . . . 162
5.10 The intersection of two linear regressions . . . 166
5.11 Non-linear calibration and calculation functions . . . 172
5.12 Calibration functions and analytical calculation functions depending on several variables . . . 175
References . . . 177

Chapter 6 Analytical Signal Detection

6.1 Introduction . . . 181
6.1.1 Analytical detection . . . 184
6.1.2 Analytical determination . . . 185
6.1.3 Analytical identification . . . 186
6.2 Detection of an analytical signal in presence of Gaussian noise . . . 187

6.3 Decision level 189
6.4 Formulation of detection by using likelihood criteria 192
6.4.1 The Bayes criterion 193
6.4.2 The minimax criterion 196
6.4.3 The criterion of the ideal observer 197
6.4.4 The criterion of maximal likelihood 197
6.4.5 The Neyman-Pearson criterion 198
6.5 The informational definition of the decision level 201
6.6 Sequential detection 202
6.7 Non-parametric detection 205
6.8 Criterion of the least-square mean error 205
6.9 The detection characteristic of an analytical signal 208
6.10 Resolution of analytical signals 214
6.11 Classical statistical tests applied to decision formulation in analytical detection 220
6.11.1 The decision level 221
6.11.2 The *t*-test in detection 223
6.11.3 Non-parametric tests in detection 226
6.11.4 Sequential probability ratio test in detection 227
6.12 Analytical applications of pattern recognition methods 231
6.12.1 The file search method 233
6.12.2 The learning machine method 240
6.12.3 Other methods of pattern recognition 247
References 251

Chapter 7 The Detection Limit in Chemical Analysis

7.1 Introduction 255
7.2 Defining the detection limit 257
7.3 The informational definition of detection limit 261
7.4 Conclusions about definition of the detection limit 262
7.5 Estimating the detection limit 266
7.5.1 Introduction 266
7.5.2 Estimation of the detection limit from a set of calibration data 268
7.5.3 Evaluation of a confidence value for the decision level and detection limit for a set of calibration data 276
7.5.4 Estimation of the detection limit from the standard deviation of analysis errors 278
7.5.5 Evaluation of the detection limit when the standard deviation of the signal varies with amount 280
7.5.6 The frequentometric method for estimating the detection limit 281

7.6 Trace analysis methods 286
7.6.1 Spectral methods 288
7.6.1.1 Emission spectroscopy methods 288
7.6.1.2 Luminescence spectroscopy methods 292
7.6.1.3 Absorption spectroscopy methods 298
7.6.1.4 Other spectral methods 302
7.6.2 Radiometric and radiochemical methods 304
7.6.2.1 Radiometric methods 304
7.6.2.2 Radiochemical methods 307
7.6.3 Electrochemical methods 309
7.6.4 Chromatographic methods 310
References 318

Chapter 8 Trace Determination
8.1 Introduction 327
8.2 The reliability of a method of analysis with respect to two error limits 333
8.3 The determination characteristic of a method 337
8.4 The determination limit 341
8.4.1 Introduction 341
8.4.2 The determination limit in terms of the coefficient of variation 342
8.4.3 The determination limit in terms of the entropy 344
8.4.4 The determination limit in terms of $p[(\Delta c)\%]$ 344
8.4.5 The relation between detection and determination limits 348
8.4.6 Lowering the determination limit by repeated determinations . 350
8.4.7 Conclusions concerning the determination limit 351
8.5 Analysis of non-homogeneous materials 353
8.5.1 The determination of the mean amount of a component in a non-homogeneous material 353
8.5.2 Experimental design for evaluating the compositional inhomogeneity of a material 356
8.5.3 The criterion for accepting the purity of a material 359
References 363

Chapter 9 Increasing the Signal-to-Noise Ratio in Analytical Chemistry
9.1 The optimization of analytical conditions 365
9.1.1 Introduction 365
9.1.2 The response surface method – RSM 372
9.1.2.1 The complete factorial experiment at two levels – the 2^n experimental design 377

9.1.2.2 The partial factorial experiment . . . 383
9.1.2.3 Factorial experiment at three levels . . . 385
9.1.3 The Box and Wilson method . . . 393
9.1.4 The simplex method . . . 397
9.2 Signal measuring techniques for increasing the signal-to-noise ratio . . . 402
9.2.1 Noise sources and their properties . . . 402
9.2.2 The noise figure . . . 407
9.2.3 Signal modulation and filtration . . . 408
9.3 Improvement of the *S/N* ratio by signal processing . . . 410
9.3.1 Signal averaging . . . 410
9.3.2 Least-squares polynomial smoothing . . . 412
9.3.3 The derivative of the signal . . . 414
9.3.4 Fourier-transform data processing . . . 416
9.3.5 Correlation method for increasing the *S/N* ratio . . . 420
References . . . 422

Appendix . . . 427

General References . . . 439

Index . . . 441

Preface

In recent years, the importance of low concentrations has tremendously increased both in science and in technology. The problems relating to pollution have extended this importance still more.

Semiconductors, nuclear materials, luminescent materials, special alloys, pharmaceutical drugs, foodstuffs and standard substances are but a few of the technical products in which control and determination of low concentrations confront the analytical chemist with special problems.

If to these technical fields we add some others of decisive importance, such as biology, medicine, oceanography, geochemistry, the chemosphere, pharmacodynamics and pollution, the need for a work treating the statistical theoretical aspects of the detection and determination of low concentrations, with very important practical implications, is strikingly evident.

In addition to problems debated in the specialized chemical literature, the the work includes to a large extent specific aspects of the statistical theory of signal detection and of information theory. The work also includes the scientific and experimental contribution of our Analytical Department as well as numerous practical examples.

Candin Liteanu
Ion Rîcă
University of Cluj-Napoca, Romania

Chapter 1

Introduction

1.1 THE CURRENT IMPORTANCE OF TRACE ANALYSIS

In the last few years a major shift of analytical chemistry towards trace element detection and determination has been stimulated by developments in science and technology, and this has become a major field of analytical research.

Though sometimes a matter of dispute, the generic name 'trace elements' will be used throughout this work to specify elements that are present in very small concentration in a substance or material; the name is kept here because of its traditional use in analytical chemistry. Trace analysis will therefore mean the detection and/or determination of small concentrations. In modern times, there is a large and rapidly increasing number of fields of activity where trace analysis is required.

Research has shown that the properties of many substances are influenced by traces of certain components, e.g. the electrical properties of semiconductors, superconductors, and ionic conductors are strongly dependent on certain impurities present in the matrix material. For this reason, silicon and germanium semiconductors have to be purified so as to contain less than $10^{-9}\%$ of impurities. The optical properties of many substances, such as absorption of radiation, photoconductivity, luminescence and phosphorescence, are also highly influenced by the presence of certain trace elements, as are other properties such as radiation effects in various solids, magnetic and mechanical properties of metals and alloys, as well as some surface properties.

It is thus clear that trace analysis is required in all the fields of activity associated with high-purity substances, including research, manufacture and utilization. Furthermore, analytical chemistry itself demands high-purity chemicals for use as reagents and standards. The resulting great demand for high-purity substances has acted as a very powerful stimulus on chemical technology.

In addition, as a result of increased demand for raw materials and the working out of the richer deposits, geological and geochemical exploration methods have undergone considerable development, and trace analysis has proved a valuable tool for location of useful raw materials. Thus, the presence of certain trace elements in river water, soils or plants can be a useful indicator for the existence and location of a natural deposit of raw materials. Again, it is now known that the presence of certain trace elements such as boron, cobalt, copper and zinc in soil and foodstuffs is essential for production of healthy crops and livestock in agriculture.

Trace analysis is, of course, fundamentally involved in space chemistry, and the development of powerful analytical methods for detection and quantitative measurement of low concentrations of biochemical compounds is a prerequisite for major advances in modern medicine.

In the field of environmental pollution, the complexity of the contemporary problem of the pollution level and its consequences for mankind and the natural equilibrium also demands analytical methods for identification and measurement of various trace substances. As is well known, health standards demand that the level of some common toxic chemicals in air should be very low, for example, the mercury or lead content in air should be less than 10^{-9} g/l. Likewise, drinking water, or water used for irrigation purposes, has to satisfy special purity requirements.

It is certain that future developments in science and technology will make still higher demands on trace analysis, so that new methods will be necessary. By developing new methods of trace analysis and by improving the performance of existing techniques analytical chemistry will make its full contribution to the general progress of contemporary science and technology.

1.2 ANALYTICAL PROBLEMS IN TRACE ANALYSIS

There are two distinct problems in analytical research: the detection and the determination of one or more components in the material investigated, i.e. qualitative and quantitative analysis.

To solve an analytical problem, an **analytical method** or **procedure** is used, i.e. a set of instructions for obtaining the required information. This set of instructions has to be complete in the sense that if it is followed rigorously, a maximal quantity of information will be acquired.

The practical application of these instructions results in an **analytical system of measurement**, which is an information-producing system; the information output concerns the qualitative and quantitative chemical composition of the material investigated and possibly the structure of the chemical compounds present.

An analytical system of measurement can be thought of as an informational chain; the investigated material is the input, and the analysis report is the output. In general an analytical system can be regarded as an ensemble of four distinct elements: sampling, physico-chemical treatment of the sample, measurement, and data-processing and calculation (Fig. 1.1).

Sampling is necessary for two reasons: (a) analysis of the whole of the material available would be wasteful and uneconomic in the case of large stocks or expensive materials, and (b) if large quantities are handled the operations become very unwieldy and slow. The whole of the material available is analysed only when it is so small in amount that it must all be used to give the necessary accuracy. The sampling has to be done in such a way that the material selected for analysis is as truly representative of the bulk composition as possible, and this is dictated by the nature of the material, characteristics of the analytical system, and the purpose of the analysis. The most important property of the material is the degree of inhomogeneity. No special care is required to take a correct sample of a homogeneous material such as a solution, any portion being representative of the whole. However, the material to be analysed is often very inhomogeneous, and in such cases correct sampling is a very complex problem in itself.

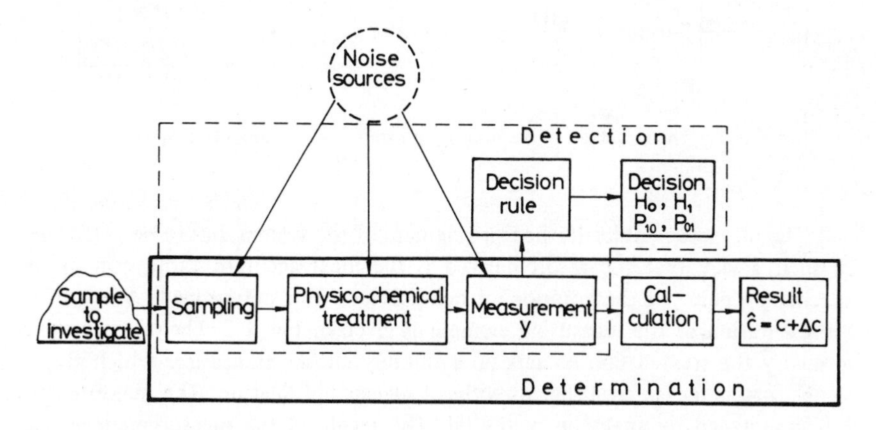

Fig. 1.1 The structure of an analytical system with the function of detection and determination. H_0, H_1 – hypotheses for decision; P_{10}, P_{01} – errors of kinds I and II; y – the signal; $\hat{c}$ – the true value; c – analysis result; Δc – the error.

With an inhomogeneous material, the analytical task can be either the estimation of the average content of one or more components, or evaluation of the variability of composition throughout the material and to obtain in this way a quantitative measure of the inhomogeneity.

When the material is an ensemble of distinct individual units we can also consider the problem of estimating the percentage of units having a composition outside a prescribed range. Sometimes the material is unstable, i.e. the composition varies with time. We then consider the material composition at a given time. For trace analysis correct sampling is an extremely complex problem: in this case, apart from the general considerations involved in sampling, care must be taken to prevent contamination of the samples during their extraction or preservation, or loss of the desired constituent by adsorption or volatilization.

The purpose of the second element of the informational chain (the physico-chemical treatment) is to bring the sample into a state in which the measurement of some property (conductivity, potential, absorption or emission of radiation, etc.) will give maximal information about the amount of the component to be determined. Usually, there are several successive analytical operations to be applied, such as dissolution, vaporization, oxidation, reduction, separation. Various ancillary measurements may be required during this treatment, for instance of mass, volume, pH, temperature. In trace analysis, the treatment may be very complex. Care has to be taken to avoid any possible loss or contamination.

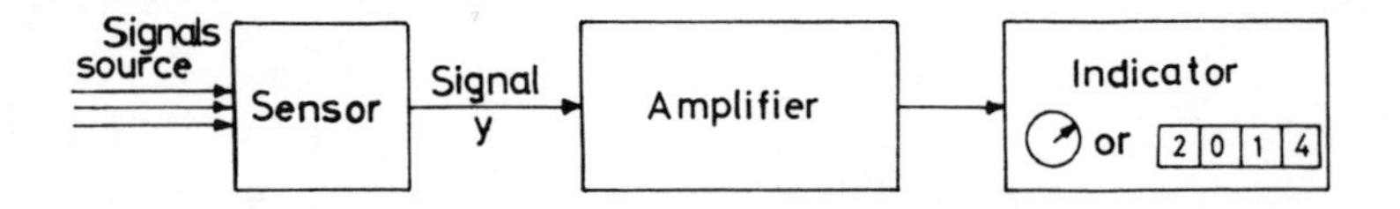

Fig. 1.2 The structure of the measuring element of the analytical system.

The measurement itself, the third element of the system, has to be performed in such a way that the result derived is the most accurate expression of the specific property of the component being detected or determined. A simplified representation of the measuring element is given in Fig. 1.2. The primary signal given by the treated sample acts on a suitably chosen transducer which gives a signal that is measured either directly or after amplification. The measurement instrument can be analogue or digital. The result of the measurement, y, can always be split into two terms:

$$y = f(c) + y_0 \tag{1.1}$$

where the first term $f(c)$ is a known function of the concentration c that is to be established. The second term, y_0, is a random variable associated with various perturbations and errors involved in the analysis process. The sampling, physico-chemical treatment, and measurement have to be accomplished and correlated so as to minimize the ratio $y_0/f(c)$.

The final step of an analytical process is the data-processing and calculation. Electronic computers are increasingly used for this purpose.

If the analysis is done for the purpose of detection, there are two possible hypotheses: H_0 that the component sought is not present in the investigated material, and H_1 that it *is* present. The result of the measurement is therefore converted into a detection decision which picks one of these hypotheses according to a previously established rule.

In an analytical determination we are interested in the value of the concentration c. This is obtained from the measured result by using the calibration function of the analytical system, which has to be already known or determined by means of a standard.

All the steps of the process are conducted and supervised by the analyst, and the correctness of the results is largely dependent upon his manipulative skill.

Two major lines have been followed in analytical research dealing with the special problems posed by trace analysis. Thus some analytical researches aim at improving the performance of established methods, others at the discovery and application of new analytical principles. Valuable results have been obtained along both lines.

Electrochemical methods, which have wide application in trace analysis, have been very much diversified, especially the polarographic methods. Emission and absorption spectroscopy also have wide application, especially atomic-absorption and X-ray fluorescence spectroscopy, and electron-probe microanalysis. Radiochemical, kinetic and enzymatic methods are very valuable in trace analysis. The remarkable progress made in trace analysis is also largely due to the advances in separation and concentration techniques.

Generally, the present advance in analytical chemistry is a consequence of the various accomplishments in physico-chemical science and of the recently developed electronic techniques of signal measurement, coupled with the use of electronic computers. These computers have taken over some of the routine operations, which is very convenient as it saves the valuable time of trained human operators and also because it may lead to more precise performance of the analytical systems. Integrated analytical systems of measurement with built-in processing computers can take over a large part of the command-control operations, all the operations of data-processing and calculation, and conveyance of the results to the user as well.

1.3 ANALYTICAL ERRORS

An analytical measurement can never be made in purely ideal conditions. Each step will always be subject to various perturbations that act as sources of noise, of which there are three classes.

(a) General noise sources: these are related to the discrete character of matter and energy, thermal agitation, and the uncertainty principle.

(b) Noise sources related to the actual conditions in which the given system is operating. They include vibration and fluctuations of working parameters such as temperature, humidity, light-sources, power supply, etc.

(c) Noise sources which are the consequence of various imperfections of the measurement system and/or imperfections related to the way in which the process is conducted. Various interferences, calibration errors and sample contamination are examples of this class. They all lead to systematic errors as well as to fluctuations inherent in analytical operations.

Because of the noise sources, there will always be errors in the analytical measurements and, hence, in the final results.

When the object of the analysis is a decision between the hypotheses H_0 (component absent) and H_1 (component present) there can be two types of error: an error of the first kind is when hypothesis H_1 is claimed to be true but, in fact, it is H_0 that is valid (i.e. false detection), and an error of the second kind occurs when hypothesis H_0 is accepted, but actually H_1 is true (i.e. failure to detect the component). A measure of these errors is given by their probability. Generally, the smaller the concentration of the component to be detected, the smaller the signal:noise ratio and, hence, the higher the probability of errors of both types: consequently, there is a limit of detection set by the degree of certainty that is required.

In an analytical determination, the goal is to find the amount of a particular component in a material. Let c designate the result obtained, and $\hat{c}$ the true value; then the error in c can be expressed as the absolute error Δc, the relative error $(\Delta c)_r$, or $(\Delta c)\%$, defined by:

$$\Delta c = c - \hat{c}$$

$$(\Delta c)_r = (c - \hat{c})/\hat{c} \tag{1.2}$$

$$(\Delta c)\% = (c - \hat{c}) \times 100/\hat{c}$$

The relative error is generally more significant than Δc, especially in trace analysis.

Errors in determination are customarily classified as random, systematic or gross. This classification is useful from a methodological point of view, but should not be taken too rigidly: there is not a sharp boundary between random and systematic errors. Some errors may first be considered as systematic and later regarded as random, or vice versa.

Random errors are inherent in the results. The easiest way to show their existence is to repeat the experiment: because of the random errors, there will invariably be a different result every time. The result of a determination is obtained after a large number of analytical operations, so its random error is a summation of the random errors involved in each step of the process. Generally, the individual random errors are unpredictable, but their set is usually assumed to conform to a well defined distribution function.

An error is said to be systematic if it causes a bias, i.e. it shifts the results from the true value by a constant value and in one direction only. However, this is not rigorously valid. In many cases, the analytical system is unstable and the systematic errors vary with time, so that they can change their value and even their sign. It is useful, then, to consider the time factor and effects related to it. This naturally leads to a classification of systematic errors into those which are constant with time, and those which are variable with time. Systematic errors can stem from various causes. There can be a single cause or several causes acting simultaneously. Such causes may be incorrect use of the measuring instruments, use of incorrect calibration functions, various interferences, reagents that are not pure enough (this is an extremely important source of errors in trace analysis).

Clearly, an important task is identification of causes of systematic errors, and their removal, for example by improving the measurement system. It is best to try to eliminate systematic errors; applying corrections for them should be a last resort.

Systematic errors can also be classified as dependent on the amount of the component to be determined, or independent of it. Thus, systematic error in titration or weighing operations is a linear function of the magnitude of the quantity measured, but the error due to contamination from impure reagents does not depend on the amount of substance to be measured (except insofar as this determines the amount of reagent used).

The classification of determination errors, given above, is illustrated in Fig. 1.3. The double arrows indicate the possibility that systematic errors can be changed into random errors and vice versa, i.e. that there is no clear demarcation between these classes of error.

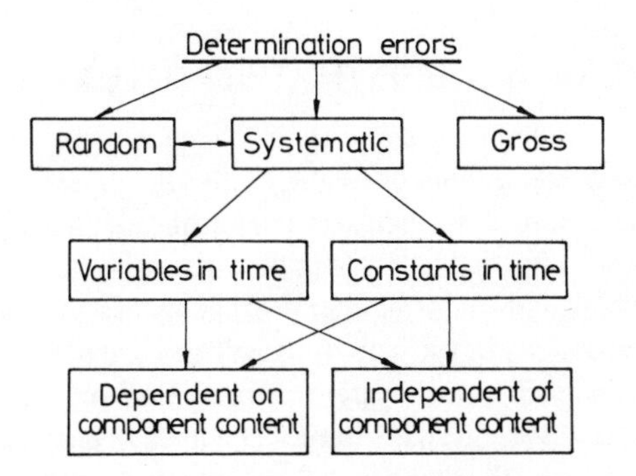

Fig. 1.3 The classification of analytical determination errors.

Closely connected with determination errors are two well known concepts in analytical practice: precision and accuracy. The accuracy is a measure of the degree to which a result (or the mean of several results) coincides with the true value. The precision is a measure of the degree to which an ensemble of results is grouped around their mean value. Good precision means good reproducibility, but not necessarily good accuracy as well.

The accuracy of the result can be expressed as the absolute value of the deviation of the result (or the mean of n results c_i) from the true value: $\Delta c = |c_i - \hat{c}|$ or $\Delta c = |\bar{c} - \hat{c}|$ where $\bar{c} = (\sum_{i=1}^{n} c_i)/n$.

The precision of a set of individual measurements is given by their standard deviation σ. The precision of the mean of the set of n individual measurements is $\sigma/\sqrt{n}$. Hence, the average result has better precision than an individual result.

Thus, results are made more correct by removing the systematic errors and increasing the precision. As a general conclusion, an analytical system of measurements will be well adjusted if there is no systematic error, and if the precision is good. In this case the average result of n determinations converges towards the real value if n is increased: this is a consequence of Glivenko's theorem. In other words, the probability P of zero error converges towards 1 on increasing n, that is:

$$\lim_{n \to \infty} [P|F(c) - F_n(c)| = 0] = 1, \tag{1.3}$$

where $F(c)$ is the theoretical distribution and $F_n(c)$ is the experimental, so we can write:

$$\lim_{n \to \infty} [P|\hat{c} - \bar{c}| = 0] = 1 \tag{1.4}$$

1.4 APPLICATIONS OF MATHEMATICAL STATISTICS

The utility of mathematical statistics in chemical analysis is already proven by the large number of publications on it. In the present work, the general application of mathematical statistics in chemical analysis will be extended by introduction of new knowledge on stability of analytical systems, and also elements of the informational treatment of analytical measurements; finally, specific statistical problems in trace analysis will be treated in detail.

By stability of an analytical system of measurement we mean the qualitative behaviour of the system with time. Obviously, because any process of analysis takes a finite time, the stability of an analytical system is very important.

Mathematical statistics give methods of estimating and analysing the stability of analytical systems, and also methods of increasing the stability.

As explained above, analytical systems are informational because the results are derived information concerning the qualitative and quantitative composition (chemical or structural) of the material investigated. As in any other field of science, in chemical analysis it is necessary to measure and know the quantity of information that has been obtained from an analytical experiment, and this can be done by means of information theory.

The two problems – the stability of analytical systems, and the informational treatment of results – are particularly important in trace analysis, because both aim at evaluating and improving the quality of performance of analytical systems.

Mathematical statistics also have specific applications in trace analysis. A rigorous theory of trace detection and determination can be obtained from mathematical statistics. The development of such a theory is a prerequisite for defining and evaluating certain quantities which would rigorously characterize the power of detection and determination of analytical systems of measurement. In this way it is possible to make an objective comparison of the performance of two or more analytical systems. It also gives the optimal approach to the problem of increasing the power of detection and determination in chemical analysis.

A problem that has raised much discussion in analytical chemistry is the limit of detection or determination; this problem can be solved completely by use of the statistical theory of signal detection.

From the statistical theory of signal detection, the rules for making detection decisions can be established. In other words, the methodology of extracting maximum information from a given set of results can be established. Many of the general notions in the statistical theory of signal detection, e.g. detection limit, detection characteristic) are basic to trace analysis.

Mathematical statistics are also used for characterizing the power of determination of an analytical system by means of such concepts as the determination characteristic, limit of determination, reliability of the method, and so on. The statistical theory of signal detection also gives the general principles for lowering the detection or determination limit (Section 8.4.6).

The optimal experimental conditions for trace detection and determination can also be established most accurately by using mathematical statistics. In order to illustrate the general problems involved in this process we shall refer to the four components of the functional structure of an analytical system (Fig. 1.1).

When the objective is determination of the average amount of one or several of the components of the material, the sampling should provide a representative portion for analysis: by representative we mean here that the sample composition should be as close as possible to the average composition of the material. In sampling, variability in the composition of the material should be considered first. The composition of inhomogeneous material can generally be represented by a random vector in multi-dimensional space. The random variation of a component as a function of mass of the material to be analysed is symbolically illustrated in Fig. 1.4; $\hat{c}$ stands for the average amount of the component. In the case of an inhomogeneous material, the sampling has to be optimized to give the best possible estimate of $\hat{c}$. For this, it is not enough simply to extract, from only one part of the material, a sample having the mass m_0 prescribed in the working instructions. Indeed, because of the inhomogeneity (and depending on its degree), the concentration in a sample taken from a small region of the material can be largely different from the average concentration $\hat{c}$. Only by chance would the concentration coincide with (or be close enough to) $\hat{c}$. Therefore, the statistical parameters characterizing the variability of the material composition must be taken into account when planning the sampling of an inhomogeneous material.

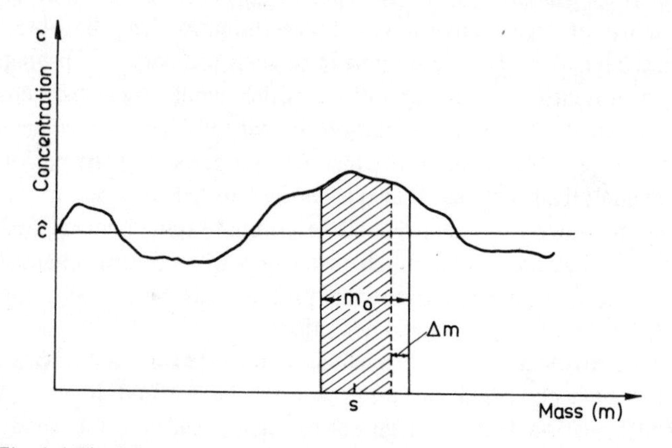

Fig. 1.4 The inhomogeneous distribution of a component in a material.

The sampling plan must define the number and positions of the points in the material at which the subsamples will be extracted, the masses of these subsamples, and the steps and particular procedure that will lead finally to the sample mass m_0 prescribed for the analysis. In the ideal case the concentration of determinand would be the same in the sample of mass m_0 and in the basic material. If the concentration is expressed in per cent, the mass of the component to be determined in an ideal sample would be $d_0 = \hat{c}m_0/100$. However, because of sample inhomogeneity and weighing errors, the amount actually taken for analysis will generally be different from the ideal quantity d_0.

Suppose that in the sample taken the average concentration of the component to be determined is $\bar{c}$ and let Δm be the weighing error. Then a mass $d = (m_0 + \Delta m)\bar{c}/100$ of this component is subjected to the analysis process.

Thus, when the purpose is the determination of the average amount of a component in a given material, the accuracy of the sampling process is measured by the quantity $\Delta d = |d_0 - d|$. The smaller Δd, the more correct the sampling. This requirement can be fulfilled by organizing the sampling on statistical principles.

An analytical procedure is generally established empirically, and is unlikely to be optimal. To establish the optimum procedure, experimental–statistical optimization methods have to be used. The physico-chemical treatment of the sample represents a succession of operations: homogenization, dissolution, separation of some components, enrichment in the component to be determined, etc. The samples to be analysed are subjected to all these operations in order that the result of the analytical determination supplies the maximum amount of information on the content of the component to be determined. The physico-chemical treatment specific to each analytical system may be outlined as in Fig. 1.5. At the input of this element of the analytical system there is the sample and at the output the property of the sample (potential, conductivity, emission or absorption of radiation etc.) which is measured to yield the analytical signal (the carrier of information regarding the component determined).

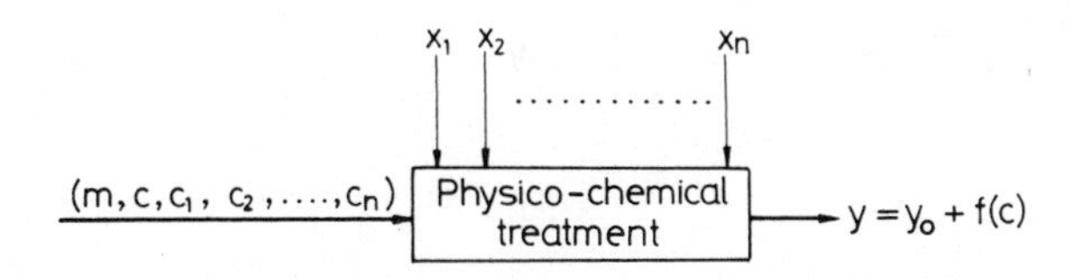

Fig. 1.5 The physico-chemical treatment of the sample to be analysed.

In a formulation made as complete as possible the response y must evidently be represented as a function of the concentration of the component to be determined c, of the concentrations of the other components in the sample $(c_1, c_2, \ldots, c_n)$ as well as of the parameters of the treatment $(x_1, x_2, \ldots, x_m)$ (e.g. the pH, temperature, duration of treatment, amounts of reagents employed, etc.), that is:

$$y = f(c, c_1, \ldots, c_n; x_1, \ldots, x_m) \tag{1.5}$$

The response function, y, may in general be split into two terms:

$$y = y_0 + f(c) \tag{1.6}$$

The first term, y_0, which is random in character, originates in interference phenomena, in the fluctuations of the treatment parameters and in the perturbations of the environment. The second term, $f(c)$, is the useful part of the response that is due to the component to be determined.

The optimization of the physico-chemical treatment aims at finding those values of the treatment parameters which yield a maximum $f(c)/y_0$ ratio. In this way conditions are established for improvement of the signal/noise ratio.

The techniques for improving the signal to noise ratio are based on various principles, such as integration and averaging of signals, signal amplification by modulation, correlation, self-correlation.

The final step is the calculation and data processing.

In the case of detection, the analytical measurement is translated into a detection decision by means of a decision rule. Generally, the measurements show variability and the decision rule can be rigorously established only on statistical criteria. This is very important in trace analysis because in this case there is always a small signal to noise ratio. In determination the result is obtained from the measurement by use of the calibration function, which is usually evaluated from an ensemble of experimental data. Again, statistical methods generally give the most correct evaluation, also very important in trace analysis. Hence, statistical methods have wide application in this field.

Chapter 2

Elements of Statistics Applied to Data Processing for Analytical Measurements

In the first chapter it has been shown that for various reasons the results of analytical measurement always show a certain variability, and statistical treatment is required in order to obtain the maximum of information from them. Consequently, statistical problems frequently arise in chemical analysis. Such problems are:

- establishing the laws of the statistical distribution of data;
- sampling problems and estimation of various parameters;
- evaluation of the confidence interval of the estimated quantities;
- testing of hypotheses.

Some of the more important elements of applied statistics for chemical analysis will be given in this chapter. There will be no insistence upon theoretical aspects; these can be found in detail in books on statistics. Only the more important theoretical results will be reviewed, and their analytical applications.

2.1 DISTRIBUTION OF RANDOM VARIABLES

Repeating an analytical determination gives a multitude of results; these can be described by a distribution function $F(x)$, which is the expression for the probability of finding the random variable at a value less than x.

The general model for the distribution function for one random variable (x) is given in Fig. 2.1.

The probability that x is less than x_1 is

$$P(x < x_1) = F(x_1) \tag{2.1}$$

The values of the distribution function fall in the interval [0, 1], i.e.

$$0 \leqslant F(x) \leqslant 1 \tag{2.2}$$

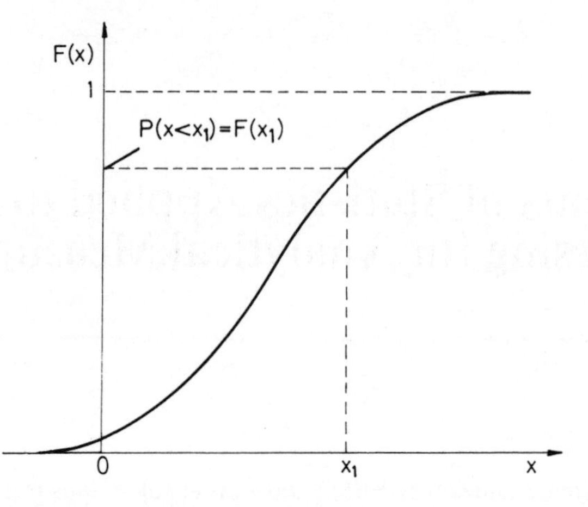

Fig. 2.1 The model of the distribution function for random variables.

Differentiation of $F(x)$ gives the corresponding probability density function $f(x)$:

$$f(x) = \frac{\mathrm{d}F(x)}{\mathrm{d}x} = F'(x) \tag{2.3}$$

The general model for $f(x)$ (the frequency function) is schematically presented in Fig. 2.2.

The probability that x is in the interval (x_1, x_2) is measured by the integral

$$P(x_1 < x < x_2) = \int_{x_1}^{x_2} f(x)\mathrm{d}x \tag{2.4}$$

This integral is represented by the dashed area in Fig. 2.2.

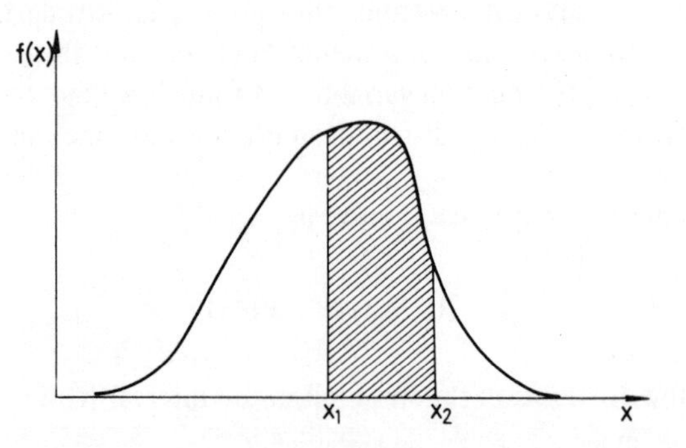

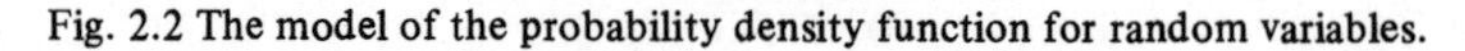

Fig. 2.2 The model of the probability density function for random variables.

The most important typical values associated with a distribution function are the mean value m and the variance σ^2, defined by

$$m = \int_{-\infty}^{+\infty} xf(x)\mathrm{d}x = M(x) \tag{2.5}$$

and

$$\sigma^2 = \int_{-\infty}^{+\infty} [x - M(x)]^2 f(x)\mathrm{d}x \tag{2.6}$$

where $M(x)$ is the expectation value for the mean.
The mean value and the variance are only particular examples of the more general characteristics known as 'moments'. Generally, the central moment of order r is defined by:

$$\mu_r = \int_{-\infty}^{+\infty} [x - M(x)]^r f(x)\mathrm{d}x \tag{2.7}$$

Obviously, the variance σ^2 is the centred moment of second order.

2.2 THE NORMAL DISTRIBUTION OF RANDOM VARIABLES

By studying the distribution of random variables, Gauss concluded that the normal distribution function can be taken as a model for probabilistic investigation of measurement processes.

This function, which is the fundamental distribution in mathematical statistics and the theory of errors, leads to the following probability density:

$$f(x; m,\sigma) = \frac{1}{\sigma\sqrt{2\pi}} \exp\left[-\frac{(x-m)^2}{2\sigma^2}\right]; -\infty < x < +\infty \tag{2.8}$$

where m is the mean value of the distribution, and σ the standard deviation. For the normal distribution function we shall use the symbolic notation $N(m, \sigma)$. The graphical representation of the function is given in Fig. 2.3.

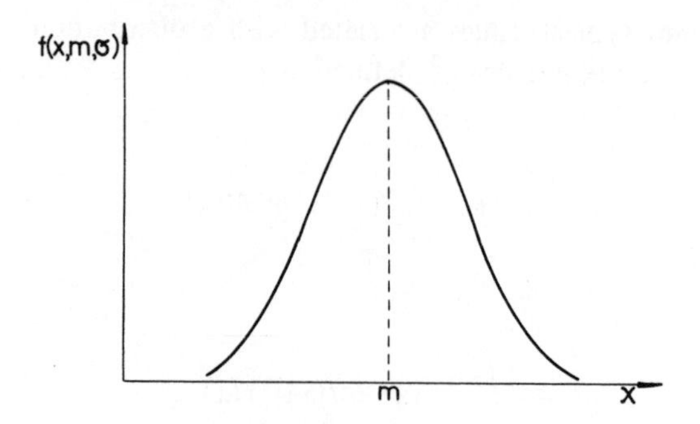

Fig. 2.3 The graphical model of the normal probability density function.

Its most important properties are the following:
- it has a maximum at $x = m$;
- it is symmetrical with respect to $x = m$;
- any change of the value of m entails a translation of the normal curve along the x-axis, but the shape of the curve is not affected;
- a modification of σ will either widen or narrow the peak, but m will be left unchanged;
- there are two points of inflexion, at $x = \pm\,\sigma$.

For the subsequent applications, the following two theorems are useful.

1. The mean value of n values pertaining to a distribution function $N(m, \sigma)$ is itself a random variable described by the distribution function $N(m, \sigma/\sqrt{n})$.

2. The probability density function of the mean value of n values pertaining to a non-normal distribution is, for large n, either the normal density function, or very close to it.

It can easily be seen that, if x is a random variable with a distribution function $N(m, \sigma)$, the normal deviate, which is the random variable z defined by

$$z = \frac{x - m}{\sigma} \quad , \tag{2.9}$$

has the distribution function $N(0, 1)$. Hence, the normal density function (2.8) can be reduced to the normal deviate function:

$$f(z) = \frac{1}{\sqrt{2\pi}} e^{-\frac{z^2}{2}} ; -\infty < z < +\infty \tag{2.10}$$

This function plays an important role, and the distribution corresponding to it is given by the equation

$$F(z) = \frac{1}{\sqrt{2\pi}} \int_{-\infty}^{z} e^{-\frac{z^2}{2}} \mathrm{d}z = F(x) = \frac{1}{2} \pm \Phi(z) \tag{2.11}$$

where $\Phi(z)$ is the Laplace function:

$$\Phi(z) = \frac{1}{\sqrt{2\pi}} \int_0^z e^{-\frac{z^2}{2}} dz \qquad (2.12)$$

Values of $\Phi(z)$ and $F(z)$ are given in Tables I and II in the Appendix. For instance, the probability of finding the variable z in the interval $(-1.5; 1.5)$ will be: $2\Phi(1.5) = 2 \times 0.433 = 0.866$.

The normal distribution function has been widely accepted as a probabilistic model for the evaluation, investigation and processing of the fluctuations arising in analytical determinations, or, generally, in any other measurements.

Generally speaking, the fluctuations in measurements are due to a complex of causes that cannot be discerned individually, but each of which contributes a certain error. The random error of a determination is the result of all the individual random errors.

In the theory of probability it is demonstrated (the central limit theorem) that if a random variable is the sum of a large number of independent random variables which each make a small contribution to the sum, the sum random variable has a normal distribution or one close to it. In the case of a well devised analytical system of measurement and a properly performed analysis, all the errors are small and the requirements of the central limit theorem are satisfied; the analytical results will be normally distributed or, at least, almost so.

Nevertheless, in trace analysis the results are frequently not normally distributed, and this may be attributed to the effect of inhomogeneity in the samples used, which will result in superposition of a non-normal distribution on on the normal error distribution.

There are some analytical determinations for which the logarithm of the results, i.e. the variable $\log x$, is normally distributed (such results are said to have a log-normal distribution). In these cases, the calculations will involve the normal distribution function (2.8) in the form:

$$f(\log x; \log m, \sigma) = \frac{1}{x\sigma\sqrt{2\pi}} e^{-\frac{(\log x - \log m)^2}{2\sigma^2}} \qquad (2.12')$$

2.3 OTHER DISTRIBUTIONS

There are other useful distribution functions which are treated in detail in statistical treatises. We shall summarize only those which will be used in solving statistical problems of chemical analysis.

2.3.1 The χ^2 distribution

It can be demonstrated that, if $x_1, x_2, \ldots, x_n$ are independent and normally distributed random variables [i.e. they are independent, and the individual distribution is $N(0, 1)$ for each of them], then the variable

$$\chi^2 = \sum_{i=1}^{n} x_i^2 \; ; 0 < \chi^2 < +\infty \tag{2.13}$$

has the probability density function:

$$f(\chi^2; \nu) = \frac{1}{2^{\frac{\nu}{2}} \Gamma\left(\frac{\nu}{2}\right)} (\chi^2)^{\frac{\nu}{2} - 1} e^{-\frac{\chi^2}{2}} \tag{2.14}$$

where $\nu = n - 1$ is the number of degrees of freedom. The distribution function $F(\chi^2)$ corresponding to this probability density is given in Table III in the Appendix.

It can be shown that for the values $x_1, x_2, \ldots, x_n$ of a random variable with a distribution function $N(m, \sigma)$, the corresponding sample variance

$$s^2 = \frac{1}{n-1} \sum_{i=1}^{n} (x_i - \bar{x})^2 \, ,$$

belongs to the set of values given by

$$s^2 \in \frac{\chi^2 \sigma^2}{n-1} \tag{2.15}$$

Here χ^2 stands for the value of the χ^2 distribution corresponding to $\nu = n - 1$ degrees of freedom.

It results that (2.15) can be used in order to obtain the distribution function of the sample variance by using the distribution function $F(\chi^2)$. This will be used subsequently for evaluating a confidence interval for the variance.

It is also possible to obtain a test of significance for comparing two sample variances by using (2.15) and considering the F-distribution of the variable (see Section 2.3.3).

The following considerations will be convenient for the purpose of using the χ^2 variable in testing hypotheses concerning the distribution functions of the random variables.

Let x be a discrete random variable with the distribution

$$X \begin{pmatrix} x_1, x_2, \ldots, x_m \\ p_1, p_2, \ldots, p_m \end{pmatrix} \tag{2.16}$$

and assume that n independent observations are made of x. Assume also that the random variable takes the value x_i in n_i cases, so that $\sum_{i=1}^{m} n_i = n$. Then it can be shown that the distribution function of the variable

$$\frac{(n_1 - np_1)^2}{np_1} + \frac{(n_2 - np_2)^2}{np_2} + \ldots + \frac{(n_m - np_m)^2}{np_m} \qquad (2.17)$$

tends towards a χ^2 distribution with the parameters $\nu = m - 1$ and $\sigma = 1$ when $n \to \infty$.

2.3.2 Student's *t*-distribution

It can be proved that given a random variable x with the distribution $N(0, 1)$ and another random variable y which is independent of x and has a χ^2 distribution with ν degrees of freedom the random variable

$$t = \frac{x}{\sqrt{y/\nu}} \qquad (2.18)$$

will have a t distribution with ν degrees of freedom; the corresponding probability density function is

$$F(t; \nu) = \frac{1}{\sqrt{\pi\nu}} \cdot \frac{\Gamma\left(\frac{\nu+1}{2}\right)}{\Gamma\left(\frac{\nu}{2}\right)} \left(1 + \frac{t^2}{\nu}\right)^{-\frac{\nu+1}{2}}; \quad -\infty < t < +\infty \qquad (2.19)$$

Values of the t distribution are given in Table IV in the Appendix.

It can be shown that for $\nu \to \infty$ the t distribution tends towards the normal distribution.

The $F(t)$ distribution function will be used for finding the confidence interval and for testing statistical hypotheses arising from chemical analysis of small numbers of samples (usually for $n < 30$).

With this in mind we shall make the following statements.

(a) If $x_1, x_2, \ldots, x_n$ is a sample of a random variable with the distribution $N(m, \sigma)$, and $\bar{x}$ and s^2 are the corresponding sample mean and variance, then the variable

$$t = \frac{\bar{x} - m}{s/\sqrt{n}} \qquad (2.20)$$

has a t distribution with $\nu = n - 1$ degrees of freedom.

(b) If $x_{11}, x_{12}, \ldots, x_{1n_1}$ and $x_{21}, x_{22}, \ldots, x_{2n_2}$ are two samples of a variable with a distribution $N(m, \sigma)$, and $\bar{x}_1, \bar{x}_2, s_1^2, s_2^2$ are the corresponding means and sample variances, then the variable

$$t = \frac{\bar{x}_1 - \bar{x}_2}{\bar{s}\sqrt{\dfrac{1}{n_1} + \dfrac{1}{n_2}}}, \tag{2.21}$$

where $\bar{s} = \sqrt{\dfrac{(n_1 - 1)s_1^2 + (n_2 - 1)s_2^2}{n_1 + n_2 - 2}}$, has a t distribution with $\nu = n_1 + n_2 - 2$ degrees of freedom.

2.3.3 The F-distribution

Given two independent random variables χ_1^2 and χ_2^2 with ν_1 and ν_2 degrees of freedom, the variable

$$F = \frac{\nu_2 \chi_1^2}{\nu_1 \chi_2^2} \tag{2.22}$$

has the probability density function given by the equation:

$$f(F; \nu_1, \nu_2) = \left(\frac{\nu_1}{\nu_2}\right)^{\frac{\nu_1}{2}} \cdot \frac{\Gamma\left(\dfrac{\nu_1 + \nu_2}{2}\right)}{\Gamma\left(\dfrac{\nu_1}{2}\right)\Gamma\left(\dfrac{\nu_2}{2}\right)} F^{\frac{\nu_1}{2} - 1} \left(1 + \frac{\nu_1}{\nu_2} F\right)^{-\frac{\nu_1 + \nu_2}{2}} ; 0 < F < \infty \tag{2.23}$$

This function can be used for testing significance hypotheses, and for this the following remarks concerning sample variances are useful.

For the samples $x_{11}, x_{12}, \ldots, x_{1n_1}$ and $x_{21}, x_{22}, \ldots, x_{2n_2}$ of the variables $N(m_1, \sigma)$ and $N(m_2, \sigma)$, according to (2.15) the corresponding sample variances are:

$$s_1^2 = \frac{\chi_1^2 \sigma^2}{\nu_1} \text{ and } s_2^2 = \frac{\chi_2^2 \sigma^2}{\nu_2} \tag{2.24}$$

where χ_1^2 and χ_2^2 correspond respectively to $\nu_1 = n_1 - 1$ and $\nu_2 = n_2 - 1$ degrees of freedom.

It results from (2.24) and (2.22) that the ratio of the two sample variances, $F = s_1^2/s_2^2$ (where $s_1^2 > s_2^2$), is a random variable with an F distribution for ν_1, ν_2 degrees of freedom.

Values of the F distribution are given in Tables V and VI in the Appendix.

2.3.4 The binomial distribution

If an event occurs with a probability p, the probability that it occurs x times in n independent experiments (trials) is given by the function:

$$f(x, n) = \frac{n!}{x!(n-x)!} p^x(1-p)^{n-x}, \tag{2.25}$$

where $x = 1, 2, 3, \ldots, n$. The distribution associated with $f(x, n)$ is called the binomial distribution.

Table 2.1

x	0	1	2	3	4	5	6	7	8	9	10
$f(x, n)$	$\frac{1}{1024}$	$\frac{10}{1024}$	$\frac{45}{1024}$	$\frac{120}{1024}$	$\frac{210}{1024}$	$\frac{252}{1024}$	$\frac{210}{1024}$	$\frac{120}{1024}$	$\frac{45}{1024}$	$\frac{10}{1024}$	$\frac{1}{1024}$

Example 2.1. Ten analytical determinations are performed on a sample. Assuming that there are no systematic errors, calculate the probability for obtaining positive random errors, for all possible outcomes of the experiment.

Obviously, when there are no systematic errors, both positive and negative random errors occur with a probability $p = 1/2$. For $n = 10$, the number of occurrences (x) of a positive error may take any integral value from 0 to 10. From (2.25), the corresponding probabilities have the values given in Table 2.1. The calculation is verified by summing the probabilities:

$$\sum_{x=1}^{10} f(x, n) = 1$$

It can be proved that, for $0.30 \leqslant p \leqslant 0.70$ and $n \geqslant 30$, the binomial distribution is well approximated by a normal distribution with mean $m = np$ and variance $\sigma^2 = np(1-p)$, so that (2.25) becomes:

$$f(x, n) = \frac{1}{\sqrt{2\pi\, np\,(1-p)}}\, e^{-\frac{(x-np)^2}{2np(1-p)}} \tag{2.26}$$

2.4 SAMPLING AND ESTIMATION

The notion of sampling is closely connected with that of estimation. Both of them are very important for chemical analysis.

A typical example of sampling analysis is given by the population of results $x_1, x_2, \ldots, x_n$ obtained from analytical determinations made on a material for evaluating the amount of a component. In the absence of systematic errors, the mean of this sampling population is the best estimation of the true amount of the component in the material.

Another example is estimation of the precision (scatter) of the results, obtained by using the standard deviation for the population:

$$s = \sqrt{\frac{1}{n-1} \sum_{i=1}^{n} (x_i - \bar{x})^2} \qquad (2.27)$$

There are many other examples of estimation in chemical analysis; we shall come back to this later.

An estimation is said to be non-biased when its 'expectance' coincides with the true value. It is positively (or negatively) biased when its 'expectance' is greater (or smaller) than the true value, and is said to be consistent when the estimation value converges towards the true value when the population size n is increased.

When there are no systematic errors, the mean is a non-biased estimation, and tends, when n is increased towards the true value x, i.e. $\lim_{n \to \infty} |\bar{x} - x| = 0$ in agreement with (1.4). The smaller the standard deviation, the faster the convergence of the estimation value towards the true value.

Example 2.2 Repeated analysis of a material gives the following percentage results:

0.0013, 0.0011, 0.011, 0.0009, 0.0010, 0.0012, 0.0010, 0.0011, 0.0008, 0.0012.

The sample mean and standard deviation are: $\bar{x} = 0.00107, s = 0.00015$.

2.5 THE CONFIDENCE INTERVAL

The notion of the confidence interval was first used by Laplace in connection with estimating the binomial distribution parameters, but development of the theory and terminology is largely due to Neyman. The notion of the confidence interval is implied by the notion of estimation. That is, to ascertain the significance of an estimation we must make a statement about the degree of confidence related to it.

Consider a random variable with the probability density function $f(x, \theta)$. The θ parameter is estimated for the sample $x_1, x_2, \ldots, x_n$ by using an estimation equation $\theta' = \phi(x_1, x_2, \ldots, x_n)$.

The estimated value θ' is itself a random variable with a frequency function which will be written as $f(\theta', \theta)$.

If the limits of a confidence interval corresponding to the probability P are (θ_1, θ_2), then, obviously, we have the equation:

$$P(\theta_1 < \theta' < \theta_2) = P = \int_{\theta_1}^{\theta_2} f(\theta', \theta) \mathrm{d}\theta' \tag{2.28}$$

The probability P is also called the confidence coefficient and it is the probability that the true value of the θ parameter will lie in the interval (θ_1, θ_2).

2.5.1 Confidence interval for the mean of the distribution $N(m, \sigma)$

There can be two distinct situations in this case.

(a) The distribution has a known standard deviation σ.

In order to estimate the mean, the sample mean $\bar{x} = (\sum_{i=1}^{n} x_i)/n$ is calculated from the population $x_1\ x_2, \ldots, x_n$. Then, in agreement with Section 2.1, the distribution of the sample mean $\bar{x}$ will be $N(m, \sigma/\sqrt{n})$, so the variable z defined by

$$\frac{\bar{x} - m}{\sigma/\sqrt{n}} = z \tag{2.29}$$

will have the distribution $N(0, 1)$. Hence, for a confidence probability P, the distribution mean m will be found within the confidence interval centred upon it and defined by:

$$(\bar{x} - |z_{\alpha/2}|\sigma/\sqrt{n}) < m < (\bar{x} + |z_{\alpha/2}|\sigma/\sqrt{n}) \tag{2.30}$$

where $\alpha = 1 - P$, and $z_{\alpha/2}$ is obtained from the condition $F(z_{\alpha/2}) = \alpha/2$.

(b) The standard deviation is unknown.

In this case the confidence interval for the distribution mean m is obtained by using the t distribution. Let $\bar{x}$ and s be the sample mean and standard deviation. The variable $t = (\bar{x} - m)\sqrt{n}/s$ has a t distribution with $\nu = n - 1$ degrees of freedom (see Section 2.3.2). For the variable t, the limits of the confidence interval (t_1, t_2) corresponding to the confidence probability P must satisfy the equation

$$P(t_1 < t < t_2) = P = \int_{t_1}^{t_2} f(t, \nu) \mathrm{d}t \tag{2.31}$$

where $f(t, \nu)$ is the frequency function for the variable t corresponding to ν degrees of freedom.

If we choose a confidence interval which is symmetrical with respect to the sample mean $\bar{x}$ we obtain $t_1 = t_2 = t_{\alpha/2}$, where $\alpha = 1 - P$ and $t_{\alpha/2}$ is obtained from the condition:

$$F(t_{\alpha/2}; \nu) = \int_{-\infty}^{t_{\alpha/2}} f(t, \nu) \mathrm{d}t = \alpha/2 \qquad (2.32)$$

Example 2.3 For the data given in example 2.2 calculate the confidence interval for the mean, corresponding to $P = 0.90$.

Because $\bar{x} = 0.00107$, $s = 0.00015$, $n = 10$, $t_{0.05;9} = 1.83$, the confidence interval of the sample mean will be:

$$0.00098 < m < 0.00116$$

2.5.2 The confidence interval of the sample variance

For estimating the variance of a random variable with a distribution $N(m, \sigma)$, we calculate the sample variance s^2 for a population $x_1, x_2, \ldots, x_n$ by (2.27). We know that the variable

$$\chi^2 = \frac{s^2 (n-1)}{\sigma^2} \qquad (2.33)$$

has a χ^2 distribution with $\nu = n - 1$ degrees of freedom; we shall write the corresponding probability density as $f(\chi^2, \nu)$.

For the χ^2 variable, the confidence interval (χ_1^2, χ_2^2) corresponding to a probability P must satisfy the equation:

$$P(\chi_1^2 < \chi^2 < \chi_2^2) = P = \int_{\chi_1^2}^{\chi_2^2} f(\chi^2, \nu) \mathrm{d}\chi^2 \qquad (2.34)$$

From (2.33) and (2.34) the confidence interval of the variance is given by

$$\frac{\nu s^2}{\chi_2^2} < \sigma^2 < \frac{\nu s^2}{\chi_1^2} \qquad (2.35)$$

The χ_1^2 and χ_2^2 values are obtained from the following conditions:

$$P(\chi^2 < \chi_1^2) = (1 - P)/2 \text{ and } P(\chi^2 < \chi_2^2) = (1 + P)/2$$

Example 2.4 For the data given in example 2.2, calculate the confidence interval of the variance corresponding to a confidence probability $P = 0.90$.

We have $n = 10$ and $s^2 = 223 \times 10^{-10}$. For $\nu = 9$ and $P = 0.90$, the condition $P(\chi^2 < \chi_1^2) = 0.05$ gives $\chi_1^2 = 3.33$, and the condition $P(\chi^2 < \chi_2^2) = 0.95$ gives $\chi_2^2 = 16.92$. Then, from (2.35) the confidence interval of the variance is:

$$119 \times 10^{-10} < \sigma^2 < 603 \times 10^{-10}$$

or, taking the square root:

$$11 \times 10^{-5} < \sigma < 25 \times 10^{-5}$$

This method of evaluating the confidence interval holds for any population, so there is no restriction on the number n. However, for $n \geqslant 30$ a simplified procedure can be used.

For a large n, it is proved that the standard deviation s is normally distributed with a mean equal to the general standard deviation σ and a standard deviation which is equal to $\sigma/\sqrt{2\nu}$. Consequently, we can assume that the variable

$$t = \frac{\sigma - s}{s/\sqrt{2\nu}}, \tag{2.36}$$

has a t distribution with $\nu = n - 1$ degrees of freedom.

Thus, according to Section 2.5.1(a), the confidence interval for the standard deviation corresponding to a probability P is

$$(s - |t_{\alpha/2}|\, s/\sqrt{2\nu}) < \sigma < (s + |t_{\alpha/2}|\, s/\sqrt{2\nu}) \tag{2.37}$$

where $\alpha = 1 - P$ and $t_{\alpha/2}$ is obtained from the condition $F(t_{\alpha/2}, \nu) = \alpha/2$.

Example 2.5 By using (2.37), for a confidence probability $P = 0.90$ and $n = 30$, the confidence interval for the standard deviation is:

$$0.78s < \sigma < 1.22s$$

By using the χ^2 distribution for solving this problem we obtain the interval:

$$0.82s < \sigma < 1.28s$$

Thus, the two methods give consistent results for the confidence interval of the general standard deviation.

2.5.3 Confidence interval of the binomial distribution parameters

We shall assume that if the experiment is repeated n times, the phenomenon of interest will take place m times. The best estimation of the probability p that the event occurs is given by the ratio $p' = m/n$.

The confidence interval which, with a probability P, contains the p parameter (a parameter of the binomial distribution) can be evaluated by several methods.

(a) In order to calculate the limits p_1 and p_2 we use the binomial distribution. These limits are obtained by solving the equations:

$$\sum_{x=m}^{n} \binom{n}{x} p_1^x (1-p_1)^{n-x} = \frac{1-P}{2} \tag{2.38}$$

$$\sum_{x=0}^{m} \binom{n}{x} p_2^x (1-p_2)^{n-x} = \frac{1-P}{2} \tag{2.39}$$

where

$$\binom{n}{x} = \frac{n!}{x!(n-x)!}$$

(b) The limits of the confidence interval can be evaluated by using the normal distribution as an approximation of the binomial distribution (see Section 2.3.4). In this case, the following equations give the limits of the confidence interval:

$$p_1 = \frac{1}{n+z_{\alpha/2}^2}\left[m - \frac{1}{2} + \frac{z_{\alpha/2}^2}{2} - |z_{\alpha/2}|\sqrt{\frac{(m-\frac{1}{2})(n-m+\frac{1}{2})}{n} + \frac{z_{\alpha/2}^2}{4}}\right] \tag{2.40}$$

$$p_2 = \frac{1}{n+z_{\alpha/2}^2}\left[m + \frac{1}{2} + \frac{z_{\alpha/2}^2}{2} + |z_{\alpha/2}|\sqrt{\frac{(m+\frac{1}{2})(n-m-\frac{1}{2})}{n} + \frac{z_{\alpha/2}^2}{4}}\right] \tag{2.41}$$

where $\alpha = 1 - P$, and $z_{\alpha/2}$ is obtained from the condition $F(z_{\alpha/2}) = \alpha/2$.

The following approximations can also be used.

$$p_1 = p' - |z_{\alpha/2}|\sqrt{\frac{p'(1-p')}{n}} \tag{2.42}$$

$$p_2 = p' + |z_{\alpha/2}|\sqrt{\frac{p'(1-p')}{n}} \tag{2.43}$$

Example 2.6 For testing a method of analysis, a material with known content of determinand has been analysed 50 times; 5 of the results had errors greater than the acceptable limits. Estimate the probability p' for a single result being satisfactory and find the corresponding confidence interval for $P = 0.95$.

The probability that the result is satisfactory is estimated by the ratio $p' = 45/50 = 0.90$.

By using the pairs of equations (2.38)-(2.39), (2.40)-(2.41) and (2.42)-(2.43) we find the following confidence intervals for the probability of obtaining a satisfactory result:

$$0.782 < p < 0.967$$

$$0.788 < p < 0.953$$

$$0.817 < p < 0.983$$

2.6 TESTING STATISTICAL HYPOTHESES

The Neyman and Pearson theory of tests used for checking hypotheses has a wide range of application in chemical analysis. In testing hypotheses, the decisions are based on samples drawn from the distributions investigated.

Consider the general case of a unidimensional distribution, where the frequency function $f(x, \theta, \ldots)$ depends on one parameter or several (the mean, variance, etc). In some cases of practical interest it must first be verified whether the parameters of the distributions investigated have certain values which are known in advance.

For example, we must test the hypothesis that the θ parameter of the distribution has the value θ_0. For testing this hypothesis, a sample is first taken from the investigated distribution. This sample is used to obtain an estimated value θ' for the θ parameter. The decision on the hypothesis is formulated by establishing the statistical significance of the estimated value θ' with respect to the θ_0 value of the θ parameter.

A problem of this kind arises in chemical analysis when establishing the significance of a systematic error. In such a case, measurements are made on samples which have a known content of the determinand, and the significance of the means with respect to the known values is established.

In other cases, we may require to know the significance of the sample variance with respect to certain values known in advance, or the significance of a calibration function with respect to a preselected value for it, or we must decide whether the parameters of two or several distributions have the same value.

For example, if there are two distributions for which we are testing the hypothesis that the θ parameter has the same value, we take a sample from each of these distributions and obtain the estimated values θ'_1 and θ'_2, then reach the decision by establishing the statistical significance of the difference between them. Such a problem arises in investigating the significance of the differences between the results obtained in two laboratories or derived by two different methods.

Two procedures can be applied for testing significance hypotheses.

(a) By the confidence interval. Consider the hypothesis that the θ parameter of the investigated distribution has the value θ_0. Let θ' be the estimated value for a sample for which the confidence interval for the θ parameter is (θ_1, θ_2) at a given conveniently chosen confidence probability P. If the θ_0 value lies within the confidence interval, there is no significant difference between the estimated value θ' and the θ_0 value of the parameter (Fig. 2.4a), so the hypothesis that θ' is equal to θ_0 cannot be rejected. However, if θ_0 lies outside the confidence interval (Fig. 2.4b) θ' is significantly different from it, and the hypothesis must be rejected.

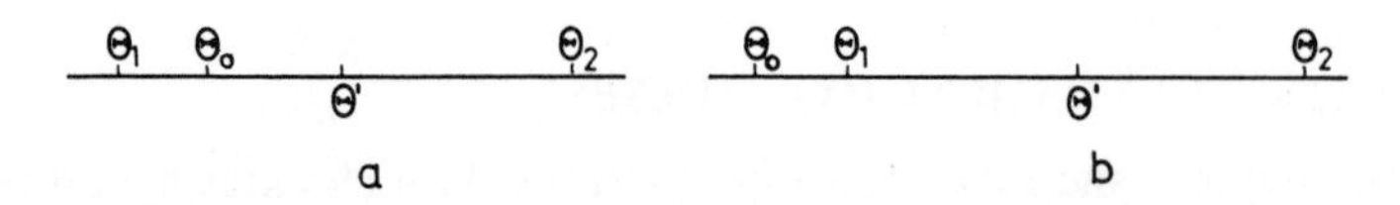

Fig. 2.4 The significance of the difference between an estimation value θ' and the known value θ_0 of a parameter when they are (a) not significantly different, (b) significantly different.

To verify the hypothesis that the parameters of two distributions have the same value a sample is taken for each distribution and the parameter and its confidence interval are evaluated for each. Let θ' and θ'', and (θ'_1, θ'_2) and (θ''_1, θ''_2), be the corresponding values. If the two confidence intervals overlap there is no significant difference between the estimated values θ' and θ'', so the hypothesis that the two distributions have the same value of the θ parameter cannot be rejected (Fig. 2.5a), but if the two confidence intervals do not overlap it must be rejected (Fig. 2.5b).

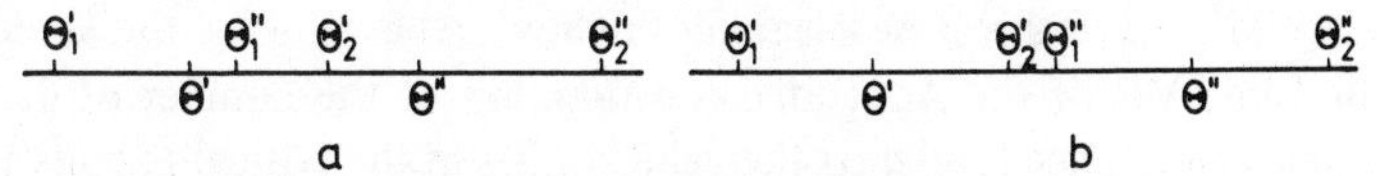

Fig. 2.5 The hypothesis that two distributions have the same Θ value (a) cannot be rejected, (b) must be rejected.

(b) From the statistics of the variables $(\theta_0 - \theta')$ and $(\theta' - \theta'')$. In this case we use the distribution function of the variables $(\theta_0 - \theta')$ and $(\theta' - \theta'')$. For a given significance level, critical regions are established such that if the variable $(\theta' - \theta_0)$ or $(\theta' - \theta'')$, lies within a critical region, the hypothesis is rejected, but it it lies outside the critical regions, the hypothesis is accepted.

If the test requires that the hypothesis be rejected, the results are said to be significantly different; otherwise the results are said not to differ significantly and, consequently, there are not sufficient reasons for rejecting the hypothesis.

If the hypothesis cannot be rejected by these tests, this means then that the results do not give enough evidence for a valid decision but they can be and are used for guiding action as if the hypothesis were true. Obviously, two types of error may be made in hypothesis testing: rejection of a true hypothesis (error of the first kind) and acceptance of a false hypothesis (error of the second kind).

To arrive at a more profound understanding of the process of decision taking we shall give various examples of hypothesis testing.

2.6.1 Testing extreme values

In a series of measurements, certain results may appear suspect. Such results should not be rejected on subjective criteria: statistical tests must be used.

(a) *Student's* t *criterion*

We shall assume that among n measurements of a property there is a doubtful result x_d. For testing the hypothesis that x_d belongs to the population of results x, two critical regions are defined by using the t distribution:

$$\frac{\bar{x} - x_d}{s\sqrt{\frac{n}{n-1}}} < -|t_{\alpha/2}|; \quad \frac{\bar{x} - x_d}{s\sqrt{\frac{n}{n-1}}} > |t_{\alpha/2}| \tag{2.44}$$

where $\bar{x}$ and s are the mean and the standard deviation for the $n - 1$ results other than the doubtful one; $t_{\alpha/2}$ is the t parameter for $(n - 1) - 1 = n - 2$ degrees of freedom (Table IV in the Appendix).

If the experimental value $(\bar{x} - x_d)/s\ \sqrt{n/(n-1)}$ lies between the critical regions defined above, the hypothesis that the result x_d belongs to the population of n results must be accepted.

(b) *Grubbs's* R *criterion*

The critical regions are defined by using the R criterion:

$$\frac{x_d - \bar{x}}{s} < -R_\alpha; \quad \frac{x_d - \bar{x}}{s} > R_\alpha \tag{2.45}$$

where $\bar{x}$ and s have the same meaning as above. The value of the variable R is given in Table VII of the Appendix as a function of the number of data n and the chosen significance level α. If the value x_d lies in the critical regions, it should be rejected,

(c) *Dixon's* Q *criterion*

This is a simple criterion for removing doubtful values. The results are first arranged in increasing order. The difference between the doubtful value x_d (which is either first or last in the series) and its neighbour x_{d-1} is divided by the difference between the first value x_1 and the last value x_n in the series, giving the ratio Q:

$$Q = \frac{x_d - x_{d-1}}{x_n - x_1} \tag{2.46}$$

If the value of Q is in either of the critical regions

$$\frac{x_d - x_{d-1}}{x_n - x_1} < -Q_\alpha \,; \frac{x_d - x_{d-1}}{x_n - x_1} > Q_\alpha \tag{2.47}$$

the doubtful value is rejected. The Q values are given in Table VIII in the Appendix.

Example 2.7 Consider the series of 10 results: 0.26, 0.21, 0.21, 0.20, 0.21, 0.19, 0.18, 0.17, 0.18, 0.19. The value $0.26 = x_d$ will be regarded as a dubious result.

The t *criterion.* From the series of $n-1=9$ values we obtain $\bar{x}=0.193$ and $s = 0.015$. For $\alpha = 0.05$ and $\nu = 10 - 2 = 8$, $t_{0.05;8} = 2.31$, so from (2.44) we have:

$$\frac{x_d - \bar{x}}{s\sqrt{n/(n-1)}} < -2.31; \frac{x_d - \bar{x}}{s\sqrt{n/(n-1)}} > 2.31$$

Thus for $x_d = 0.26$, $(x_d - \bar{x})/s\sqrt{n/(n-1)} = 4.24$ which is in a critical region, so the value $x_d = 0.26$ should be rejected.

The R *criterion.* For $\alpha = 0.05$, $R_{0.05;10} = 3.54$. From (2.45) it results that

$$\frac{x_d - \bar{x}}{s} < -3.54 \,; \frac{x_d - \bar{x}}{s} > 3.54$$

Taking $x_d = 0.26$ gives $(x_d - \bar{x})/s = 4.46$ which is in a critical region, so the result $x_d = 0.26$ must be rejected.

The Q *criterion.* For $\alpha = 0.05$, the table gives $Q_{0.05;10} = 0.412$, so from (2.47):

$$\frac{x_d - x_{d-1}}{x_n - x_1} < -0.412 \; ; \; \frac{x_d - x_{d-1}}{x_n - x_1} > 0.412$$

For $x_d = 0.26$, $(x_d - x_{d-1})/(x_n - x_1) = 0.556$ which is within a critical region, so the result $x_d = 0.26$ is again rejected.

2.6.2 Testing the mean

2.6.2.1 *Testing the mean with respect to a known value*

There can be two distinct situations for such a statistical problem.

(a) The standard deviation σ of the distribution to which the sample belongs has a known value (the z test).

Consider that the sample population $x_1, x_2, \ldots, x_n$ belongs to a distribution $N(m, \sigma)$. The quantity $z = (\bar{x} - m)\sqrt{n}/\sigma$ has a distribution $N(0, 1)$; by choosing a convenient significance level α, we can find a value $z_{\alpha/2}$ such that

$$P(-|z_{\alpha/2}| < \frac{\bar{x} - m}{\sigma}\sqrt{n} < |z_{\alpha/2}|) = 1 - \alpha \tag{2.48}$$

The value $z_{\alpha/2}$ is obtained from the condition $F(z_{\alpha/2}) = \alpha/2$. In this way, the following critical regions are defined:

$$\frac{\bar{x} - m}{\sigma}\sqrt{n} < -|z_{\alpha/2}| \; ; \; \frac{\bar{x} - m}{\sigma}\sqrt{n} > |z_{\alpha/2}| \tag{2.49}$$

Now, if $z = (\bar{x} - m)\sqrt{n}/\sigma$ is in a critical region, the hypothesis that the mean of the distribution from which the experimental data were obtained is equal to m is rejected. In other words, the sample mean $\bar{x}$ is significantly different from m.

(b) The standard deviation of the distribution is unknown (the t test).

Let $\bar{x}$ and s be the mean and standard deviation of the sample data $x_1, x_2, \ldots, x_n$, which belong to a distribution $N(m, \sigma)$. The variable $t = (\bar{x} - m)\sqrt{n}/s$ has a t distribution with $\nu = n - 1$ degrees of freedom. Hence, for the function $f(t, \nu)$ we can define a critical region such that if $t = (\bar{x} - m)\sqrt{n}/s$ falls within it, the hypothesis that the distribution from which the sample is extracted has a mean value m should be rejected. Hence, by conveniently choosing a significance level α, we can write

$$P(-|t_{\alpha/2\,;\,\nu}| < \frac{\bar{x} - m}{s}\sqrt{n} < |t_{\alpha/2\,;\,\nu}|) = 1 - \alpha \tag{2.50}$$

Consequently, the critical regions for rejection of the hypothesis will be

$$\frac{\bar{x}-m}{s}\sqrt{n} < -|t_{\alpha/2\,;\,\nu}|\,;\;\frac{\bar{x}-m}{s}\sqrt{n} > |t_{\alpha/2\,;\,\nu}| \tag{2.51}$$

The problem of testing a sample mean $\bar{x}$ with respect to a preselected value m has many applications in chemical analysis. For example, to establish the significance of a systematic error, a reference material is analysed several times. From the known amount of determinand in the reference material, and the collection of results, the experimental value of t is calculated. If this value falls in a critical region the hypothesis that there is a systematic error should be accepted.

Example 2.8 For the data given in example 2.2, at a significance level $\alpha = 0.10$ test the hypothesis that the results belong to a distribution which has a mean $m = 0.0010\%$.

The mean and standard deviation of the 10 results are $\bar{x} = 0.00107$ and $s = 0.00015$. Hence the experimental value of t is -1.47.

For $\alpha/2 = 0.05$ and $\nu = 9$ we obtain $t_{0.05;9} = -1.83$ from the t distribution table. The hypothesis is rejected if the experimental value of t lies in the following critical regions:

$$t < -1.83\,;\; t > 1.83$$

Because the experimental value $t = -1.47$ lies outside the critical regions the hypothesis cannot be rejected.

2.6.2.2 *Testing two means*

There are three distinct situations in this case:

(a) The two distributions to which the sample populations belong have known standard deviations σ_1 and σ_2 (the z test).

Let $x_{11}, x_{12}, \ldots, x_{1n_1}$ and $x_{21}, x_{22}, \ldots, x_{2n_2}$ be taken from the distributions $N(m_1, \sigma_1)$ and $N(m_2, \sigma_2)$. The hypothesis to be verified is that $m_1 = m_2$ (the standard deviations σ_1 and σ_2 are already known). It can be demonstrated that if $m_1 = m_2$, the variable:

$$z = \frac{\bar{x}_1 - \bar{x}_2}{\sigma_{(\bar{x}_1 - \bar{x}_2)}} \tag{2.52}$$

has a distribution $N(0, 1)$. Here $\bar{x}_1$ and $\bar{x}_2$ are the means of the two populations, and

$$\sigma_{(\bar{x}_1 - \bar{x}_2)} = \sqrt{\frac{\sigma_1^2}{n_1} + \frac{\sigma_2^2}{n_2}} \tag{2.53}$$

is the standard deviation corresponding to the difference of the means of the populations of sizes n_1 and n_2.

For a given significance level α, the critical regions for rejection of the hypothesis that the distributions from which the two populations were taken have equal means m_1 and m_2 are

$$\frac{\bar{x}_1 - \bar{x}_2}{\sigma_{(\bar{x}_1-\bar{x}_2)}} < -|z_{\alpha/2}|\,;\quad \frac{\bar{x}_1 - \bar{x}_2}{\sigma_{(\bar{x}_1-\bar{x}_2)}} > |z_{\alpha/2}| \tag{2.54}$$

(b) The standard deviations of the two distributions are unknown, but are assumed to be equal (the t test).

Let $x_{11}, x_{12}, \ldots, x_{1n_1}$ and $x_{21}, x_{22}, \ldots, x_{2n_2}$ be drawn from the distributions $N(m_1, \sigma)$ and $N(m_2, \sigma)$, which are assumed to have equal (though unknown) standard deviations.

It is known that if the means and the standard deviations both have equal values for the two distributions then the variable

$$t = \frac{\bar{x}_1 - \bar{x}_2}{\bar{s}\sqrt{\dfrac{1}{n_1} + \dfrac{1}{n_2}}} \tag{2.55}$$

has a t distribution with $\nu = n_1 + n_2 - 2$ degrees of freedom (see Section 2.3.2). Here $\bar{s}$ is the weighted standard deviation of the two values s_1 and s_2, defined by

$$\bar{s} = \sqrt{\frac{(n_1-1)s_1^2 + (n_2-1)s_2^2}{n_1 + n_2 - 2}} \tag{2.56}$$

From the comment just made about the t test, it follows that the critical regions (associated with a significance level α) for rejection of the hypothesis that the two distributions have equal means will be

$$\frac{\bar{x}_1 - \bar{x}_2}{\bar{s}\sqrt{\dfrac{1}{n_1}+\dfrac{1}{n_2}}} < -|t_{\alpha/2\,;\,\nu}|\,;\quad \frac{\bar{x}_1 - \bar{x}_2}{\bar{s}\sqrt{\dfrac{1}{n_1}+\dfrac{1}{n_2}}} > |t_{\alpha/2\,;\,\nu}| \tag{2.57}$$

In other words, if the experimental value of t (i.e. the value calculated from (2.55)) lies in a critical region, we shall reject the hypothesis that the two sample means $\bar{x}_1$ and $\bar{x}_2$ are valid estimations for the one mean value.

Example 2.9 By using the t test and a significance level $\alpha = 0.05$, establish the significance of the difference between the mean values of the following two series of analysis results:

I. 0.019; 0.023; 0.022; 0.015; 0.017; 0.016; 0.022; 0.017; 0.019; 0.017; 0.019; 0.020; 0.019; 0.019; 0.019; 0.015; 0.021; 0.020; 0.022; 0.024

II. 0.019; 0.019; 0.023; 0.020; 0.020; 0.020; 0.025; 0.023; 0.019; 0.019; 0.020; 0.021; 0.020; 0.017; 0.023; 0.023; 0.020; 0.020; 0.021; 0.017

The means and the standard deviations are:

$\bar{x}_1 = 0.0192; \bar{x}_2 = 0.0205; s_1 = 0.0026; s_2 = 0.0022$

From these we obtain the experimental value $t = 1.71$. For a significance level $\alpha = 0.05$ and $\nu = 38$ degrees of freedom we obtain $t_{0.025;38} = -2.00$. Hence, the hypothesis that the two distributions have the same mean will be rejected within the critical regions

$$t < -2.00; t > 2.00$$

Because the experimental result $t = 1.71$ does not lie in a critical region, we decide that the two series of results are not significantly distinguishable, so that both $\bar{x}_1$ and $\bar{x}_2$ are estimations of the same mean value.

(c) The two distributions for which the means are to be tested have non-equal standard deviations (the approximate t test). We consider two samples taken from the distributions $N(m_1, \sigma_1)$ and $N(m_2, \sigma_2)$ with $\sigma_1 \neq \sigma_2$. It has been proved that if $m_1 = m_2$ the variable

$$t = \frac{\bar{x}_1 - \bar{x}_2}{\sqrt{\dfrac{s_1^2}{n_1} + \dfrac{s_2^2}{n_1}}} \tag{2.58}$$

has a t distribution with ν degrees of freedom with

$$\nu = \frac{1}{c^2/(n_1-1) + (1-c)^2/(n_2-1)} \tag{2.59}$$

where

$$c = \frac{s_1^2/n_1}{s_1^2/n_1 + s_2^2/n_2} \tag{2.60}$$

From the discussion of the comparison of two samples by using the t test, it results that the hypothesis that the two distribution means are equal will be rejected if t lies in the critical regions which, for a given significance level α, are defined by

$$\frac{\bar{x}_1 - \bar{x}_2}{\sqrt{\dfrac{s_1^2}{n_1} + \dfrac{s_2^2}{n_2}}} < -|t_{\alpha/2\,;\,\nu}|\,; \quad \frac{\bar{x}_1 - \bar{x}_2}{\sqrt{\dfrac{s_1^2}{n_1} + \dfrac{s_2^2}{n_2}}} > |t_{\alpha/2\,;\,\nu}| \tag{2.61}$$

2.6.3 Testing sample variances

Another problem in chemical analysis is that of testing sample variances. This is done in order to evaluate and compare the reproducibility of analytical methods and also to identify and evaluate the factors causing errors.

In what follows, we shall consider a few such problems, limiting ourselves to the most important examples in chemical analysis.

2.6.3.1 *Testing a sample variance with respect to a known value*

Let $x_1, x_2, \ldots, x_n$ be a sample population of a distribution $N(m, \sigma)$, and verify the hypothesis that the variance of the distribution to which the sample belongs is $\sigma^2 = \sigma_0^2$.

Let s^2 be the sample variance, evaluated from the sample data. If the hypothesis were true (i.e. if the sample were taken from a distribution with variance σ_0^2), the variable

$$\chi^2 = \frac{(n-1)s^2}{\sigma_0^2} \tag{2.62}$$

would have a χ^2 distribution with $\nu = n - 1$ degrees of freedom.

For a significance level α, this hypothesis is rejected for χ^2 values in the following critical regions.

$$\frac{(n-1)s^2}{\sigma_0^2} < \chi^2_{\alpha/2;\,\nu}\,; \quad \frac{(n-1)s^2}{\sigma_0^2} > \chi^2_{(1-\alpha/2);\,\nu} \tag{2.63}$$

The quantities $\chi^2_{\alpha/2;\,\nu}$ and $\chi^2_{(1-\alpha/2);\,\nu}$ are obtained from the conditions:

$P(\chi^2 < \chi^2_{\alpha/2;\nu}) = \alpha/2$ and $P(\chi^2 < \chi^2_{(1-\alpha/2);\nu}) = (1 - \alpha/2)$.

Hence, if the experimental variable $\chi^2 = (n-1)s^2/\sigma_0^2$ is in a critical region, the hypothesis will be rejected and we shall decide that the sample variance s^2 is significantly different from σ_0^2.

Example 2.10 For a significance level $\alpha = 0.10$ assess the hypothesis that the analytical results given in example 2.2 belong to a distribution with variance $\sigma_0^2 = 100 \times 10^{-10}$.

From the 10 experimental results the sample variance s^2 is 223×10^{-10}. The conditions $P(\chi^2 < \chi^2_{0.05;9}) = 0.05$ and $P(\chi^2 < \chi^2_{0.95;9}) = 0.05$ give $\chi^2_{0.05;9} = 3.33$ and $\chi^2_{0.95;9} = 16.9$.

Hence, the critical regions where the hypothesis is rejected are:

$$\frac{(n-1)s^2}{\sigma_0^2} < 3.33\,; \quad \frac{(n-1)s^2}{\sigma_0^2} > 16.9$$

The experimental value $(n-1)s^2/\sigma_0^2 = 20.07$ is within the upper region, and we say that the sample variance $s^2 = 223 \times 10^{-10}$ is significantly different from $\sigma_0^2 = 100 \times 10^{-10}$. Hence, the hypothesis must be rejected.

2.6.3.2 *Testing two sample variances (the* F *test)*

Assume that two random variables with the distribution functions $N(m_1, \sigma_1)$ and $N(m_2, \sigma_2)$ are given, and two population samples $x_{11}, x_{12}, \ldots, x_{1n_1}$, and $x_{21}, x_{22}, \ldots, x_{2n_2}$ are taken from the two distributions. The problem is to verify that the two variances are equal, i.e. $\sigma_1^2 = \sigma_2^2$.

It was proved in Section 2.3.3 that if two sample variances s_1^2 and s_2^2 obtained from n_1 and n_2 independent experiments are estimating the same common variance σ^2, their ratio $F = s_1^2/s_2^2$ has an F distribution with $\nu_1 = n_1 - 1$ and $\nu_2 = n_2 - 1$ degrees of freedom.

Consequently, by using the F distribution for a significance level α, we can evaluate the critical regions:

$$\frac{s_1^2}{s_2^2} < F_{\alpha/2;\,\nu_1,\,\nu_2}\,;\quad \frac{s_1^2}{s_1^2} > F_{(1-\alpha/2);\,\nu_1,\,\nu_2} \tag{2.64}$$

such that, if the experimental variable $F = s_1^2/s_2^2$ falls in either region, the hypothesis that the two sample variances s_1^2 and s_2^2 estimate the same common variance must be rejected.

The values $F_{\alpha/2;\,\nu_1,\,\nu_2}$ and $F_{(1-\alpha/2)\,;\nu_1,\,\nu_2}$ are obtained from the conditions:

$$P(F_{\nu_1,\,\nu_2} < F_{\alpha/2;\,\nu_1,\,\nu_2}) = \alpha/2$$

and

$$P(F_{\nu_1,\,\nu_2} < F_{(1-\alpha/2);\,\nu_1,\,\nu_2}) = (1-\alpha/2)$$

Example 2.11 By using the F test and a significance level $\alpha = 0.10$ verify the hypothesis that the two series of data given in example 2.9 belong to distributions with equal variances.

We obtain $s_1^2 = 676 \times 10^{-8}$ and $s_2^2 = 484 \times 10^{-8}$. As $s_1^2 > s_2^2$, the experimental F value will be $s_1^2/s_2^2 = 1.40 = F$. The critical region for the significance level $\alpha = 0.10$ and $\nu_1 = \nu_2 = 19$ degrees of freedom will be:

$$\frac{s_1^2}{s_2^2} > F_{0.05;\,19,19}$$

As $F_{0.05;19,19} = 2.11$ the experimental value $s_1^2/s_2^2 = 1.40$ is outside the critical region, so that the hypothesis that the two sample variances are estimating the same general variance is accepted.

2.6.3.3 *Testing several sample variances (Bartlett's test)*

Suppose that there is an arbitrary number of sample variances $s_1^2, s_2^2, \ldots, s_n^2$ with $\nu_1, \nu_2, \ldots, \nu_n$ degrees of freedom. The hypothesis that all these sample variances estimate the same general variance can be checked by using Bartlett's test.

In this case the variable is

$$B = \frac{1}{c}\left(\nu \ln \bar{s}^2 - \sum_{i=1}^{n} \nu_i \ln s_i^2\right) \tag{2.65}$$

where

$$\nu = \nu_1 + \nu_2 + \ldots + \nu_n$$

$$\bar{s}^2 = \frac{\sum_{i=1}^{n} \nu_i s_i^2}{\nu}$$

$$c = 1 + \frac{\left(\sum_{i=1}^{n} \frac{1}{\nu_i} - \frac{1}{\nu}\right)}{3(n-1)}$$

For $\nu_i \geqslant 5$, the variable B has a χ^2 distribution with $\nu - 1$ degrees of freedom. Consequently, by using the χ^2 distribution we can evaluate, for a significance level α, critical regions such that, if they do not contain the experimental value of B, the hypothesis that all the sample variances $s_1^2, s_2^2, \ldots, s_n^2$ estimate the same common variance will be accepted.

Obviously, the critical regions are

$$B < \chi^2_{\alpha/2;\,(\nu-1)};\, B > \chi^2_{(1-\alpha/2);\,(\nu-1)} \tag{2.66}$$

2.6.4 Testing sampling probabilities

Some analytical problems require examination of the sampling probabilities resulting from investigation of various phenomena.

2.6.4.1 *Testing sampling probability with respect to a known value*

We shall assume that the phenomenon of interest occurs m' times in n repetitions of the experiment. For large values of n, taking the normal distribution as approximating the binomial distribution, we find that the variable $p' = m'/n$, which estimates the probability p for the occurrence of the phenomenon in any one experiment, has a distribution $N(m, \sigma)$, with:

$$m = p \text{ and } \sigma = \sqrt{\frac{p(1-p)}{n}}$$

Consequently, the variable

$$z = \frac{p' - p}{\sqrt{\frac{p(1-p)}{n}}}$$

has a distribution $N(0, 1)$.

Hence, for a significance level α, we can evaluate critical regions such that, if z falls within either, then the hypothesis that the sampling probability p' estimates the probability p, must be rejected. The critical regions will be

$$\frac{p' - p}{\sqrt{\frac{p(1-p)}{n}}} < -|z_{\alpha/2}|\,; \quad \frac{p' - p}{\sqrt{\frac{p(1-p)}{n}}} > |z_{\alpha/2}| \tag{2.67}$$

Example 2.12 Of 40 repetitions of the analysis of a reference material 25 gave results exceeding the true value. For a significance level $\alpha = 0.10$, test the hypothesis that there is no systematic error in the analysis.

Obviously, in the absence of systematic errors, the probability of obtaining a positive error is equal to the probability for a negative error, so $p = 1/2$. Then, by (2.67), the critical regions in which the hypothesis is rejected will be

$$\frac{p' - p}{\sqrt{\frac{p(1-p)}{n}}} < -1.67; \quad \frac{p' - p}{\sqrt{\frac{p(1-p)}{n}}} > 1.67$$

The experimental value $\dfrac{p' - p}{\sqrt{\frac{p(1-p)}{n}}} = 1.56$ is outside the critical regions,

so the hypothesis that systematic errors are absent cannot be rejected.

2.6.4.2 *Testing two sampling probabilities*

We shall now consider two independent series of experiments. In the first series, after n_1 experiments, the phenomenon investigated has occurred m_1 times, while in the second series, after n_2 experiments, the phenomenon has occurred m_2 times. Let us assess whether the probabilities $p_1 = m_1/n_1$ and $p_2 = m_2/n_2$ estimate the same common probability.

For solving this problem, the binomial distribution will be approximated by the normal distribution, for large values of n_1 and n_2.

It has been proved that when the hypothesis is true, the variable $(p_1 - p_2)$ will have a distribution $N(0, \sigma)$, where

$$\sigma = \sqrt{p(1-p)(1/n_1 + 1/n_2)} \text{ and } p = \frac{m_1 + m_2}{n_1 + n_2}$$

Hence, the variable

$$z = \frac{p_1 - p_2}{\sqrt{p(1-p)(1/n_1 + 1/n_2)}} \tag{2.68}$$

has a distribution $N(0, 1)$.

The approximation can be improved by taking into account continuity corrections. Thus, if $p_1 \geqslant p_2$, the equation (2.68) can be written as

$$z = \frac{(p_1 - 1/2n_1) - (p_2 + 1/2n_2)}{\sqrt{p(1-p)(1/n_1 + 1/n_2)}} \tag{2.69}$$

Consequently, in order to test the hypothesis that the sampling probabilities p_1 and p_2 estimate the same common probability, for a significance level α, the following critical regions are defined:

$$\frac{(p_1 - 1/2n_1) - (p_2 + 1/2n_2)}{\sqrt{p(1-p)(1/n_1 + 1/n_2)}} < -|z_{\alpha/2}|; \tag{2.70}$$

$$\frac{(p_1 - 1/2n_1) - (p_2 + 1/2n_2)}{\sqrt{p(1-p)(1/n_1 + 1/n_2)}} > |z_{\alpha/2}|$$

If the experimental value evaluated from (2.69) falls in either critical region, the hypothesis that p_1 and p_2 estimate the same common probability will be rejected.

Example 2.13 In order to compare the performance of two methods of analysis, a sample of reference material is repeatedly analysed, 34 times (n_1) by the first method and 30 times (n_2) by the second. When two error limits are set (one each side of the true amount) 8 results (m_1) exceed them in the case of the first method and 5 (m_2) in the case of the second. Test the hypothesis that the results outside the admissible error limits occur with equal probability in the two methods.

We have the estimations $p_1 = 8/34$ and $p_2 = 5/30$ for the corresponding probabilities. From (2.70), with a significance level $\alpha = 0.10$, the hypothesis that the probabilities are equal will be rejected in the following critical regions:

$$\frac{(p_1 - 1/2n_1) - (p_2 + 1/2n_2)}{\sqrt{p(1-p)(1/n_1 + 1/n_2)}} < -1.67; \quad \frac{(p_1 - 1/2n_1) - (p_2 + 1/2n_2)}{\sqrt{p(1-p)(1/n_1 + 1/n_2)}} > 1.67$$

Because the experimental value $\dfrac{(p_1 - 1/2n_1) - (p_2 + 1/2n_2)}{\sqrt{p(1-p)(1/n_1 + 1/n_2)}} = 0.36$ lies outside the critical regions, the hypothesis cannot be rejected.

2.6.5 Testing hypotheses about the type of the distribution function

For extracting the desired information from a set of analytical measurements by statistical methods, a prerequisite is knowledge of the nature of their distribution function. Hence hypotheses concerning the type of distribution function must be tested. We shall give here two statistical tests used for this purpose.

2.6.5.1 *The* χ^2 *test*

In Section 2.3.1 it was shown that for a random variable x for which the possible values x_i are taken with probabilities p_i, if $n_1, n_2, \ldots, n_m$ are the absolute frequencies of these values appearing in $n = n_1 + n_2 + \ldots + n_m$ experiments, the variable:

$$\chi^2 = \sum_{i=1}^{m} \frac{(n_i - np_i)^2}{np_i} \tag{2.71}$$

has a χ^2 distribution with $\nu = m-1$ degrees of freedom.

Consequently, for testing the hypothesis that the experimental frequencies n_i estimate the theoretical frequencies np_i we can evaluate, by using the χ^2 distribution, the following critical region corresponding to a significance level α:

$$\sum_{i=1}^{m} \frac{(n_i - np_i)^2}{np_i} > \chi^2_{(1-\alpha);\nu} \tag{2.72}$$

Example 2.14 The results given in Table 2.2 (the first two columns) were obtained for a sample of steel with an X-ray fluorescence spectrometer. Test for a significance level $\alpha = 0.05$, the hypothesis that the results belong to a normal distribution.

The calculations are done as shown in Table 2.2.

Because the experimental value $\chi^2 = 1.877 < 9.49 = \chi^2_{0.95;4}$, the hypothesis that the results are normally distributed cannot be rejected.

Table 2.2 Testing the hypothesis of the normality of the measurement results by using the χ^2 test

Measured result (digits)	Absolute frequency	Frequency within intervals (n_i)	$z_i = \frac{x_i - \bar{x}}{\sigma}$	$F(z_i) = F(x_i)$	Probability within intervals (p_i)	$\frac{(n_i - np_i)^2}{np_i}$
1759	0					
1760	1					
1761	0	10			0.100	0.999
1762	2					
1763	5					
1764	2					
			−1.288	0.100		
1765	2					
1766	1					
1767	5	14			0.242	0.761
1768	1					
1769	5					
			−0.405	0.342		
1770	5					
1771	6					
1772	5	25			0.342	0.000
1773	7					
1774	2					
			+0.479	0.684		
1775	5					
1776	3					
1777	3	18			0.229	0.098
1778	4					
1779	3					
			+1.362	0.913		
1780	2					
1781	0					
1782	2	6			0.087	0.019
1783	0					
1784	2					
1785	0					
						$\chi^2_{\text{exp.}} = 1.877$

2.6.5.2 *Kolmogorov's test*

Let $F(x)$ be the theoretical distribution function of a variable, and $F_n(x)$ the the empirical distribution function of the same variable, obtained by repeating the experiment. It has been proved that

$$\lim_{n \to \infty} P[\sup|F_n(x) - F(x)| \leqslant \frac{\lambda}{\sqrt{n}}] = K(\lambda) = \sum_{\nu=-\infty}^{+\infty} (-1)^{\nu} e^{-2\nu^2\lambda^2} \quad (2.73)$$

For sufficiently large values of n, there is a test which can be used for verifying the hypothesis that the empirical function $F_n(x)$ estimates the theoretical function $F(x)$.

The critical region where the hypothesis is rejected is given by

$$\sup |F_n(x) - F(x)| > \frac{\lambda}{\sqrt{n}} \quad (2.74)$$

The values of the Kolmogorov function $K(\lambda)$ are given in Table IX in the Appendix.

Example 2.15 For the data of the previous example use Kolmogorov's test at a significance level $\alpha = 0.05$, to verify the hypothesis that the results in Table 2.3 belong to a normal distribution.

The mean of the results is $\bar{x} = 1771.7$ digits, and the standard deviation is $\sigma = 5.66$ digits. Hence, we have to verify the hypothesis that the results belong to the distribution $N(1771.7;5.66)$.

The values of the distribution function $F(x)$, calculated by using the intermediary variable $z = (x - \bar{x})/\sigma$ and the tabulated values $F(z)$ (Table II in the Appendix), are given in Table 2.3 along with the empirical values $F_n(x)$ and the differences $|F_n(x) - F(x)|$. For the significance level $\alpha = 1 - K(\lambda) = 0.05$, $\lambda = 1.36$. Thus, for $n = 73$, $\sup|F(x) - F_n(x)| = 0.053 < 0.159 = \lambda/\sqrt{n}$, so the hypothesis that the tabulated results belong to a normal distribution cannot be rejected.

2.6.6 Variance analysis

Variance analysis was introduced into mathematical statistics by Fisher, and has a wide range of applications in biology, chemistry, sociology, psychology, etc., and also a wide range of technical applications.

In chemical analysis, the methods of variance analysis are successfully used for identification and evaluation of errors and their causes.

Variance analysis is based on a very simple idea, namely that for a given random variable, comparison of its variances estimated in the presence and absence of a factor acting on it will give information about the effect of the factor. If the variance in presence of the factor is significantly the greater we

Table 2.3 Testing the hypothesis of the normality of measurement results by Kolmogorov's test

Measurement results (digits)	Frequency of the results	$z = \dfrac{x-\bar{x}}{\sigma}$	$F(z) = F(x)$	$F_n(x)$	$\lvert F(x)-F_n(x)\rvert$
1760	1	−2.07	0.019	0.014	0.005
1761	0	−1.86	0.031	0.014	0.017
1762	2	−1.71	0.044	0.041	0.003
1763	5	−1.54	0.062	0.109	0.047
1764	2	−1.36	0.087	0.137	0.050
1765	2	−1.18	0.119	0.164	0.045
1766	1	−1.01	0.156	0.177	0.021
1767	5	−0.83	0.203	0.246	0.043
1768	1	−0.65	0.258	0.260	0.002
1769	5	−0.48	0.316	0.328	0.012
1770	5	−0.30	0.382	0.400	0.018
1771	6	−0.12	0.452	0.479	0.027
1772	5	0.05	0.520	0.548	0.028
1773	7	0.23	0.591	0.644	*0.053*
1774	2	0.41	0.659	0.671	0.012
1775	5	0.58	0.719	0.739	0.020
1776	3	0.76	0.776	0.780	0.004
1777	3	0.94	0.826	0.822	0.004
1778	4	1.11	0.867	0.877	0.010
1779	3	1.29	0.901	0.926	0.025
1780	2	1.47	0.929	0.945	0.016
1781	0	1.64	0.949	0.945	0.004
1782	2	1.82	0.966	0.972	0.006
1783	0	2.00	0.977	0.972	0.005
1784	2	2.17	0.985	1.000	0.015

decide that the factor has an influence on the random variable, the size of the effect being measured by the difference between the two variances.

More generally, variance analysis can be used to assess the relative importance of several simultaneously acting factors.

2.6.6.1 *Single-factor variance analysis*

Single-factor variance analysis can be used to compare statistically r sample means ($r > 2$) to assess the influence of one factor.

We shall assume that the r populations from which the samples of sizes $n_1, n_2, \ldots, n_r$ have been taken (to test the hypothesis of the equality of their means, i.e. $m_1 = m_2 = \ldots = m_r$) have the distributions $N(m_i, \sigma)$, where $1 \leqslant i \leqslant r$. Consequently, the data to be processed belong to the model

$$x_{ij} = m_i + e_{ij} \tag{2.75}$$

where $1 \leqslant i \leqslant r$ and $1 \leqslant j \leqslant n_r$, and e_{ij} has a distribution $N(0, \sigma)$.

When the r samples are of equal size, i.e. $n_1 = n_2 = \ldots = n_r = n$, the data to be processed have the structure given in Table 2.4. Thus, the factor acting upon the variable is given r different values, and we wish to investigate the effect of this on the variable considered.

Table 2.4 The data to be treated by single-factor variance analysis

j	i				
	1	2	— — —	$r-1$	r
1	x_{11}	x_{21}	— — —	$x_{(r-1)1}$	x_{r1}
2	x_{12}	x_{22}	— — —	$x_{(r-1)2}$	x_{r2}
—	—	—	— — —	—	—
—	—	—	— — —	—	—
—	—	—	— — —	—	—
$n-1$	$x_{1(n-1)}$	$x_{2(n-1)}$	— — —	$x_{(r-1)(n-1)}$	$x_{r(n-1)}$
n	x_{1n}	x_{2n}	— — —	$x_{(r-1)n}$	x_{rn}

The procedure and method of establishing the significance of the factor are given concisely in Table 2.5. The following notations are used in the table:

$$\bar{x}_{i.} = \frac{\sum_{j=1}^{n_j} x_{ij}}{n_i}$$ — the mean of the i groups of measurements (a sample characteristic)

$$\bar{x}_{..} = \frac{\sum_{i=1}^{r}\sum_{j=1}^{n_i} x_{ij}}{\sum_{i=1}^{r} n_i}$$ — the mean of all the measurements (a sample characteristic)

Table 2.5 Single-factor variance analysis

Source of variance	Sums of squares	Degrees of freedom	Mean of the sum of squares	Estimated value	F
Between the groups	$S_2 = \sum_{i=1}^{r} n_i(\bar{x}_{i.} - \bar{x}_{..})^2$	$r-1$	$s_2^2 = \dfrac{S_2}{r-1}$	$\sigma^2 + \dfrac{1}{r-1} \sum_{i=1}^{r} n_i (m_i - m)^2$	$\dfrac{s_2^2}{s_1^2}$
Within the groups	$S_1 = \sum_{i=1}^{r} \sum_{j=1}^{n_i} (x_{ij} - \bar{x}_{i.})^2$	$\left(\sum_{i=1}^{r} n_i\right) - r$	$s_1^2 = \dfrac{S_1}{\sum_{i=1}^{r} n_i - r}$	σ^2	
Total	$S = \sum_{i=1}^{r} \sum_{j=1}^{n_i} (x_{ij} - \bar{x})^2$	$\left(\sum_{i=1}^{r} n_i\right) - 1$		$F_{\alpha; r-1, \sum_{i=1}^{r} n_i - r}$	

$$m = \frac{\sum_{i=1}^{r} m_i}{r}$$ – the mean of the means of the r distributions (unknown quantity)

m_i – the means of the r distributions investigated (unknown quantities)

The difference $(m_i - m)$ measures the effect of the variation of the investigated factor on the observed variable.

If the hypothesis $m_1 = m_2 = \ldots = m_r = m$ is true, i.e. if variation of the factor has no influence on the observed variable, then both s_1^2 and s_2^2 will be two independent estimations of σ^2, so the variable $F = s_2^2/s_1^2$ will have an F distribution with $(r-1)$ and $(\sum_{i=1}^{r} n_i) - r$ degrees of freedom.

Consequently, for a given significance level α, the critical region for rejection of the hypothesis of equal means of the r distributions will be

$$s_2^2/s_1^2 > F_{\alpha;\,(r-1),\,(\sum_{i=1}^{r} n_i) - r} \tag{2.76}$$

Hence, if the experimental value $F = s_2^2/s_1^2$ is outside the critical region, it will be concluded that the factor investigated has no influence on the observed variable.

Example 2.16 In an emission spectral analysis by densitometry, the difference in density (six samples) of the line-pair $\lambda_{Mn} = 2576.1$ Å; $\lambda_{Co} = 2583.1$ Å as a result of variation in the amount of Fe, Cr, and Ni present gave the data in Table 2.6.

Table 2.6 Determination of the significance of the influence of the amount of iron, chromium, and nickel on the line pair $\lambda_{Mn} = 2576.1$ Å and $\lambda_{Co} = 2583.1$ Å

	Difference in density $\Delta S \times 10^2$					
Sample \ Matrix	1	2	3	4	5	6
1	14	17	13	15	15	17
2	14	13	14	16	17	15
3	15	13	13	19	18	14
4	17	17	15	15	17	14
5	17	17	15	16	16	17
6	17	17	17	14	17	16

To conclude that the amount of Fe, Cr, and Ni does not affect the results (ΔS), we have to verify the hypothesis that the means of the six groups of results estimate the same general mean, i.e. they are equal. This can be done by single-factor variance analysis, as shown in Table 2.7.

Table 2.7 Single-factor variance analysis of the measurements given in Table 2.6

Source of variance	Sum of squares	Degrees of freedom	Mean of the sum of squares	F
Between the groups	$S_2 = 14.3$	5	$s_2^2 = 2.86$	$s_2^2/s_1^2 = 1.1$
Within the groups	$S_1 = 74$	30	$s_1^2 = 2.47$	$F_{0.05;5,30} = 2.53$

Because the experimental value $s_2^2/s_1^2 = 1.1$ is smaller than the tabulated value $F_{0.05;5,30} = 2.53$, it follows that at this significance level, the hypothesis must be accepted. In other words, all 36 results can be considered as belonging to a homogeneous set.

2.6.6.2 *Two-factor variance analysis*

We shall consider now the case of a balanced two-factor experiment in which the two factors under study, A and B, take respectively t and r levels. Thus, we have a two-factor experiment with rt cells and each cell contains n

Table 2.8 The structure of the observed data in a two-factor experiment

B	A				
	1	2	— — —	t–1	t
1	x_{11}	x_{12}	— — —	$x_{1(t-1)}$	x_{1t}
2	x_{21}	x_{22}	— — —	$x_{2(t-1)}$	x_{2t}
—	—	—	— — —	—	—
—	—	—	— — —	—	—
—	—	—	— — —	—	—
r–1	$x_{(r-1)1}$	$x_{(r-1)2}$	— — —	$x_{(r-1)(t-1)}$	$x_{(r-1)t}$
r	x_{r1}	x_{r2}	— — —	$x_{r(t-1)}$	x_{rt}

elements. In such a case the experimental data have the structure given in Table 2.8. Hence, the following model corresponds to the experimental data:

$$x_{ijv} = m_{ij} + e_{ijv} \tag{2.77}$$

where $1 \leqslant i \leqslant r$; $1 \leqslant j \leqslant t$; $1 \leqslant v \leqslant n$; and e_{ijv} has the distribution $N(0, \sigma^2)$.

To understand the significance of the parameters involved in the calculation we shall use Table 2.9, where the symbols have the following meaning;

m_{ij}– the cell means;
$m_{i.}$ – the row means;
$m_{.j}$– the column means:
ζ_j– the effect of factor A;
η_i– the effect of factor B;
m– the general mean of the data.

Table 2.9 The significance of the influence of the factors A and B upon the observed data

B	A				Mean	Deviation
	1	2	– – –	t		
1	m_{11}	m_{12}	– – –	m_{1t}	$m_{1.}$	$(m_{1.} - m) = \eta_1$
2	m_{21}	m_{22}	– – –	m_{2t}	$m_{2.}$	$(m_{2.} - m) = \eta_2$
–	–	–	– – –	–	–	–
–	–	–	– – –	–	–	–
–	–	–	– – –	–	–	–
r	m_{r1}	m_{r2}	– – –	m_{rt}	$m_{r.}$	$(m_{r.} - m) = \eta_r$
Mean	$m_{.1}$	$m_{.2}$	– – –	$m_{.t}$	m	
Deviation	$m_{.1}-m=\zeta_1$	$m_{.2}-m=\zeta_2$	– – –	$m_{.t}-m=\zeta_t$		

The effect of interaction of factors A and B (the interaction constant) is given by the equation:

$$\theta_{ij} = m_{ij} - (m + \zeta_j + \eta_i) \tag{2.78}$$

Interaction is the phenomenon in which the effect of one of the factors depends on the level of the other. Here we consider only two factors, but there can be interactions between several factors.

The calculations are concisely given in Table 2.10.

Table 2.10 Two-factor variance analysis

Source of variance	Sum of squares	Degrees of freedom	Mean of the sum of squares	Estimates of the quantity	F
Rows	$S_4 = nt \sum_{i=1}^{r} (\bar{x}_{i..} - \bar{x}_{...})^2$	$r-1$	$s_4^2 = \dfrac{S_4}{r-1}$	$\sigma^2 + \dfrac{nt \sum_{i=1}^{r} \eta_i^2}{r-1}$	$F_B = \dfrac{s_4^2}{s_1^2}$
Columns	$S_3 = nr \sum_{j=1}^{t} (\bar{x}_{.j.} - \bar{x}_{...})^2$	$t-1$	$s_3^2 = \dfrac{S_3}{t-1}$	$\sigma^2 + \dfrac{nr \sum_{j=1}^{t} \zeta_j^2}{t-1}$	$F_A = \dfrac{s_3^2}{s_1^2}$
Interactions	$S_2 = n \sum_{i=1}^{r} \sum_{j=1}^{t} (\bar{x}_{ij.} - \bar{x}_{i..} - \bar{x}_{.j.} + \bar{x}_{...})^2$	$(r-1)(t-1)$	$s_2^2 = \dfrac{S_2}{(r-1)(t-1)}$	$\sigma^2 + \dfrac{r \sum_{i=1}^{r} \sum_{j=1}^{t} \theta_{ij}}{(r-1)(t-1)}$	$F_{AB} = \dfrac{s_2^2}{s_1^2}$
Within the cells	$S_1 = \sum_{i=1}^{r} \sum_{j=1}^{t} \sum_{\nu=1}^{n} (x_{ij\nu} - \bar{x}_{ij.})^2$	$rt(n-1)$	$s_1^2 = \dfrac{S_1}{rt(n-1)}$	σ^2	
Total	$S = \sum_{i=1}^{r} \sum_{j=1}^{t} \sum_{\nu=1}^{n} (x_{ij\nu} - \bar{x}_{...})^2$	$rtn-1$	$s = \dfrac{S}{rtn-1}$		$F_{\alpha;\,(r-1),\,rt(n-1)}$ $F_{\alpha;\,(t-1),\,rt(n-1)}$ $F_{\alpha;\,(r-1)(t-1),\,rt(n-1)}$

From the analysis of the parameters given in Table 2.10 we conclude the following.

(a) If the interaction constant θ_{ij} is equal to zero, i.e. there is no interaction between A and B, the s_2^2 parameter will be an estimation of the variance σ^2 and independent of s_1^2. Hence, in this case, the ratio $F_{AB} = s_2^2/s_1^2$ will have an F distribution with $(r-1)$ $(t-1)$ and $rt(n-1)$ degrees of freedom. Consequently, for a significance level α, the hypothesis that there is no interaction of A and B will be accepted if

$$F_{AB} < F_{\alpha;(r-1)(t-1),\, rt(n-1)}$$

(b) If $\eta_i = 0$, i.e. there is no effect by factor B, the s_4^2 parameter will be an estimation of σ^2, so the ratio $F_B = s_4^2/s_1^2$ will have an F distribution with $(r-1)$ and $rt(n-1)$ degrees of freedom. Hence, for a significance level α, the hypothesis that factor B has no effect on the data is accepted if

$$F_B < F_{\alpha;\,(r-1),\, rt(n-1)}$$

(c) If $\zeta_j = 0$, i.e. factor A has no effect, the s_3^2 parameter estimates the variance σ^2, so the ratio $F_A = s_3^2/s_1^2$ has an F distribution with $(t-1)$ and $rt(n-1)$ degrees of freedom. Hence, for a significance level α, the hypothesis that factor A does not affect the data will be accepted if

$$F_A < F_{\alpha;\,(t-1),\, rt(n-1)}$$

Example 2.17 We have to establish the influence of the amount of manganese and aluminium in a steel sample on the density difference (ΔS) for the line-pair $\lambda_{Mo} = 2816.2$ Å, $\lambda_{Fe} = 2813.6$ Å. Six samples are prepared with various amounts of aluminium and manganese, but the same amount of molybdenum (the element to be determined), iron (the standard element) and a constant amount of all the other elements.

The six samples will be designated Mn_0Al_0, Mn_0Al_1, Mn_0Al_2, Mn_1Al_0, Mn_1,Al_1, and Mn_1Al_2. These notations have the following meanings.

Mn_0Al_0 indicates that these elements are not present in the sample.

Mn_1 indicates 2% manganese.

Al_1 indicates 0.3% aluminium.

Al_2 indicates 0.8% aluminium.

Hence, one of the factors (the amount of manganese) has two levels and the other (the amount of aluminium) has three.

The results are given in Table 2.11 which contains 6 cells, with 5 measurements in each. The calculations are given in Table 2.12.

Table 2.11 Investigation of the influence of manganese and aluminium on the analytical pair $\lambda_{Mo} = 2816.2$ Å; $\lambda_{Fe} = 2813.6$ Å

	Difference of density $\Delta S \times 10^2$		
	Al_0	Al_1	Al_2
Mn_0	−43 −41 −42 −40 −43	−40 −42 −41 −43 −43	−37 −36 −34 −36 −38
Mn_1	−44 −43 −43 −41 −44	−42 −43 −38 −40 −40	−38 −35 −35 −39 −36

Table 2.12 Variance analysis of the measurements in Table 2.11

Source of variance	Sum of squares	Degrees of freedom	Mean of the squares	F
Between rows (Mn)	$S_4 = 0.51$	$r-1 = 1$	$s_4^2 = 0.51$	$F_{Mn} = 0.23$ not significant
Between columns (Al)	$S_3 = 208.70$	$t-1 = 2$	$s_3^2 = 104.30$	$F_{Al} = 47.40$ significant
Interactions (Al-Mn)	$S_2 = 3.40$	$(r-1)(t-1) = 2$	$s_2^2 = 1.70$	$F_{Al\text{-}Mn} = 0.77$ not significant
Within cells	$S_1 = 53.40$	$rt(n-1) = 24$	$s_1^2 = 2.20$	$F_{0.05;1,24} = 4.25$ $F_{0.05;2,24} = 3.40$

Considering the data in Table 2.12, as $F_{Mn} < F_{0.05;1,24}$, $F_{Al} > F_{0.05;2,24}$, and $F_{Al\text{-}Mn} < F_{0.05;2,24}$, for a significance level $\alpha = 0.05$ it is evident that neither the manganese level nor the Al–Mn interaction has a significant effect on the data, but the aluminium level does. The influence of the aluminium is explained by the overlap between its spectral line $\lambda_{Al} = 2816$ Å with the line $\lambda_{Mo} = 2816.1$ Å which was used throughout the measurements.

2.6.7 Order statistics

The object of order statistics is investigation of observed values of random variables in terms of order relationships. The great advantage of order statistics is that the results obtained do not depend on the nature of the distribution of the random variable. Usually, these methods are known as non-parametric methods, because there is no distribution parameter involved in using them for testing hypotheses.

Let x and y be two independent random variables, with a continuous distribution function. Two samples of sizes n_1 and n_2, i.e. the samples $x_1, x_2, \ldots, x_{n_1}$ and $y_1, y_2, \ldots, y_{n_2}$ are considered. The problem is to verify that the two samples have the same distribution.

2.6.7.1 *Smirnov's test*

This test is valid for large values of the sum $n_1 + n_2$. In order to reject the hypothesis of identity of the two distribution functions, the following critical region is evaluated for a significance level α:

$$D_{n_1 n_2} > \lambda_\alpha \sqrt{\frac{1}{n_1} + \frac{1}{n_2}} \tag{2.79}$$

where $D_{n_1 n_2} = \sup |F_{n_1}(x) - F_{n_2}(y)|$. Here $F_{n_1}(x)$ and $F_{n_2}(y)$ are the empirical distribution functions associated with the two samples.

A good approximation for the value of λ_α is given by the equation

$$\lambda_\alpha = \sqrt{-\frac{1}{2} \ln \alpha} \tag{2.80}$$

Because Smirnov's test is valid only for large values of $n_1 + n_2$, it can seldom be applied in chemical analysis. The tests which are convenient for testing hypotheses in chemical analysis are those which do not impose any restriction on the order of magnitude of the sum $(n_1 + n_2)$. Two tests of this kind will now be given.

2.6.7.2 *Wilcoxon's test*

Assume that for the two samples the observed values, x_i and y_j (where $1 \leqslant i \leqslant n_i$, and $1 \leqslant j \leqslant n_j$) are arranged in increasing order. In this way we obtain a series of the form:

$$y\,y\,y\,x\,y\,x\,x\,y\,y\,x\,x\,x \tag{2.81}$$

In such a series, for any x succeeding any y, we shall say that we have an inversion. For example, there are 29 inversion in the series (2.81) (3 for the first x, 4 each for the second and third, and 6 for each of the last three).

Let u be the total number of inversions from the experimental series. For $n_1, n_2 \geqslant 4$ and $n_1 + n_2 \geqslant 20$ it can be proved that, if the two samples belong to the same distribution, the random variable u is normally distributed with the mean $\bar{u}$ and the standard deviation σ_u given by

$$\bar{u} = \frac{n_1 n_2}{2}$$

and

$$\sigma_u = \sqrt{\frac{n_1 n_2}{12}(n_1 + n_2 + 1)}$$

Obviously, the distribution of the variable $z = (u - \bar{u})/\sigma_u$ will be the $N(0, 1)$ distribution. Hence, the critical regions for rejecting the hypothesis of identity of the two distributions will be, for a significance level α:

$$\frac{u - \bar{u}}{\sigma_u} < -|z_{\alpha/2}|; \frac{u - \bar{u}}{\sigma_u} > |z_{\alpha/2}| \tag{2.82}$$

For small n_1 and n_2, there are tables in which the probabilities $P(u \leqslant u_\alpha) = \alpha$ are given (Table X in the Appendix). Then for a significance level α the critical regions for rejecting the hypothesis will be

$$u < u_\alpha;\ u > n_1 n_2 - u_\alpha \tag{2.83}$$

where u_α is obtained from the equation $P(u \leqslant u_\alpha) = \alpha$.

2.6.7.3 *Van der Waerden's test*

In this case the two samples of sizes n_1 and n_2, will also be arranged in increasing order; a series similar to (2.81) will thus be obtained.

Let r_i be the order number of value x_i in the series, and let a new random variable be defined:

$$X = \sum_{i=1}^{n_1} \psi\left(\frac{r_i}{n+1}\right) \tag{2.84}$$

where $n = n_1 + n_2$, and ψ is the inverse function of the distribution function $N(0, 1)$.

For $n \to \infty$, it can be shown that the distribution of X tends towards the distribution $N(0, \sigma_X^2)$, where

$$\sigma_X^2 = \frac{n_1 n_2}{n(n-1)} \sum_{k=1}^{n} \left[\psi\left(\frac{k}{n+1}\right)\right]^2 \tag{2.85}$$

Consequently, for $n \to \infty$, the distribution of the variable X/σ_X tends towards the $N(0, 1)$ distribution. From these considerations, for a significance level α we can evaluate the critical regions

$$X < -X_{\alpha/2}\,;\; X > X_{\alpha/2} \tag{2.86}$$

for rejecting the hypothesis of the identity of the two distributions.

The theoretical values of $X_{\alpha/2}$ are tabulated for various significance levels α and various numbers n, and $n_1 - n_2$ in Table XI in the Appendix.

Example 2.18 With the data of example 2.9 and Wilcoxon's test at a significance level $\alpha = 0.05$, verify the hypothesis that the two series of analytical results I and II belong to the same distribution function.

Denoting by x each of the values in series I and by y each of those in series II, we obtain, after arranging all the data in increasing order, the series $x\,x\,x\,y\,x\,y\,x\,x\,y\,x\,y\,x\,y\,x\,y\,x\,x\,x\,x\,y\,x\,y\,y\,y\,y\,y\,y\,y\,x\,y\,x\,x\,x\,y\,x\,y\,y\,y\,x\,y$ (the equal values in the two series must be alternated).

By using Wilcoxon's test we find $u = 142, \bar{u} = 200, \sigma_u = 37$, so $z = -1.57$. The critical regions for rejecting the hypothesis that the two series of results belong to a common distribution function are, for $\alpha = 0.05$:

$$z < -1.96;\; z > 1.96$$

Because the experimental value $z = -1.57$ lies outside these regions, the hypothesis must be accepted.

By using van der Waerden's X test with the same data, we find the experimental value $X = -4.68$. From the corresponding Table, for $n = 40$, $n_1 = n_2$ and $\alpha/2 = 0.025$ we find $X = 5.75$. Hence, the critical regions for rejecting the hypothesis are

$$X < -5.75;\; X > 5.75$$

Because the experimental value lies outside these region, the hypothesis must again be accepted.

2.6.8 The sequential probability ratio test

In all the hypothesis testing problems given above, samples of well-defined size have been used. The sequential method of hypothesis testing operates on the successive terms of the series of observed data, as soon as they are obtained. For testing hypotheses, sequential analysis gives a procedure by which one of the following three decisions can be taken at each step of the experiment: (1) the acceptance of the hypothesis; (2) the rejection of the hypothesis; (3) the continuation of the experiment in order to consider one more observation.

One of the sequential analysis tests, namely the sequential probability ratio test, will be presented concisely here.

Let x be a random variable with a probability density $f(x, \theta)$, where θ is a parameter which can take the values θ_0 and θ_1. Thus, the hypotheses which can be tested are the H_0 hypothesis, i.e. $\theta = \theta_0$, and the H_1 hypothesis, i.e. $\theta = \vartheta_1$.

In order to decide between the two possible hypotheses, we make use of the probability ratio:

$$\frac{P_{1n}}{P_{0n}} = \frac{\prod_{i=1}^{n} f(x_i, \theta_1)}{\prod_{i=1}^{n} f(x_i, \theta_0)} \tag{2.87}$$

where $n = 1, 2, \ldots$

In all the tests already described (using a defined sample size) a critical region was defined which allowed formulation of the decision. The sequential probability ratio test is built up by extending the method used for a defined sample size, so as to include also the region for which the sampling is to be continued, i.e. the region for which neither hypothesis can be accepted. This region is defined as follows:

$$A_1 < \frac{P_{1n}}{P_{0n}} < A_2 \tag{2.88}$$

If $\dfrac{P_{1n}}{P_{0n}} \leqslant A_1$, the H_0 hypothesis is accepted, and if $\dfrac{P_{1n}}{P_{0n}} \geqslant A_2$, the H_1 hypothesis is accepted.

Accepting the values α and β for the errors of the first and second kind respectively, we have:

$$A_1 = \frac{\beta}{1 - \alpha} \text{ and } A_2 = \frac{1 - \beta}{\alpha} \tag{2.89}$$

2.6.8.1 *The sequential probability ratio test for the normal distribution*

We consider the distribution $N(m, \sigma)$ and the following hypotheses about the parameter m: the H_0 hypothesis (i.e. $m = m_0$) and the H_1 hypothesis (i.e. $m = m_1$). The normal probability density is:

$$f(x, m) = \frac{1}{\sigma\sqrt{2\pi}} \exp\left[\frac{-(x-m)^2}{2\sigma^2}\right]$$

so the ratio (2.87) will be:

$$\frac{P_{1n}}{P_{0n}} = \frac{\prod_{i=1}^{n} \exp\left[-\frac{(x_i-m_1)^2}{2\sigma^2}\right]}{\prod_{i=1}^{n} \exp\left[-\frac{(x_i-m_0)^2}{2\sigma^2}\right]} = \exp\left[-\frac{m_1-m_0}{\sigma^2}\right] \sum_{i=1}^{n} x_i + \frac{n}{2\sigma^2}(m_0^2 - m_1^2)$$

From this, by using the inequalities in (2.88) we obtain, after taking the logarithm:

$$\ln\frac{\beta}{1-\alpha} + \frac{n}{2\sigma^2}(m_1^2-m_0^2) < \frac{(m_1-m_0)}{\sigma^2} \sum_{i=1}^{n} x_i < \ln\frac{1-\beta}{\alpha} + \frac{n}{2\sigma^2}(m_1^2-m_0^2)$$

If $m_1 > m_0$ we can write:

$$\frac{\sigma^2}{m_1-m_0}\ln\frac{\beta}{1-\alpha} + \frac{n}{2}(m_1+m_0) < \sum_{i=1}^{n} x_i < \frac{\sigma^2}{m_1-m_0}\ln\frac{1-\beta}{\alpha} + \frac{n}{2}(m_1+m_0)$$

Consequently, the three decisions which can be taken at each step of the experiment are the following:

(a) if $\sum_{i=1}^{n} x_i \leqslant \frac{\sigma^2}{m_1-m_0}\ln\frac{\beta}{1-\alpha} + \frac{n}{2}(m_1+m_0)$, the H_0 hypothesis is accepted;

(b) if $\sum_{i=1}^{n} x_i \geqslant \frac{\sigma^2}{m_1-m_0}\ln\frac{1-\beta}{\alpha} + \frac{n}{2}(m_1+m_0)$, the H_1 hypothesis is accepted;

(c) if neither inequality is satisfied, the sampling will be continued.

Example 2.19 In an X-ray fluorescence spectrometric method for phosphorus in steel, a standard sample containing 0.055% phosphorus is used as a performance monitor. The variance corresponding to the random error fluctuations is known to be $\sigma^2 = 1 \times 10^{-6}\%^2$. The analytical system is regarded as running normally if the systematic error is less than ± 0.002%. The object of the monitoring is to ascertain which of the three possible states below applies to the analytical system at any particular moment. The three states are

(a) no systematic error,
(b) giving positive systematic error,
(c) giving negative systematic error.

In the absence of systematic errors, the analysis errors have the distribution $N(m_0 = 0.000; \sigma = 0.001)$; in the presence of a negative systematic error equal to -0.002%, the analysis errors have the distribution $N(m_1 = -0.002; \sigma = 0.001)$; in the presence of a positive systematic error equal to $+0.002$, the analysis errors have the distribution $N(m_2 = +0.002; \sigma = 0.001)$.

To ascertain the state of the system at a given moment, the sequential probability ratio test will be used with $\alpha = \beta = 0.025$. The hypothesis of the absence of a positive systematic error will be tested by using the distributions $N(m_0 = 0.000; \sigma = 0.001)$ and $N(m_2 = +0.002; \sigma = 0.001)$.

Let x_i be the result of a determination, and n the order number of this result. From the application of the sequential probability ratio test theory to these two distributions, the following decision rules are obtained:

(a) if $\sum_{i=1}^{n} (x_i - 0.055) \leqslant -0.0019 + 0.001n$, we decide that the systematic error is either absent or negative (the H_0 hypothesis);

(b) if $\sum_{i=1}^{n} (x_i - 0.055) \geqslant 0.0019 + 0.001n$, we decide that there is a positive systematic error (the H_1 hypothesis);

(c) if neither inequality is satisfied, the experiment will be continued.

To verify the hypothesis of the absence of a negative systematic error, we apply the same test to the distributions $N(m_0 = 0.000; \sigma = 0.001)$ and $N(m_1 = -0.002; \sigma = 0.001)$. In this case the following decision rules are obtained:

(a′) if $\sum_{i=1}^{n} (x_i - 0.055) \geqslant 0.0019 - 0.001n$, we decide that the systematic error is either absent, or positive (the H_0 hypothesis);

(b′) if $\sum_{i=1}^{n} (x_i - 0.055) \leqslant -0.0019 - 0.001n$, we decide that the systematic error is negative (the H_1 hypothesis);

(c′) if neither condition is fulfilled, the experiment will be continued.

Obviously, when conditions (a) and (a′) are simultaneously fulfilled the conclusion is that systematic errors are absent.

The results are given in columns 3 and 4 of Table 2.13 and in Fig. 2.6 where it can be seen that after the 11th determination it could be decided that the analytical system had a negative systematic error, i.e. the accepted hypothesis is H_1, that is $m_1 \geqslant -0.002$.

Table 2.13 Formulation of a decision on the sequential probability ratio test, for the normal distribution (columns 3 and 4), and the binomial distribution (columns 5 and 6). The hypotheses considered: H_0, $m = 0.000$; H_1, $m = 0.002$ or alternatively $m = -0.002$).

n	x_i	$\sum_{i=1}^{n} (x_i-0.055)$	The decision level for the hypothesis H_1 (i.e. the presence of the systematic error)	m	The decision level for the hypothesis H_1 (i.e. the presence of the systematic error)
1	2	3	4	5	6
1	0.054	--0.001	–0.0029	1	1.741
2	0.054	–0.002	–0.0039	2	2.418
3	0.055	–0.002	–0.0049	2	3.095
4	0.053	–0.004	–0.0059	3	3.772
5	0.054	–0.005	–0.0069	4	4.449
6	0.054	–0.006	–0.0079	5	5.126
7	0.054	–0.007	–0.0089	6*	5.803
8	0.053	–0.009	–0.0099	7	6.480
9	0.053	–0.011	–0.0109	8	7.157
10	0.055	–0.011	–0.0119	8	7.834
11*	0.053	–0.013	–0.0129	9	8.511
12	0.054	–0.014	–0.0139	10	9.188
13	0.054	–0.015	–0.0149	11	9.865
14	0.053	–0.017	–0.0159	12	10.543
15	0.053	–0.019	–0.0169	13	11.220

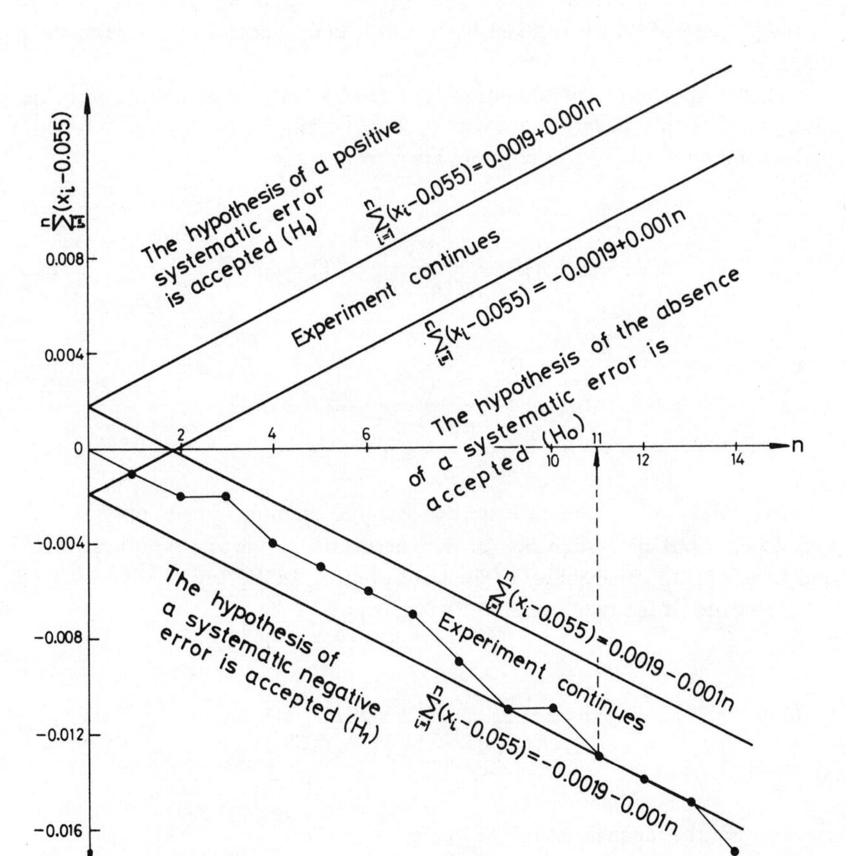

Fig. 2.6 The application of the sequential probability ratio test as a decision method (for detecting a systematic error) in the case of a normal distribution.

2.6.8.2 *The sequential probability ratio test for the binomial distribution*

We shall consider the case of a phenomenon which in a single experiment is observed with a probability p, and for which there are two distinct values p_0 and p_1, one of which is taken by p. Thus, two hypotheses can be made; H_0, that $p = p_0$, and H_1 that $p = p_1$.

These hypotheses can be rigorously tested by using the sequential probability ratio test. For this, in agreement with (2.25), the binomial distribution functions corresponding to the two hypotheses are considered:

$$f_0(m, n) = \frac{n!}{m!(n-m)!} p_0^m(1-p_0)^{n-m} \tag{2.90}$$

$$f_1(m, n) = \frac{n!}{m!(n-m)!} p_1^m(1-p_1)^{n-m} \tag{2.91}$$

Here $f_0(m, n)$ is the probability that the phenomenon is observed in m experiments out of a sequence of n experiments if the H_0 hypothesis is true, and $f_1(m, n)$ is the probability of the same event if the H_1 hypothesis is true.

The ratio of the two probabilities is given by:

$$\frac{f_1(m, n)}{f_0(m, n)} = \frac{p_1^m(1-p_1)^{n-m}}{p_0^m(1-p_0)^{n-m}} \tag{2.92}$$

In this case, the inequalities in (2.88) are:

$$\ln\frac{\beta}{1-\alpha} < m\ln\frac{p_1}{p_0} + (n-m)\ln\frac{(1-p_1)}{(1-p_0)} < \ln\frac{1-\beta}{\alpha} \tag{2.93}$$

from which we obtain

$$A_{0n} = \frac{\ln\dfrac{\beta}{1-\alpha} - n\ln\dfrac{1-p_1}{1-p_0}}{\ln\dfrac{p_1}{p_0} - \ln\dfrac{1-p_1}{1-p_0}} < m < \frac{\ln\dfrac{1-\beta}{\alpha} - n\ln\dfrac{1-p_1}{1-p_0}}{\ln\dfrac{p_1}{p_0} - \ln\dfrac{1-p_1}{1-p_0}} = A_{1n} \tag{2.94}$$

Consequently if $m \leqslant A_{0n}$, the H_0 hypothesis is accepted; if $m \geqslant A_{1n}$, the H_1 hypothesis is accepted; if $A_{0n} < m < A_{1n}$, the experiment will be continued.

In order to calculate A_{0n} and A_{1n} a reference value x_r is chosen, the random variable being taken as the 0-values (with probability p_0) when $x < x_r$, and the 1-values (with probability p_1) when $x > x_r$. In the case of the variable 0 the distribution is considered to be without systematic error $[N(m_0, \sigma)]$, and in the case of the variable 1, the distribution is regarded as having systematic error $[N(m, \sigma)]$.

Example 2.20 For the case given in example 2.19, by using the sequence of measurements given in Table 2.13, establish the significance of a negative systematic error.

In Fig. 2.7, we give the schematic model for the distribution of the analysis results obtained with the standard sample (which has phosphorus concentration 0.055%) for two cases – when systematic errors are absent, and when there is a negative systematic error equal to –0.002%.

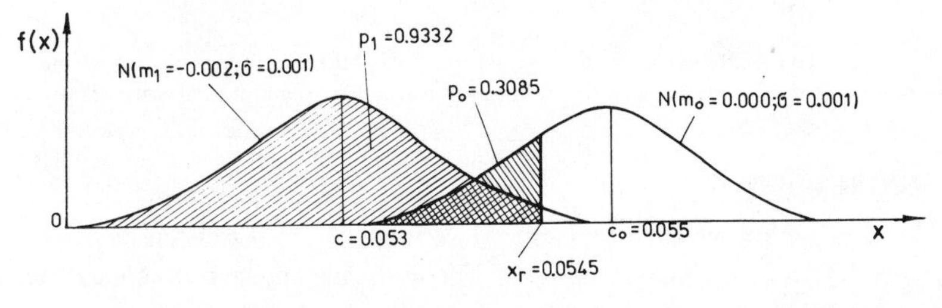

Fig. 2.7 The distribution of the analysis results: without systematic errors (c_0 = 0.055); with negative systematic errors (c = 0.053).

Choosing $x_r = 0.0545$, we have $z_0 = (0.0545\text{-}0.055)/0.001 = -0.5$, so $p_0 = F(z_0) = 0.309$, and further $z_1 = (0.0545\text{-}0.053)/0.001 = +1.5$, so $p_1 = F(z_1) = 0.933$.

For $\alpha = \beta = 0.025$, the sequential probability ratio test leads to the following decision rules.

1. If $m \geqslant 1.064 + 0.677n$, the H_1 hypothesis (i.e. that there is a negative systematic error) is accepted.

2. If $m \leqslant -1.064 + 0.677n$, the H_0 hypothesis (i.e. that negative systematic errors are absent) is accepted.

3. If $-1.064 + 0.677n < m < 1.064 + 0.677n$, the experiment is continued.

Here n is the total number of experiments, and m is the number of experiments in which the value is less than 0.0545.

The results are given in the columns 5 and 6 of Table 2.13 and in Fig. 2.8. After the 7th determination it can be decided that there is a negative systematic error.

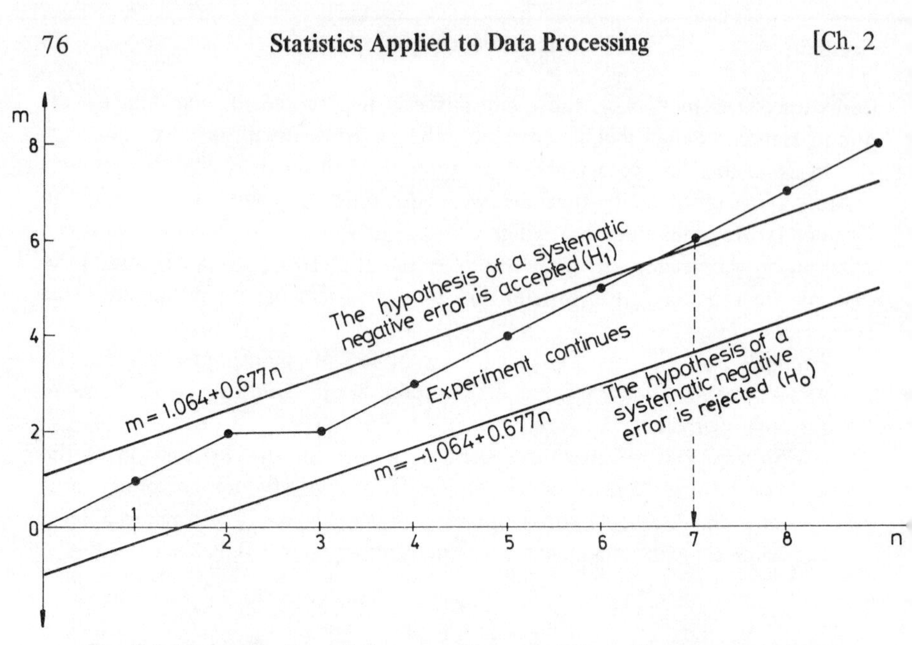

Fig. 2.8 The application of the sequential probability ratio test as a decision method (for detecting a systematic error) in the case of the binomial distribution.

2.7 REGRESSION ANALYSIS

In this section we shall limit ourselves to topics of regression analysis that are more relevant to chemical analysis. The following applications are extremely important.

(a) Evaluation of the calibration functions of analytical systems.
(b) Discovery and correction of interference phenomena.
(c) Finding the optimal working conditions for analytical measurements.
(d) Evaluation of quality of performance.

Detailed treatment of theoretical aspects of regression analysis can be found in specialist books on statistics.

2.7.1 Regression analysis for a single independent variable

2.7.1.1 *Linear regression*

The simplest case of regression analysis, from both the practical and the mathematical point of view, is that of two linearly dependent variables. A large part of this section will be devoted to the linear regression analysis of two variables, because of its wide application in analytical chemistry. Also, a detailed treatment will illustrate the most important topics raised by regression analysis in general.

In the following we shall consider the classical model in which the independent variable x is known exactly (i.e. without error), and the dependent variable y is normally distributed in such a way that its mean $\bar{y}$ is linearly dependent on x, and its standard deviation σ_y is constant, which we shall write as

$$y \Rightarrow N(\bar{y} = a + bx; \sigma_y)$$

This model is shown graphically in Fig. 2.9.

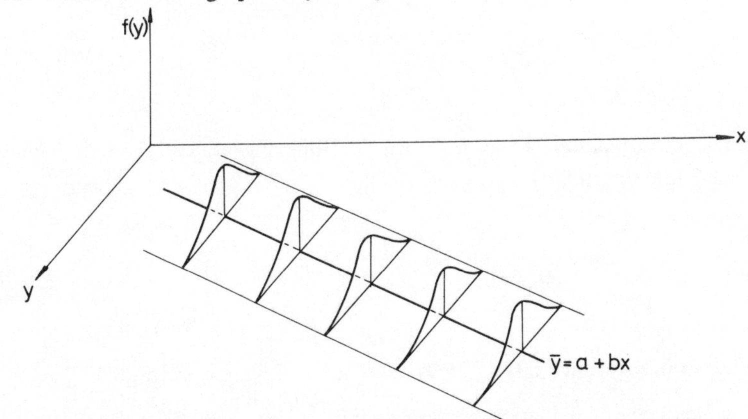

Fig. 2.9 Graphical method for linear dependence between two variables.

By use of a set of pairs of values (x_i, y_i) belonging to this model, the model parameters a, b and σ_y can be estimated and their confidence intervals evaluated. The significances of the estimated values of these parameters with respect to certain reference values can also be established.

(a) *Estimation of the model parameters*

The classical procedure for estimating the parameters a and b is the least-squares method, which consists in summing the squares of the deviations of y_i from a general line $\bar{y} = a + bx$ and finding the a and b values for which the sum is a minimum.

The sum of the squares of the deviations will be written as:

$$\text{SSD} = \sum_{i=1}^{n} [y_i - (a + bx_i)]^2 \tag{2.95}$$

Taking the partial derivatives of this expression with respect to a and b, and equating to zero, we obtain the equations

$$\begin{aligned} an + b\sum_{i=1}^{n} x_i &= \sum_{i=1}^{n} y_i \\ a\sum_{i=1}^{n} x_i + b\sum_{i=1}^{n} x_i^2 &= \sum_{i=1}^{n} x_i y_i \end{aligned} \tag{2.96}$$

from which the values of a and b are easily found.

Obviously, only estimation values for the model parameters will be obtained for a finite ensemble of pairs (x_i, y_i), given by

$$a' = (\bar{y} - b'\bar{x}) \tag{2.97}$$

$$b' = \frac{S_{xy}}{S_{xx}} \tag{2.98}$$

The variance σ_y^2 associated with the fluctuations of y_i around the x-dependent mean values* is estimated by the quantity:

$$s_y^2 = \frac{S_{yy} - b'S_{xy}}{n - 2} \tag{2.99}$$

The symbols given in (2.97)-(2.99) have the following meaning:

$$S_{xx} = \sum_{i=1}^{n} x_i^2 - \frac{\left(\sum_{i=1}^{n} x_i\right)^2}{n} \qquad S_{xy} = \sum_{i=1}^{n} x_i y_i - \frac{\sum_{i=1}^{n} x_i \sum_{i=1}^{n} y_i}{n}$$

$$S_{yy} = \sum_{i=1}^{n} y_i^2 - \frac{\left(\sum_{i=1}^{n} y_i\right)^2}{n} \qquad \bar{x} = \frac{\sum_{i=1}^{n} x_i}{n} \text{ and } \bar{y} = \frac{\sum_{i=1}^{n} y_i}{n}$$

(b) *The confidence intervals for the linear regression parameters*
We can prove that the variable a' has an $N(a, \sigma_{a'}^2)$ distribution, where:

$$\sigma_{a'}^2 = \frac{\sigma_y^2 \sum_{i=1}^{n} x_i^2}{n \sum_{i=1}^{n} (x_i - \bar{x})^2} \tag{2.100}$$

It follows that the variable $t = (a-a')/s_{a'}$ has a t distribution with $n-2$ degrees of freedom. Here $s_{a'}$ is the sample standard deviation for the a' variable, and is obtained from (2.100) by using the sample value s_y instead of σ_y. Obviously,

* We shall define by 'dependence function' the function which describes the variation of the mean values of y with respect to x.

corresponding to a probability P, the confidence interval of a is given by:

$$a' - |t_{\alpha/2;n-2}|s_{a'} < a < a' + |t_{\alpha/2;n-2}|s_{a'} \qquad (2.101)$$

where $\alpha = 1 - P$.

We can also prove that the variable b' estimating the parameter b has the distribution $N(b, \sigma_{b'}^2)$, where:

$$\sigma_{b'}^2 = \frac{\sigma_y^2}{\sum_{i=1}^{n} (x_i - \bar{x})^2} \qquad (2.102)$$

Hence, the variable $t = (b' - b)/s_{b'}$ has a t distribution with n—2 degrees of freedom. Here $s_{b'}$ is the estimated standard deviation for b' obtained by using Eq. (2.102) with the sample value s_y instead of σ_y. Hence, the confidence interval of b, calculated for a confidence probability P, is given by:

$$b' - |t_{\alpha/2;n-2}|s_{b'} < b < b' + |t_{\alpha/2;n-2}|s_{b'} \qquad (2.103)$$

According to Section 2.5.2, the confidence interval of the variance of y with respect to the regression line, calculated for a probability P, will be:

$$\frac{s_y^2(n-2)}{\chi_2^2} < \sigma_y^2 < \frac{s_y^2(n-2)}{\chi_1^2} \qquad (2.104)$$

where the values χ_1^2 and χ_2^2 are obtained from the conditions

$$P(\chi^2 < \chi_1^2) = (1-P)/2 \text{ and } P(\chi^2 < \chi_2^2) = (1+P)/2$$

(c) *The confidence interval of the regression line*

From the discussion above it is clear that with a finite number of data pairs (x_i, y_i), the true equation for the regression line remains unknown.

From the estimation values of a' and b' and the corresponding standard deviations ($s_{a'}$ and $s_{b'}$), the confidence interval of the regression line for a confidence probability P will be given by:

$$a' + b'x - |t_{\alpha/2;n-2}|s_y\sqrt{\frac{1}{n} + \frac{(x - \bar{x})^2}{S_{xx}}} <$$

$$a + bx < a' + b'x + |t_{\alpha/2;n-2}|s_y\sqrt{\frac{1}{n} + \frac{(x - \bar{x})^2}{S_{xx}}} \qquad (2.105)$$

(d) *The confidence interval of* y

We can also evaluate the confidence interval for y at a fixed value of x. In this way we obtain the so-called 'width of standard deviation', often used in chemical analysis, which is given by:

$$a' + b'x - |t_{\alpha/2;n-2}|s_y\sqrt{1 + \frac{1}{n} + \frac{(x - \bar{x})^2}{S_{xx}}} < y <$$

$$a' + b'x + |t_{\alpha/2;n-2}|s_y\sqrt{1 + \frac{1}{n} + \frac{(x - \bar{x})^2}{S_{xx}}} \quad (2.106)$$

The significance of the confidence intervals for the regression line and for the values of y is represented graphically in Fig. 2.10.

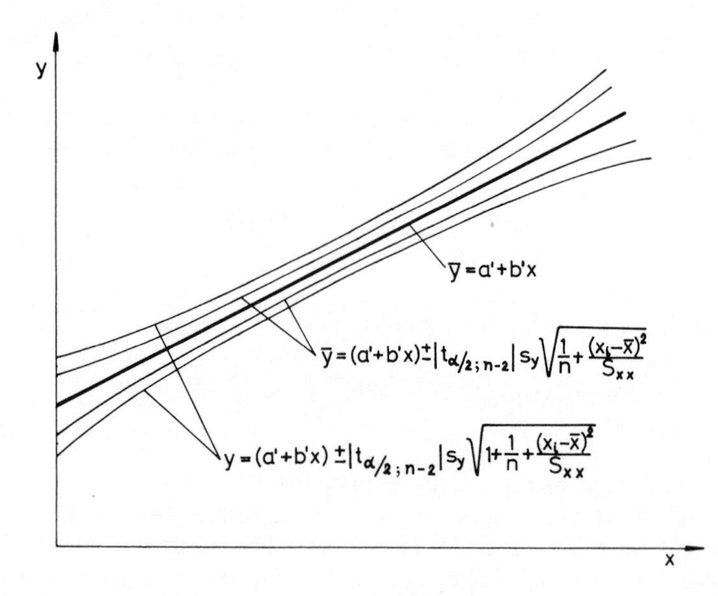

Fig. 2.10 The significance of the confidence interval for the regression line and the variable y.

Example 2.21 The results obtained by X-ray fluorescence spectrometry for 9 steel samples with a known amount of molybdenum are given in Table 2.14. Verify the hypothesis that systematic errors are absent within the range of concentrations investigated, and evaluate the precision of the method.

Table 2.14 Analysis results for nine samples of steel, with a known concentration of molybdenum

Sample number	Concentration c_i of the standard sample, %	Analysis results c'_i, %
1	0.068	0.069
2	0.045	0.045
3	0.075	0.075
4	0.035	0.034
5	0.015	0.015
6	0.039	0.037
7	0.085	0.083
8	0.011	0.012
9	0.014	0.016

Obviously, in the absence of systematic errors, the coefficients of the line $c'_i = f(c_i)$, describing the relation between the results and the true amounts, must have the values $a = 0$, and $b = 1$.

However, from the data pairs given in Table 2.14, the regression line coefficients are $a' = 0.000819$ and $b' = 0.97836$, with standard deviations $s_{a'} = 0.000787$ and $s_{b'} = 0.049$.

In order to verify the hypothesis that systematic errors are absent we shall establish, by using the t test, the significance of the value $a' = 0.000819$ with respect to $a = 0.000$ and the value $b' = 0.97836$ with respect to $b = 1.000$.

Thus, $t = 0.000819/0.000787 = 1.04 < 2.37 = t_{0.05;7}$, and $t = (1-0.97836)/0.049 = 0.44 < 2.37 = t_{0.05;7}$. Hence the hypothesis that there is no systematic error in the two parameters of the regression line must be accepted.

The same data lead to the value $s^2_{c'} = 1.5 \times 10^{-4}$ for the variance of the results, i.e. the standard deviation (which measures the precision of the results) is equal to $s_{c'} = 1.22 \times 10^{-2}$.

The hypothesis that systematic errors are absent can also be verified by using the confidence interval of the line which describes the relation between the result and the true amount. For a confidence probability $P = 0.90$, this line will be inside the confidence interval within the hyperbolic branches:

$$y = 0.000819 + 0.9784\,c \pm 2.885 \times 10^{-3} \sqrt{\frac{1}{9} + \frac{(c - 0.043)^2}{0.006146}}$$

The regression line for $a = 0.000$ and $b = 1.000$ lies in this confidence interval, so the hypothesis that systematic errors are absent must be accepted.

2.7.1.2 *Non-linear regression*

There are many cases where the relation between the variables x and y cannot be regarded as linear over the interval investigated. In such a case there are two distinct situations.

(a) The relation between x and y, though non-linear, can be linearized by a suitable mathematical transformation. In this case, the coefficients can still be calculated by using the linear regression equations.

For example, the relation between a spectral line intensity I and the sample concentration c is given by $I = mc^b$. If logarithms are taken, the equation will be linearized:

$$\log I = \log m + b \log c$$

The regression coefficients, $a = \log m$ and b, can be obtained by using the method already described.

(b) The relation between the two variables cannot be linearized. For example, this is the case for the second-order equation:

$$y = b_0 + b_1 x + b_2 x^2 \tag{2.107}$$

In this case the least-squares method is also used for the determination of the parameters b_0, b_1, and b_2. Generalizing, we can say that in order to estimate the parameters $b_0, b_1, b_2, \ldots$ from function $y_i = f(x_i, b_0, b_1, b_2, \ldots)$ we have to calculate the deviations of the variable y from the generalized function, and then to sum the squares of these deviations, obtaining

$$\sum_{i=1}^{n} [y_i - f(x_i, b_0, b_1, b_2, \ldots)]^2 \tag{2.108}$$

Then, the partial derivatives of this equation with respect to the coefficients $b_0, b_1, b_2, \ldots$ are calculated and equated to zero. The resulting system of equations is then solved. Thus, the estimation values of the correlation function parameters are finally obtained.

For the parabolic function (2.107), the system of equations is

$$nb_0 + b_1 \sum_{i=1}^{n} x_i + b_2 \sum_{i=1}^{n} x_i^2 = \sum_{i=1}^{n} y_i$$

$$b_0 \sum_{i=1}^{n} x_i + b_1 \sum_{i=1}^{n} x_i + b_2 \sum_{i=1}^{n} x_i^2 = \sum_{i=1}^{n} y_i x_i \tag{2.109}$$

$$b_0 \sum_{i=1}^{n} x_i^2 + b_1 \sum_{i=1}^{n} x_i^3 + b_2 \sum_{i=1}^{n} x_i^4 = \sum_{i=1}^{n} y_i x_i^2$$

Obviously, the higher the degree of the polynomial, the more difficult the calculations, the required powers of x being twice the degree of the polynomial. Such problems of regression analysis are solved by using electronic computers.

2.7.2 Multiple regression

There are many cases in chemical analysis where several factors are involved in the phenomenon investigated. Phenomena influenced by several factors can be treated by using a regression function with several independent variables. The simplest case of multiple regression is, of course, the linear case, which has the following general equation for the regression function:

$$y = b_0 + b_1x_1 + b_2x_2 + \ldots + b_nx_n \tag{2.110}$$

where y is the dependent variable, $x_1, x_2, \ldots, x_n$ are the independent variables, and $b_0, b_1, \ldots, b_n$ the regression coefficients.

As in the case of single-variable regression analysis, the regression coefficients are estimated by minimizing the sum

$$\Sigma[y - (b_0 + b_1x_1 + b_2x_2 + \ldots + b_nx_n)]^2 \tag{2.111}$$

For this we take the partial derivatives with respect to $b_0, b_1, \ldots, b_n$, and equate each to zero. The parameters are obtained by solving the system of equations obtained in this way. For (2.110), we arrive at the system

$$\begin{aligned}
&nb_0 + b_1\,\Sigma x_1 + b_2\,\Sigma x_2 + \ldots + b_n\,\Sigma x_n = \Sigma y \\
&b_0\,\Sigma x_1 + b_1\,\Sigma x_1x_1 + b_2\,\Sigma x_1x_2 + b_3\,\Sigma x_1x_3 + \ldots + b_n\,\Sigma x_1x_n = \Sigma yx_1 \\
&b_0\,\Sigma x_2 + b_1\,\Sigma x_2x_1 + b_2\,\Sigma x_2x_2 + b_3\,\Sigma x_2x_3 + \ldots + b_n\,\Sigma x_2x_n = \Sigma yx_2 \\
&\ldots\ldots\ldots\ldots\ldots\ldots\ldots\ldots\ldots\ldots\ldots\ldots \\
&b_0\Sigma x_m + b_1\Sigma x_mx_1 + b_2\Sigma x_mx_2 + b_3\Sigma x_mx_3 + \ldots + b_n\,\Sigma x_mx_n = \Sigma yx_m
\end{aligned} \tag{2.112}$$

2.8 CORRELATION

Analytical systems of measurement have been improved in quality by using the general principles of correlation theory.

The determination of the regression equation has enabled us to establish the correlation between the variables and to know the change in one variable entailed by a change in the other. The degree of correlation between variables is measured by the 'correlation coefficient'.

Let x and y be two random variables, m_1 and m_2 their means, and σ_x and σ_y the standard deviations. The theoretical correlation coefficient of x and y is defined as:

$$\rho = \frac{M(x-m_1)(y-m_2)}{\sigma_x \sigma_y} \; ; \; -1 < \rho < +1 \tag{2.113}$$

If ρ is equal to either -1 or $+1$, there is a strict linear relationship between x and y of the form

$$y = a \pm \frac{\sigma_y}{\sigma_x} x \tag{2.114}$$

If we write (x_i, y_i), with $1 \leqslant i \leqslant n$, for a sample of the two-dimensional random vector (x, y), the sample correlation coefficient will be given by:

$$r = \frac{\sum_{i=1}^{n} (x_i-\bar{x})(y_i-\bar{y})}{\sqrt{\sum_{i=1}^{n} (x_i-\bar{x})^2 \sum_{i=1}^{n} (y_i-\bar{y})^2}} \tag{2.115}$$

The sample value r is an estimation for ρ, that is $r \to \rho$ when $n \to \infty$.

For the same system, if the function which relates y and x is $y = a + bx$, the estimation values of the coefficients are:

$$b' = r \frac{s_y}{s_x} \tag{2.116}$$

$$a' = \bar{y} - b'\bar{x} \tag{2.117}$$

where $\bar{x} = \dfrac{\sum_{i=1}^{n} x_i}{n}$, $\bar{y} = \dfrac{\sum_{i=1}^{n} y_i}{n}$, $s_x^2 = \dfrac{\sum_{i=1}^{n} (x_i-\bar{x})^2}{n-1}$ and $s_y^2 = \dfrac{\sum_{i=1}^{n} (y_i-\bar{y})^2}{n-1}$.

Consequently, the sample regression line of the two variables will be:

$$(y-\bar{y}) = r \frac{s_y}{s_x} (x-\bar{x}) \tag{2.118}$$

Fisher has shown that the sample correlation coefficient r can be transformed into a new random variable l having a distribution function that is normal or close to it:

$$l = \frac{1}{2}\ln\frac{1+r}{1-r} \tag{2.119}$$

The mean and variance of this variable are:

$$M(l) = \frac{1}{2}\ln\frac{1+\rho}{1-\rho} + \frac{\rho}{2(n-1)} \tag{2.120}$$

$$\sigma_l^2 = \frac{1}{n-3} \tag{2.121}$$

The Fisher transformation can be used for the following purposes:

(a) to establish, with respect to a value ρ_0, the significance of a sample value r of the correlation coefficient;

(b) to establish the significance of two sample values r_1 and r_2 of the correlation coefficient;

(c) to evaluate a confidence interval for the correlation coefficient.

In order to establish the significance of a sample value r with respect to a value ρ_0, the critical region is defined by:

$$\left(l - \frac{1}{2}\ln\frac{1+\rho_0}{1-\rho_0}\right)\sqrt{n-3} < -|z_{\alpha/2}|;\left(l - \frac{1}{2}\ln\frac{1+\rho_0}{1-\rho_0}\right)\sqrt{n-3} > |z_{\alpha/2}| \tag{2.122}$$

where l is the variable obtained by transforming the sample value r evaluated from n pairs of data (x_i, y_i).

If the experimental value $\left(l - \frac{1}{2}\ln\frac{1+\rho_0}{1-\rho_0}\right)\sqrt{n-3}$ falls in either critical region, the hypothesis that r is an estimation for ρ_0 will be rejected.

In order to establish the significance of two sample values r_1 and r_2, the critical regions are defined by:

$$\frac{l_1 - l_2}{\sqrt{\frac{1}{n_1-3} + \frac{1}{n_2-3}}} < -|z_{\alpha/2}|;\frac{l_1 - l_2}{\sqrt{\frac{1}{n_1-3} + \frac{1}{n_2-3}}} > |z_{\alpha/2}| \tag{2.123}$$

where l_1 and l_2 are the quantities obtained from r_1 and r_2 by using the transformation (2.119), and n_1 and n_2 are the numbers of pairs of data from which the values r_1 and r_2 have been evaluated. If the experimental value $(l_1 - l_2)/\sqrt{\dfrac{1}{n_1-3} + \dfrac{1}{n_2-3}}$ falls in either critical region, the hypothesis that r_1 and r_2 estimate the same value r will be rejected.

From the fact that the random variable $\left(l - \dfrac{1}{2}\ln\dfrac{1+\rho}{1-\rho}\right)\sqrt{n-3}$ has a distribution $N(0, 1)$, the confidence interval of the correlation coefficient ρ, for a confidence probability $P = 1 - \alpha$, will be given by:

$$\frac{\exp\left[2\left(l - \dfrac{z_{\alpha/2}}{\sqrt{n-3}}\right)\right] - 1}{\exp\left[2\left(l - \dfrac{z_{\alpha/2}}{\sqrt{n-3}}\right)\right] + 1} < \rho < \frac{\exp\left[2\left(l + \dfrac{z_{\alpha/2}}{\sqrt{n-3}}\right)\right] - 1}{\exp\left[2\left(l + \dfrac{z_{\alpha/2}}{\sqrt{n-3}}\right)\right] + 1} \tag{2.124}$$

Example 2.22 Table 2.15 gives density measurements for the line-pair molybdenum $\lambda_{Mo} = 2816.2$ Å and iron $\lambda_{Fe} = 2813.6$ Å. These results were obtained from repeated photographic recordings of the spectrum of a steel sample. Estimate the correlation coefficient between the densities of the two spectral lines, and for a confidence probability $P = 0.90$, estimate the corresponding confidence interval.

From (2.115) the estimation value for the correlation coefficient is $r = 0.942$. Fisher's transformation (2.119) then gives $l = 1.75$. From (2.121) the variance associated with this variable is $\sigma_l^2 = 1/57$.

For $P = 0.90$, the confidence interval of the correlation coefficient of the two variables is, from (2.124):

$$0.910 < \rho < 0.968$$

Generally, the higher the correlation coefficient of the density of the two spectral lines (i.e. the smaller its deviation from unity), the more reproducible is the density difference (i.e. the more reproducible the analytical signal). Consequently, a high value of the correlation coefficient will indicate a good choice of the analytical pair.

Table 2.15 Results of measurements of spectral density, corresponding to the lines $\lambda_{Mo} = 2816.2$ Å and $\lambda_{Fe} = 2813.6$ Å, for a steel sample.

Spectrum number	Density × 10²		Spectrum number	Density × 10²		Spectrum number	Density × 10²	
	S_{Mo}	S_{Fe}		S_{Mo}	S_{Fe}		S_{Mo}	S_{Fe}
1	30	60	21	30	62	41	21	50
2	35	67	22	38	69	42	15	44
3	29	58	23	32	63	43	16	47
4	40	74	24	32	63	44	27	56
5	33	60	25	30	58	45	22	52
6	50	81	26	35	66	46	39	69
7	31	65	27	33	65	47	23	49
8	40	66	28	34	64	48	37	67
9	38	68	29	26	54	49	24	51
10	31	62	30	32	62	50	31	62
11	35	68	31	27	56	51	24	51
12	19	52	32	45	74	52	32	62
13	31	60	33	30	57	53	15	44
14	33	64	34	27	56	54	36	64
15	30	57	35	36	63	55	24	53
16	40	73	36	34	64	56	37	66
17	27	57	37	33	62	57	28	56
18	42	74	38	30	60	58	36	64
19	20	51	39	31	59	59	29	56
20	51	83	40	28	56	60	26	55

Chapter 3

Informational Processing of Results

3.1 INTRODUCTION

The fundamental landmarks in information theory were given by Shannon [1] in his fundamental work *A Mathematical Theory of Communication.* This work was published in 1948 and was followed by such a rapid development that information theory became, around 1960, a well-established mathematical theory. Since 1960, the emphasis in works in this field has moved towards the field of application, including the use of information theory in analytical chemistry.

As in any other field of knowledge, in analytical chemistry we should know the amount of information obtained by performing an experiment, and be able to measure it. This can be done only by using the concepts of information theory.

The aim of this new technique in chemical analysis is to increase the quality of performance both of analytical systems and of methods of processing and judging the results. There are already many publications [2-50] on applications of these concepts in analytical chemistry, but this is only a beginning; there are many fields which have never been considered from this point of view. When problems of trace detection and determination are considered, the rigorous evaluation of the quality of performance of the analytical systems is a matter of first importance. A review of some elements of informational processing of analytical results is also justified because in this way we can rigorously define the quality characteristics of analytical systems.

Because of the analogies existing between information transmission systems and analytical systems of measurement, the concepts of information theory (coding, decoding, perturbation, entropy, quantity of information, redundancy) can readily be transposed to chemical analysis.

3.2 ANALOGIES BETWEEN INFORMATION TRANSMISSION AND ANALYTICAL SYSTEMS

The transmission of information can be regarded as based on the following elements: source of information, coder, channel, decoder, and receiver (Fig. 3.1), or in more condensed form, an assembly consisting of an information source, a channel and a receiver (the coder and decoder are included in the channel) (Fig. 3.1).

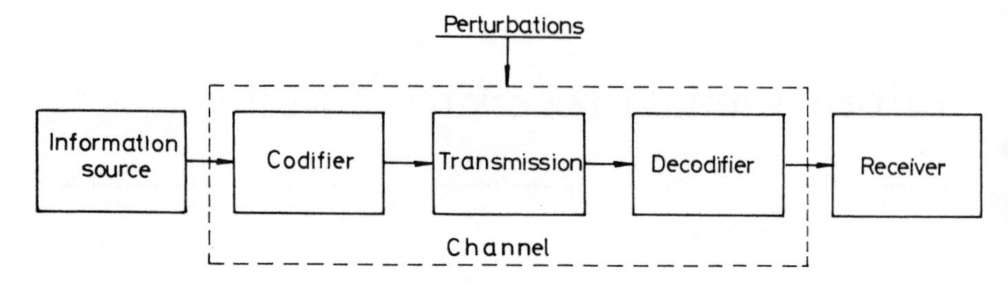

Fig. 3.1 A system of information transmission.

The analogy with an analytical system, itself an informational system, is obvious from Fig. 3.1 and the model in Fig. 3.2.

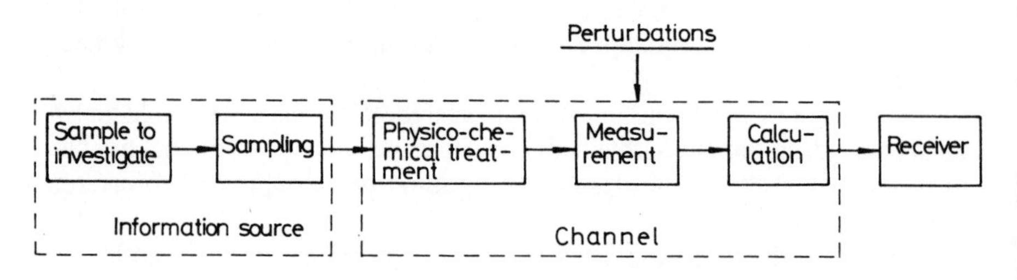

Fig. 3.2 The analogy between an analytical system and a system of information transmission.

In analysis the source of information is the set of materials which can be analysed correctly under well-specified conditions. Unfortunately, a universally applicable analytical method is an unattained (and perhaps unattainable) ideal. A sample for analysis must therefore be regarded as an element of a well-defined set, i.e. its composition must belong to the set of measurable values.

The physico-chemical treatment of the sample is intended to achieve the conditions required for obtaining a signal specific for the component to be determined, and it is by this analytical operation that the detection or determination is made possible. The physico-chemical treatment is done, essentially, for the same purpose as the coding operation in information transmission, i.e. to facilitate the transmission and to remove the effect of perturbations.

The operation of calculation and data-processing leads from the analytical signal to the analytical result. This operation is similar to the signal decoding process used in the information transmission systems.

Both analytical measurement and the transmission of information are subject to various perturbations. Consequently, both the signals received and the results of analysis will always include a degree of uncertainty.

An analytical measurement system can be said to be defined if, for certain working conditions, we know the materials which can be analysed (the alphabet of the source of information) as well as the statistical distribution of the results with respect to the true values.

Adequate analytical investigation must be made in order to define the materials which can be analysed under specified conditions; only in this way can it be guaranteed that the results will be free from errors caused by interferences and matrix effects.

Because of various perturbations, a range of analysis results will be obtained for a given value of the amount of determinand. To evaluate the perturbations arising in the process of analysis (analogous to the perturbations in the information transmission channel), we must know the statistical distribution of the analytical results.

This is shown schematically in Fig. 3.3, where $c \in C$ designates the set of samples which can be analysed, $c' \in C'$ the set of results, and $p(c'/c)$ the set of probable values for each element $c' \in C'$ corresponding to each element $c \in C$.

As we can easily see in Fig. 3.3, for a given input value c there is a corresponding range of results (due to the action of perturbations) and this range is characterized by a probability density.

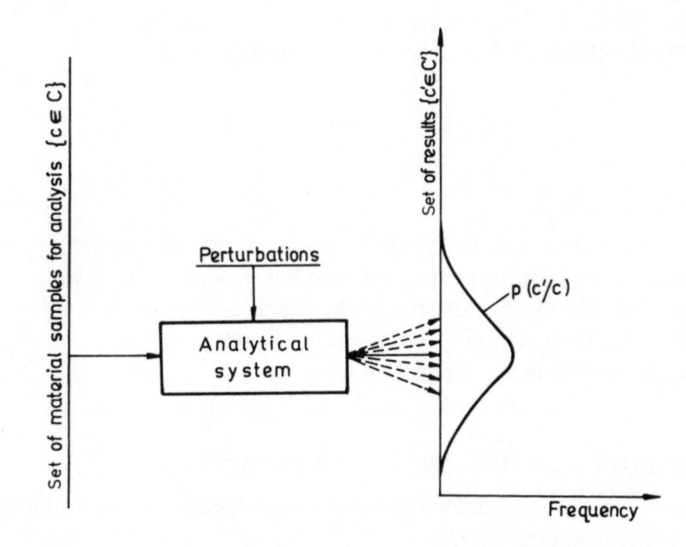

Fig. 3.3 The method for defining an analytical system in terms of probability.

Consequently, an analytical system of measurements is completely defined by the three parameters $c \in C$, $c' \in C'$ and $p(c'/c)$ and can be designated as:

$$\{c \in C, p(c'/c), c' \in C'\} \tag{3.1}$$

This definition of an analytical system of measurement on the basis of probability corresponds completely to the corresponding definition of a system of information transmission (a channel): the meaning of the symbols is $c \in C$ the set of input signals (the source alphabet), $c' \in C'$ the set of signals received, $p(c'/c)$ the probability density characterizing the perturbations on the given channel.

A more elaborate definition can be given by taking into account the complexity of the analytical systems.

We may designate by $\{c \in C, p\}$ the set of samples to be analysed, where p (the probability) arises from the fact that selecting a sample containing a certain amount of determinand is a random action.

In what follows we shall use the symbol ϕ' to designate the transformations undergone by the sample during the physico-chemical treatment (the analytical coding). Consequently, in the physico-chemical treatment, the set $\{c \in C, p\}$ of samples to be analysed produces the set $\{y \in Y, p'\}$ of analytical signals. The transformation of the set of analytical signals into the set $\{c' \in C', p''\}$ of results (i.e. the analytical decoding) will be designated by ϕ''.

Thus we have correspondence of three fields of probability:

I – the field of probability for the samples;

II – the field of probability for the analytical signals;

III – the field of probability for the analytical results:

$$\underset{\text{I}}{\{c \in C, p\}} \xrightarrow{\phi'} \underset{\text{II}}{\{y \in Y, p'\}} \xrightarrow{\phi''} \underset{\text{III}}{\{c' \in C', p''\}} \tag{3.2}$$

This is analogous to a more complete representation of a system of information transmission in which the coding and decoding processes are taken into account.

The use of information theory in chemical analysis is justified by these analogies, but it must be clearly recognized from the very beginning that specific particularities arise in the application itself.

3.3 ENTROPY AS A MEASURE OF UNCERTAINTY

The chemical applications of information theory require an understanding of the concept of entropy.

Suppose we have to do an experiment (not necessarily analytical), the outcome of which is unknown to us in advance but can be one of several possible results (events), which have a discrete distribution designated as:

$$X = \begin{bmatrix} x_1, x_2, \ldots, x_n \\ \\ p_1, p_2, \ldots, p_n \end{bmatrix} \tag{3.3}$$

We shall assume that the system of events $x_1, x_2, \ldots, x_n$ is a complete system, so that $\sum_{i=1}^{n} p_i = 1$; this signifies that the result obtained must certainly be one of these events, which removes a certain degree of uncertainty.

Obviously, the question arises: what is the measure of the degree of uncertainty? It is evident that the uncertainty depends essentially on the probabilities of the events. Indeed, the distribution

$$X = \begin{bmatrix} x_1, & x_2 \\ \\ 0.5 & 0.5 \end{bmatrix} \tag{3.4}$$

associated with two equally probable events contains a greater uncertainty than the distribution:

$$X = \begin{bmatrix} x_1, & x_2 \\ \\ 0.95, & 0.05 \end{bmatrix} \tag{3.5}$$

where the same events are involved, but with different probabilities.

As a measure of the uncertainty of a discrete field of probabilities (3.3), Shannon [1] introduced the entropy, defined by an expression of the form:

$$H(p_1, p_2, \ldots, p_n) = -\sum_{i=1}^{n} p_i \log p_i \tag{3.6}$$

To begin with, logarithms to the base 2 were used in the definition of entropy, but the choice of base is immaterial because the logarithms can be transformed from one base to another.

When logarithms to base 2 are used, the unit of uncertainty is called a 'bit'. Thus, the entropies corresponding to the distributions (3.4) and (3.5) are equal to 1 and 0.32 bit respectively.

For a continuous distribution with a probability density $p(x)$, the entropy is defined by:

$$H[p(x)] = -\int_{+\infty}^{+\infty} p(x) \log p(x) \, dx \tag{3.7}$$

In the particular case where the probability density is uniformly distributed in the interval (a, b), we obtain from (3.7) the following equation for the entropy:

$$H[p(x)] = -\int_a^b \frac{1}{a-b} \log \frac{1}{b-a} \, dx = \log(b-a) \tag{3.8}$$

Further, the entropy corresponding to a normal distribution of the event, with a standard deviation σ, is given by:

$$H[p(x)] = -\int_{-\infty}^{+\infty} \frac{1}{\sigma\sqrt{2\pi}} \exp\left[-\frac{x^2}{2\sigma^2}\right] \log\left(\frac{1}{\sigma\sqrt{2\pi}} \exp\left[-\frac{x^2}{2\sigma^2}\right]\right) dx \tag{3.9}$$

$$= \log(\sigma\sqrt{2\pi e})$$

The entropy has the following properties.

(a) It is always a non-negative quantity, i.e. $H(p_1, p_2, \ldots, p_n) \geqslant 0$.

(b) If only one state is possible (the event is certain), i.e. when the value is $p_i = 1$, the entropy is equal to zero.

(c) The entropy has its maximum value for $p_1 = p_2 = \ldots = p_n$, i.e. when all the events have the same probability; hence $H(p_1, p_2, \ldots, p_n) \leqslant H(1/n, 1/n, \ldots, 1/n)$. In order to exemplify this property we shall consider the case of a complete system of two events with probabilities p_1 and p_2. According to (3.6), the entropy of this system will be:

$$H(p_1, p_2) = -p_1 \log p_1 - p_2 \log p_2 = -p_1 \log p_1 - (1-p_1) \log(1-p_1) \tag{3.10}$$

$$= -(1-p_2) \log(1-p_2) - p_2 \log p_2$$

The variation of the entropy for this system of two events with respect to the probabilities of the events is presented in Fig. 3.4. The minimum value of the entropy (zero) occurs for $p_1 = 1$, $(p_2 = 0)$ or for $p_2 = 1$, $(p_1 = 0)$. The maximum value, $H(p_1, p_2) = 1$ bit, is reached for $p_1 = p_2 = 0.5$.

(d) For two arbitrary systems of events X_1 and X_2, the entropy of the combined systems is given by:

$$H(X_1, X_2) = H(X_1) + H(X_2|X_1) \tag{3.11}$$

where $H(X_2|X_1)$ is the entropy of the system X_2 conditioned by the system X_1.

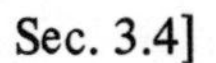

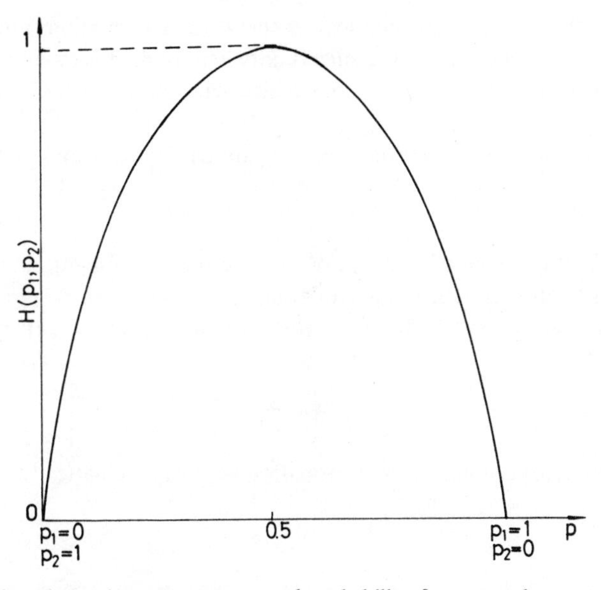

Fig. 3.4 The relation between entropy and probability for a complete system of two events.

When the events of the two systems are independent, we have:

$$H(X_1, X_2) = H(X_1) + H(X_2)$$

Obviously, $H(X_2|X_1) \leqslant H(X_2)$.

Generally, for a combination of n systems, we can write:

$$H(X_1, X_2, \ldots, X_n) = H(X_1) + H(X_2|X_1) + H(X_3|X_1, X_2) + \ldots + H(X_n|X_1, X_2, \ldots, X_{n-1}) \quad (3.12)$$

When the n systems are independent, we have:

$$H(X_1, X_2, \ldots, X_n) = H(X_1) + H(X_2) + \ldots + H(X_n) \quad (3.13)$$

The properties of the entropy, as given above for discrete variables, are equally valid for continuous variables.

3.4 INFORMATION

Information is a very general concept; we can say that we obtain information from an experiment only when the result removes some of the uncertainty existing before the experiment was performed.

Thus, there is a close connection between information and uncertainty. The information is equal to the uncertainty removed. It is obvious that because of the effect of perturbation on the results, we cannot generally remove all the uncertainty.

The amount of information I given by an experiment can be written as:

$$I = H_0 - H' \tag{3.14}$$

where H_0 is the degree of uncertaintly before the experiment, and H' that after it. The maximum amount of information, I_{max}, which can be obtained is equal to the uncertainty existing before the experiment, i.e. to the entropy before the experiment:

$$I_{max} = H_0 \tag{3.15}$$

The procedure of obtaining the information is given schematically in Fig. 3.5.

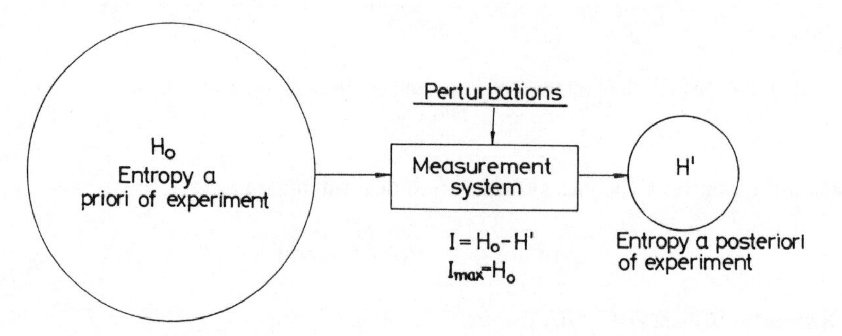

Fig. 3.5 The procedure of obtaining information from an experiment.

3.5 THE AMOUNT OF INFORMATION OBTAINED IN DETECTION AND IDENTIFICATION

Detection (or qualitative) analysis means establishing the presence or absence of one or several of the components in the sample. *Identification* means establishing the structural configuration (or nature) of a substance or one or more of its components.

Consider the simplest case of analytical detection, establishment of the presence (state 1) or absence (state 0) of a component X_1.

In the absence of any information prior to the experiment, the distribution corresponding to the two states is

$$X = \begin{bmatrix} 1 & 0 \\ \frac{1}{2} & \frac{1}{2} \end{bmatrix} \tag{3.16}$$

The uncertainty which corresponds to this distribution is $H_0 = 1$ bit. Hence, the maximum amount of information which can be obtained is $I_{max} = 1$ bit.

Now consider experiment(s) made in order to establish the presence or absence of three independent components X_1, X_2 and X_3. If the symbols 0 and 1 denote absence and presence, the eight possible states of the system will be 000, 100, 010, 001, 110, 101, 011, and 111. When no information is available prior to the detection experiments the following distribution will correspond to these equi-probable states:

$$X_1, X_2, X_3 = \begin{bmatrix} 000 & 100 & 010 & 001 & 110 & 101 & 011 & 111 \\ \frac{1}{8} & \frac{1}{8} & \frac{1}{8} & \frac{1}{8} & \frac{1}{8} & \frac{1}{8} & \frac{1}{8} & \frac{1}{8} \end{bmatrix} \tag{3.16'}$$

Consequently the uncertainty to be removed is:

$$H_0 = \log_2 2^3 = 3 \text{ bits} = I_{max}$$

Generally, when n components are to be detected in a material the uncertainty to be removed is:

$$H_0 = \log_2 2^n = n \text{ bits} = I_{max} \tag{3.17}$$

In the case of identification, the problem is formulated similarly. If the possible structural configurations are written as s_i, the experiment has to establish which of them $(s_1, s_2, \ldots, s_n)$ is the actual one. In other words, we have to remove the uncertainty which corresponds to the distribution:

$$S = \begin{bmatrix} s_1, s_2, \ldots, s_n \\ p_1, p_2, \ldots, p_n \end{bmatrix} \tag{3.18}$$

If we have no *a priori* information, it is obvious that $p_1 = p_2 = \ldots = p_n = 1/n$, so the maximum amount of information obtainable is $I_{max} = n$ bits.

The total number of known chemical elements is 104, so the number of chemical elements contained in any material will be smaller than or equal to 104. Hence, the *a priori* uncertainty in any experiment for detection of the elements contained in a material cannot exceed 104 bits ($H_0 \leqslant 104$ bits). For example, in the qualitative elemental analysis of a lunar material, about which nothing is known in advance, the hypothesis of the presence of all the chemical elements must be examined. Hence for this investigation the uncertainty prior to the experiment will be $H_0 = 104$ bits, the maximum possible uncertainty. In practice of course the uncertainty will be less than this because some elements do not occur naturally. Though the number of chemical elements is limited and relatively small, the number of known chemical compounds is very large. Hence, in the detection of chemical compounds, if no *a priori* information exists, the uncertainty is much larger than in the case of detection of elements.

Without going into detail, we should mention that the detection function of an analytical system cannot be arbitrarily extended into the domain of increasingly small concentrations, because of the effect of the perturbations. Generally, a detection experiment for a single component can give an amount of information equal to 1 bit only if the amount of the component to be detected is greater than a certain critical level.

3.6 THE AMOUNT OF INFORMATION GIVEN BY A DETERMINATION

The amount of information obtained through the determination of one or more of the components of the sample is again evaluated from the decrease in the uncertainty. Consider the simplest case, the determination of a single component. Obviously, when no *a priori* information is available, the range of uncertainty for the component concentration is (0%, 100%). When there is some information prior to the experiment (and this is usually the case), the range of uncertainty is

$$(c'\%, c''\%) < (0\%, 100\%)$$

The concentration of the component to be determined is a continuous variable. However, in practice, it will be quantized. Let Δc be the smallest difference in concentration that can be detected, i.e. the quantization interval. Then, in the concentration interval (c', c'') there will be $N = (c''-c')/\Delta c$ discrete concentration values. The object of a determination is to establish which of these corresponds to the real value, i.e. to remove an uncertainty [51] equal to:

$$H_0(c) = \log \frac{c'' - c'}{\Delta c} \tag{3.19}$$

Generally, because of perturbations, the experiment cannot remove the whole of this uncertainty. Hence, to evaluate the amount of information, we have also to evaluate the uncertainty which remains after the determination.

As shown in Chapter 2, the normal distribution is a plausible model for analytical errors, so we shall accept it for evaluating the entropy of the determination errors. When there are no systematic errors in the case of a normal distribution with a standard deviation σ for the random errors, the uncertainty remaining after the determination, as given by (3.9), will be

$$H'(c) = \log \frac{\sigma\sqrt{2\pi e}}{\Delta c} \tag{3.20}$$

From (3.14), (3.19) and (3.20), the amount of information obtained under these conditions will be given by an expression of the form:

$$I = H_0(c) - H'(c) = \log \frac{c'' - c'}{\sigma\sqrt{2\pi e}} \ . \tag{3.21}$$

The equations for the residual entropy, the amount of information, and the entropy error, all contain the standard deviation σ. However, this value is not known, and has to be estimated experimentally. With a limited sample size, an estimation value s is obtained for the standard deviation. Consequently, estimative values will be obtained for the entropy and the amount of information, because they are evaluated by using the estimated standard deviation s. By use of a confidence interval for the standard deviation (Section 2.5.2) a confidence interval of the following form can be attributed to both entropy error and the amount of information:

$$\log \frac{c'' - c'}{\sqrt{\dfrac{s^2\nu}{\chi^2_{(1-P)/2}}}\sqrt{2\pi e}} < I < \log \frac{c'' - c'}{\sqrt{\dfrac{s^2\nu}{\chi^2_{(1+P)/2}}}\sqrt{2\pi e}} \tag{3.22}$$

For example, consider an analysis for which it is known *a priori* that the concentration is in the interval 0.005–0.2% and for which there are 10 determinations, with $s = 0.002\%$. Taking the confidence probability $P = 0.90$ for $\nu = 9$ we obtain $\chi^2_{0.05} = 3.33$, $\chi^2_{0.95} = 16.32$, so from (2.35) the confidence interval for σ will be $1.49 \times 10^{-3} < \sigma < 3.29 \times 10^{-3}$. Hence the confidence interval of the amount of information is:

$$3.8 \text{ bits} < I < 5.0 \text{ bits}$$

The significance of the amount of information, (3.21), given by a determination is represented graphically in Fig. 3.6.

Equation (3.21) reveals the informational essence of an analytical determination: the numerator represents the length of the uncertainty interval before the determination and the denominator is the weighted value of the uncertainty interval after it.

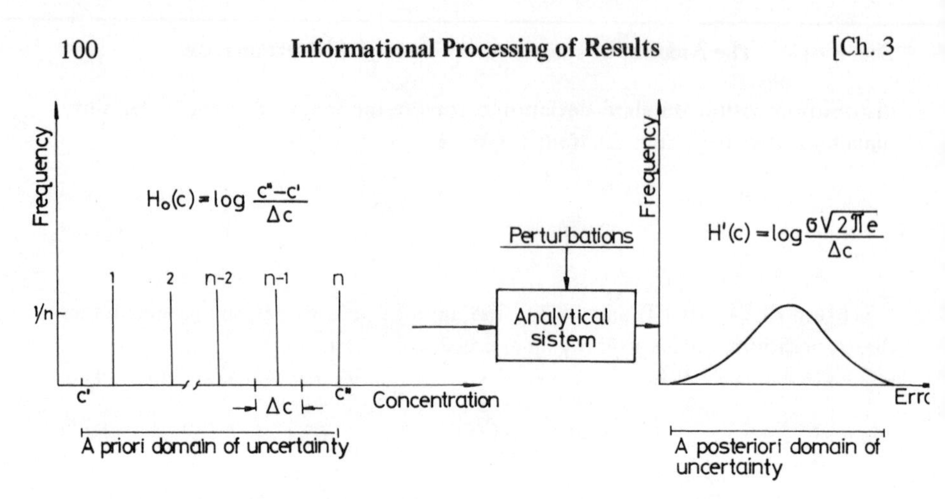

Fig. 3.6 The quantity of information given by a determination.

When n determinations are performed on the sample and the result is given as their mean value, the standard deviation of the mean will be $\sigma/\sqrt{n}$, so instead of (3.21), the following equation will give the amount of information:

$$I = \log \frac{c'' - c'}{\sqrt{2\pi e}\ \sigma/\sqrt{n}} \tag{3.23}$$

As can be seen from (3.23), the practical method for increasing the amount of information is to decrease the standard deviation σ for the individual results, both by diminishing the perturbations, and by increasing the number of determinations.

For the amount of information given by a determination, the following equation is also used in the analytical literature [9-11,14,15]:

$$I = \log \frac{c'' - c'}{2z\sigma} \tag{3.24}$$

where the denominator is a measure of the magnitude of the confidence interval for the analytical errors. Hençe, this equation takes the probability density of the analytical errors as a uniform distribution over the interval $2z\sigma$, which is actually not a rigorously valid assumption.

From (3.9) the entropy corresponding to a normal distribution with standard deviation σ is equivalent to the entropy of a uniform distribution of the event within an uncertainty interval of length $\sigma\sqrt{2\pi e}$. From this it follows that the error of analysis, which is distributed according to an arbitrary law, can be replaced by an equivalent entropy error, as calculated for a uniform distribution: this equivalent value will be called the *entropy value of the analysis errors* [52].

Consequently, the entropy value of the normally distributed analysis errors will be:

$$\Delta = 1/2\,\sigma\sqrt{2\pi e} \tag{3.25}$$

In terms of the entropy value of analysis errors, Eq. (3.21) for the amount of information is written as

$$I = \log \frac{c'' - c'}{2\Delta} \tag{3.26}$$

In what follows the expression of the amount of information will be obtained for the case of several independent components. Consider first the case of determination of three components X_1, X_2, X_3.

We accept that before the experiment we know the ranges of uncertainty for the concentrations of the three components to be (c_1', c_1''), (c_2', c_2''), (c_3', c_3'') where $c_1'' + c_2'' + c_3'' \leqslant 100\%$.

Under these conditions, the composition of the material can be represented, in three-dimension space, by a random vector contained in the rectangular parallelepiped with edges (c_1', c_1''), (c_2', c_2'') and (c_3', c_3''), as represented in Fig. 3.7.

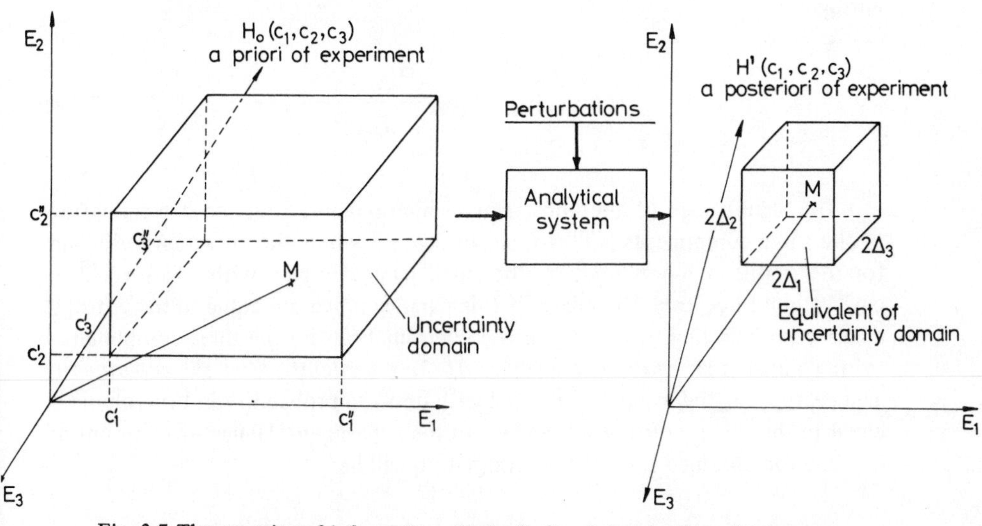

Fig. 3.7 The quantity of information obtained after determination of three components in a material. M designates a particular value from the set of possible values.

Without other information available prior to the experiment, we must accept that the concentrations of the three components are uniformly distributed in the uncertainty domain (the large parallelepiped in Fig. 3.7).

If the quantization interval Δc is the same for each of the three components, it results that the entropy before the determination is

$$H_0(c_1,c_2,c_3) = \log \frac{(c_1''-c_1')(c_2''-c_2')(c_3''-c_3')}{(\Delta c)^3} \tag{3.27}$$

To obtain the equation for the entropy after the determination, we shall consider the case where the determinations of the three components are independent and without systematic errors, and the random errors are normally distributed with standard deviations σ_1, σ_2, and σ_3. Hence, the determination error of the three components is represented by a three-dimensional random vector with a normal distribution.

Consequently, the residual entropy after the determination is:

$$H'(c_1,c_2,c_3) = \log \frac{(\sigma_1\sqrt{2\pi e})(\sigma_2\sqrt{2\pi e})(\sigma_3\sqrt{2\pi e})}{(\Delta c)^3} \tag{3.28}$$

Introducing into (3.28) the entropy values of the analysis errors for the three components, Δ_1, Δ_2, and Δ_3, we obtain the following value for the residual entropy:

$$H'(c_1,c_2,c_3) = \log \frac{2^3\,\Delta_1\,\Delta_2\,\Delta_3}{(\Delta c)^3}; \tag{3.29}$$

The significance of the uncertainty which remains after the determination of the three components is illustrated in Fig. 3.7, where the uncertainty domain for the results is represented by the small parallelepiped, with edges $\sigma_1\sqrt{2\pi e}$, $\sigma_2\sqrt{2\pi e}$, and $\sigma_3\sqrt{2\pi e}$. The edges of this parallelepiped are equal to the entropy errors corresponding to the respective determinations for the three components, so its volume is an *equivalent measure of the uncertainty after the experiment.* The entropy of the events distributed with uniform probability in this volume is equal to the actual entropy subsequent to the experiment. Hence, the amount of information obtained from the determination will be:

$$I(c_1,c_2,c_3) = \log \frac{(c_1''-c_1')(c_2''-c_2')(c_3''-c_3')}{\sigma_1\,\sigma_2\,\sigma_3\,(\sqrt{2\pi e})^3} \tag{3.30}$$

By generalization, it follows that for determination of n independent components in a material, the material composition as known before the experiment can be represented by an n-dimensional random vector, uniformly distributed within an $S^{(n)}$ volume from n-dimensional space (Fig. 3.8). Taking into account the uncertainty intervals (c_i', c_i'') and the standard deviations σ_i of the errors of analysis for all the species ($i = 1,2, \ldots, n$), the information obtained after the determination of the n components will be given by the following equations which generalize (3.29) and (3.30):

$$I(c_1, c_2, \ldots, c_n) = \log \frac{(c_1''-c_1')(c_2''-c_2') \ldots (c_n''-c_n')}{\sigma_1 \sigma_2 \ldots \sigma_n (\sqrt{2\pi e})^n} \tag{3.31}$$

$$I(c_1, c_2, \ldots, c_n) = \log \frac{(c_1''-c_1')(c_2''-c_2') \ldots (c_n''-c_n')}{2^n \Delta_1 \Delta_2 \ldots \Delta_n} = \frac{S^{(n)}}{s^{(n)}} \tag{3.32}$$

The significance of this is illustrated graphically in Fig. 3.8 which expresses the informational essence of research on an analytical determination, which assumes decrease of the uncertainty interval from a value $S^{(n)}$ to a value $s^{(n)} < S^{(n)}$.

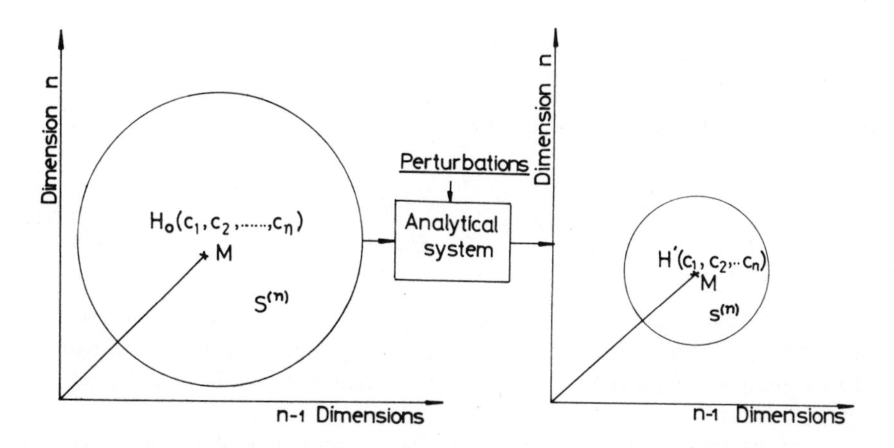

Fig. 3.8 The process of obtaining information after an analytical determination. M designates a particular value from the set of possible values.

Example. Establish the amount of information (calculated in bits) obtained by determination of the concentration of three components in a material. Before the experiment, we know that the uncertainty interval for each of the three concentrations, is (0.01%, 0.10%) and also that the standard deviations for the determination errors are $\sigma_1 = 0.005\%$, $\sigma_2 = 0.008\%$ and $\sigma_3 = 0.007\%$

From (3.30) we obtain:

$$I = \log_2 \frac{(0.10 - 0.01)^3}{0.005 \times 0.008 \times 0.007(\sqrt{2\pi e})^3} = 7 \text{ bits}$$

3.7 REDUNDANCY OF AN ANALYSED MATERIAL

This concept is introduced by analogy with the concept of redundancy, as used in the theory of information transmission, where it is a characteristic of the source of information.

The redundancy of a material which has been analysed is a measure of the deviation of the entropy (which, in its turn, measures the uncertainty specific to the investigated material) from its maximum value (which would be attained if all the possible states had equal probability, i.e. if no knowledge about the material investigated could become available).

Thus the redundancy is defined as the difference between the maximum attainable value of the entropy, $H_{max}(c)$, and the real value of the entropy, $H_0(c)$:

$$R = H_{max}(c) - H_0(c) \tag{3.33}$$

The relative redundancy is defined by:

$$\rho = 1 - \frac{H_0(c)}{H_{max}(c)} \tag{3.34}$$

From (3.33) and 3.34) it is obvious that the redundancy is a measure of the degree of information available, before the determination, about the material to be analysed. To clarify this concept, we shall consider both detection and determination.

Let us take detection first. If all 104 known chemical elements may be present, with equal probability, in the given material, it follows that the maximum entropy is:

$$H_{max} = \log_2 2^{104} = 104 \text{ bits}$$

If, however, information available prior to the experiment reduces the uncertainty concerning the elemental composition to only n elements where $n < 104$, the number of equally probable states for the investigated material is 2^n, so that the real entropy is:

$$H_0 = n \text{ bits} < 104 \text{ bits}$$

Hence the redundancy of the material is:

$$R = (104 - n) \text{ bits} \tag{3.35}$$

For example, when the elementary composition is to be established for a material which, prior to the experiment, is known to contain only $n = 26$ elements, the material redundancy is, from (3.33) and (3.34), $R = 78$ bits, and $\rho = 0.75$ bits.

In the case of determination, the problem of material redundancy is similarly formulated.

When the result of the determination is expressed as per cent of the component in the sample, the maximum entropy will correspond to the uncertainty domain (0%, 100%), so that:

$$H_{max}(c) = \log \frac{100}{\Delta c} \tag{3.36}$$

Because, in general, the real uncertainty domain (c'%, c''%) is smaller than the domain (0%, 100%), the real entropy value is:

$$H_0(c) = \log \frac{c''-c'}{\Delta c} < H_{max}(c) \tag{3.37}$$

Hence in this case the redundancy is:

$$R = \log \frac{100}{c''-c'} \tag{3.38}$$

For example, if the uncertainty domain for the component concentration is (0%, 25%) the redundancy will be:

$$R = \log_2 \frac{100}{25} = 2 \text{ bits}$$

The redundancy appears not only as a result of a restriction of the uncertainty domain, but also on going from a uniform distribution (for which the entropy is maximum) to another distribution. Thus, for the uniform distribution of the event in the uncertainty interval (c', c''), the entropy is:

$$H'_{max} = \log \frac{(c''-c')}{\Delta c}$$

but if the event has a normal distribution over this interval, i.e. a standard deviation $\sigma = (c''-c')/6$, (for $P = 0.997$), the entropy will be given by:

$$H_0' = \log \frac{\sigma\sqrt{2\pi e}}{\Delta c} = \log \frac{(c''-c')\sqrt{2\pi e}}{6\Delta c} \tag{3.39}$$

Consequently, the redundancy associated with the normal distribution for the same uncertainty interval is:

$$R' = \log_2 \frac{6}{\sqrt{2\pi e}} \text{bits}$$

3.8 INFORMATIONAL CHARACTERIZATION OF ANALYTICAL SYSTEMS

A matter of some importance is the introduction of informational criteria for characterizing the quality of performance of analytical instruments and systems of measurement.

3.8.1 Informational power of an analytical instrument

This informational criterion for performance of analytical instruments was introduced by Kaiser [7], especially in connection with spectral methods.

In the case of a simple classical analytical instrument, which measures only the amplitude of a signal (temperature, mass, concentration, etc.), the informational power is calculated from the equation:

$$P_{\text{inf}} = \log S_{\text{A}} \tag{3.40}$$

where S_{A} is the number of values which are distinguishable over the entire measuring scale of the instrument. Thus, for example, an instrument which has a measuring scale with $S_{\text{A}} = 1000$ will have an informational power $P_{\text{inf}} = \log_2 1000 \sim 10$ bits.

For many modern analytical instruments there is the possibility of changing, within a certain range, some parameter of the measured signal. For example, spectrometers are used for absorption, emission, fluorescence or phosphorescence measurements within a certain spectral region, i.e. the frequency of the radiation being measured can be modified within certain limits. If $(x_{\text{a}}, x_{\text{b}})$ denotes this range of variation, $R(x) = (x_{\text{b}}-x_{\text{a}})/\Delta x$ is the resolving power of the instrument within the range $(x_{\text{a}},x_{\text{b}})$ (i.e. Δx is the smallest difference that can be discriminated on the scale of the instrument), and $S_{\text{A}}(x)$ is the number of distinct values on the scale, the informational power is:

$$P_{\text{inf}} = \int_{x_{\text{a}}}^{x_{\text{b}}} R(x) \log S_{\text{A}}(x) \frac{\text{d}x}{x} \tag{3.41}$$

If the resolving power is almost constant over the whole range and the instrument has a uniform scale, we can substitute the means $\bar{R}(x)$ and $\bar{S}_A$ for the corresponding integrals to obtain

$$P_{inf} = \bar{R}(x) \log \bar{S}_A \ln\left(\frac{x_b}{x_a}\right) \tag{3.42}$$

For example, an ordinary absorption spectrometer working in the region from 200 to 1200 nm (that is $x_b/x_a = 1200/200$) with a resolution $\Delta\lambda = 1$ nm and a mean spectral resolving power $\bar{R}(x) = 700$*, and fitted with a transmittance scale graduated from 0 to 100% at 0.5% intervals (i.e. $\bar{S}_A = 200$), has the informational power

$$P_{inf} = 700 \log_2 200 \ln \frac{1200}{200} = 9580 \text{ bits}$$

Generally, modern instruments have a much larger informational power than classical ones. It follows from (3.42) that we can increase the informational power of an instrument by enlarging the interval (x_a, x_b) (i.e. by increasing the factor $\ln (x_b/x_a)$), by increasing the discrimination (i.e. the factor $\log \bar{S}_A$), and also by increasing the resolving power $R(x)$.

Equation (3.42) for the informational power of an instrument has been extended by Fitzgerald and Winefordner [53] for the case of an instrument provided with an auxiliary device for resolving the signal parameter, and also for an instrument in which two signal parameters are resolved. For example, if there are two devices for resolving parameter x, with mean resolving powers $\bar{R}_1(x)$ and $\bar{R}_2(x)$, the informational power will be:

$$P_{inf} = [\bar{R}_1(x) + \bar{R}_2(x)] \log \bar{S}_A \ln\left(\frac{x_b}{x_a}\right) \tag{3.43}$$

Thus for a phosphorimeter with an excitation monochromator with mean resolving power $\bar{R}_{ex}(\lambda)$, and an emission monochromator with mean resolving power $\bar{R}_{em}(\lambda)$, working in the range (λ_a, λ_b), the informational power will be:

$$P_{inf} = [\bar{R}_{ex}(\lambda) + \bar{R}_{em}(\lambda)] \log \bar{S}_A \ln\left(\frac{\lambda_b}{\lambda_a}\right) \tag{3.44}$$

When the measuring instrument is provided with two accessories for resolving the signal parameters x and y, the equation for the informational power is:

$$P_{inf} = \bar{R}(y) \bar{R}(x) \log \bar{S}_A \ln \frac{x_b}{x_a} \ln \frac{y_b}{y_a} \tag{3.45}$$

$$* \bar{R} = \frac{1}{1200-200} \int_{200}^{1200} \frac{\lambda}{\Delta\lambda} d\lambda = \frac{1}{2000} (1200^2-200^2) = 700.$$

For example, if the phosphorimeter also has an accessory for time-resolution of the signal, the equation for the informational power of the instrument becomes:

$$P_{inf} = [\bar{R}_{ex}(\lambda) + \bar{R}_{em}(\lambda)]\bar{R}(t) \log \bar{S}_A \ln \frac{\lambda_b}{\lambda_a} \ln \frac{t_2}{t_1} \tag{3.46}$$

where $\bar{R}(t)$ is the mean time-resolving power for the signal, and (t_1, t_2) the time interval within which the discrimination is being made.

By generalization of (3.46) for the analytical signal treated as an n-dimensional random vector, we obtain the following equation for the informational power of the instrument:

$$P_{inf} = \sum_{i=1}^{n} \bar{R}(x_i) \log \bar{S}_A \sum_{i=1}^{n} \ln \frac{x_{ib}}{x_{ia}} \tag{3.47}$$

A large value for the informational power indicates that the instrument has a wide range of application for solving complex analytical problems. Trace analysis usually involves such problems, so high informational power instruments are required.

It must be mentioned, however, that in real conditions, owing to the effect of various noise sources, the informational power of an instrument is correspondingly diminished.

3.8.2 Informational capacity of an analytical system

The discussion in Chapter 1 revealed that the investigation of a material to solve a given analytical problem requires the use of an analytical system. Analytical instruments are only elements of the informational chain that constitutes an analytical system of measurement, and it has been found useful to introduce informational criteria for the complete analytical system.

The informational capacity of an analytical system expresses the maximum quantity of uncertainty which can be removed by carrying out a complete analytical cycle. Let us first consider detection. If the analytical system can detect at most n components its maximum informational power for detection will be

$$(I_d)_{max} = \log_2 2^n = n \text{ bits} \tag{3.48}$$

Of all the detection systems available, spectroscopic and chromatographic methods probably have the largest informational capacity. For example, an arc spectrograph with photographic recording can detect 30-40 elements in a single exposure, giving an informational capacity of $(I_d)_{max}$ = 30–40 bits. Similarly in chromatography as many as 80–100 chemical components may be detectable in a single run, so an informational detection capacity of 80–100 bits may be attained.

To clarify the concept of informational determination capacity, we shall first consider an analytical system with a linear calibration function with 0–S units for the signal parameter (mass, voltage, intensity, transmittance, etc.). Corresponding to this measurement range $(0,S)$, there is a range of concentrations to be determined $(0,c_{max}\%)$. Under optimum conditions the system should discriminate S distinct concentration values uniformly distributed in the range $(0,c_{max}\%)$. Consequently, the maximum uncertainty which can be removed by the analytical system, or in other words, the maximum informational capacity of the system, is:

$$(I_D)_{max} = \log S = \log \frac{c_{max}}{\Delta c} \tag{3.49}$$

where $\Delta c = c_{max}/S$.

Because of the effect of perturbations and, hence, of the existence of a remaining uncertainty after the experiment, the real informational capacity $(I_D)_{real}$ will always be smaller than the maximum informational capacity $(I_D)_{max}$.

Generally, if the errors are normally distributed with standard deviation σ, the residual uncertainty is:

$$H'(c) = \log \frac{\sigma\sqrt{2\pi e}}{\Delta c} \tag{3.50}$$

and the real informational capacity will be:

$$(I_D)_{real} = (I_D)_{max} - H'(c) = \log \frac{c_{max}}{\sigma\sqrt{2\pi e}} \tag{3.51}$$

For analytical systems with several channels, i.e. those which can detect several components simultaneously, the informational capacity is:

$$\begin{aligned}(I_D)_{real} &= \log \frac{c_{1\,max}\, c_{2\,max} \ldots c_{n\,max}}{\sigma_1\, \sigma_2 \ldots \sigma_n(\sqrt{2\pi e})^n} \\ &= \log \frac{c_{1max}\, c_{2\,max} \ldots c_{n\,max}}{2^n\, \Delta_1\, \Delta_2 \ldots \Delta_n}\end{aligned} \tag{3.52}$$

where there are n channels, $c_{1\,max}, c_{2\,max}, \ldots, c_{n\,max}$ are the maximum concentration ranges for each component and channel, the corresponding standard deviations are $\sigma_1, \sigma_2, \ldots \sigma_n$ and the entropy values of the errors are $\Delta_1, \Delta_2, \ldots, \Delta_n$.

For example, the real informational capacity of a two-channel system with $c_{1\,max} = c_{2\,max} = 2\%$ and standard deviations $\sigma_1 = \sigma_2 = 0.015\%$ will be:

$$(I_D)_{real} = \log_2 \frac{2 \times 2}{0.015 \times 0.015(\sqrt{2\pi e})^2} = 10 \text{ bits}$$

3.8.3 Redundancy of an analytical system

In terms of application it is important to know the degree to which the informational capacity of an analytical system is used. The redundancy will be:

$$R = I_{real} - I'_{real} \tag{3.53}$$

and measures that part of the informational capacity which remains unutilized in the process of analysis.

Obviously, in practice the redundancy should be made as small as possible.

For example, if an arc spectrograph can detect 16 components of a sample in a single exposure it will have an informational capacity $I_{real} = 16$ bits [Eq. (3.17)]. If it is used for the detection of only two components, i.e. if it is asked to supply an information of only $I'_{real} = 2$ bits, the redundancy of the system (the unused part of its informational capacity) will be $R = 14$ bits.

Redundancy is similarly evaluated for determination systems. Again take the two-channel system with informational capacity $I_{real} = 10$ bits, but use it to determine the concentrations of two components for which the range of uncertainty is $c'_{1\,max} = c'_{2\,max} = 1\%$. Hence, the part of the informational capacity utilized is $I'_{real} = 2.5$ bits. The redundancy (the unused part of the informational capacity) will be $R = 7.5$ bits.

3.8.4 Information flow of an analytical system

From the practical point of view it is very important to have a characteristic giving a measure of the quantity of information given by the analytical system in unit time; the information rate is such a characteristic.

The information rate i_{real} of an analytical system is the ratio between the informational capacity used (I'_{real}) and the time Δt required for the analytical cycle:

$$i_{real} = \frac{I'_{real}}{\Delta t} \tag{3.54}$$

The achievement of analytical systems with large informational capacity and small cycle time is of great interest, because such systems are required in many branches of industry and research. There are several such systems available, for example X-ray fluorescence and emission spectrometers which can determine 40 components in a sample in less than a minute.

To illustrate the concept we shall consider the two-channel system with operational informational capacity $I'_{\text{real}} = 2.5$ bits, and assume that a complete analytical cycle requires 2.5 minutes. Then the information rate is $i_{\text{real}} = 1$ bit/minute.

3.8.5 Informational power of an analytical system

By this we understand the amount of information obtained in a determination when, prior to the experiment, the range of uncertainty of the concentration to be determined is 1 unit. Hence, from Eq. (3.21) for the amount of information, the informational power for the determination of one component is

$$P_{\text{I}} = \log \frac{1}{\sigma\sqrt{2\pi e}} \tag{3.55}$$

Hence the smaller the standard deviation, the higher the informational power, and the more adequate the analysis system for trace analysis.

For example, in the photometric determination of phosphorus in steel, if the standard deviation is $\sigma = 0.0002\%$, the informational power is:

$$P_{\text{I}} = \log_2 \frac{1}{0.0002\sqrt{2\pi e}} = \log_2 1250 = 10.3 \text{ bits}$$

REFERENCES

[1] Shannon, C. E., *Bell. Syst. Techn. J.*, **27**, 379, 623 (1948).
[2] Stakheev, Yu. I., *Zavodsk. Lab.*, **28**, 831 (1962).
[3] Liteanu, C. and Crişan, I. Al., *Rev. Roumaine Chim.*, **12**, 1475 (1967).
[4] Malissa, H. and Jellinek, G., *Z. Anal. Chem.*, **238**, 81 (1968).
[5] Kienitz, H. and Kaiser, R., *Method. Phys. Anal.*, **5**, 274 (1969).
[6] Doerffel, K. and Hildebrandt, W., *Wiss. Z. Tech. Hochschule für Chemie. Leuna-Merseburg*, **11**, 30 (1969).
[7] Doerffel, K., *Chem. Tech.*, **25**, 94 (1973).
[8] Kaiser, H., *Anal. Chem.*, **42** (2), 24A; (4), 26A (1970).
[9] Eckschlager, K., *Collection Czech. Chem. Commun.*, **36**, 3016 (1971).
[10] Eckschlager, K., *Collection Czech. Chem. Commun.*, **37**, 137 (1972).
[11] Eckschlager, K., *Collection Czech. Chem. Commun.*, **38**, 1486 (1972).
[12] Eckschlager, K., *Collection Czech. Chem. Commun.*, **38**, 1330 (1973).
[13] Eckschlager, K., and Vajda, J., *Collection Czech. Chem. Commun.*, **39**, 3076 (1974).

[14] Eckschlager, K., *Collection Czech. Chem. Commun.*, **40**, 3627 (1975).
[15] Eckschlager, K., *Chem. Listy,* **69**, 810 (1975).
[16] Eckschlager, K., *Z. Chem.*, **16**, 111 (1976).
[17] Eckschlager, K., *Collection Czech. Chem. Commun.*, **41**, 1875 (1976).
[18] Griepink, B. F. A. and Dijkstra, G., *Z. Anal. Chem.*, **257**, 269 (1971).
[19] Belyaev, Yu. I. and Koveshnikova, T. A., *Zh. Analit. Khim.*, **27**, 429 (1972).
[20] Danzer, K., *Z. Chem.*, **13**, 20 (1973).
[21] Danzer, K., *Z. Chem.*, **13**, 69 (1973).
[22] Danzer, K., *Z. Chem.*, **13**, 229 (1973).
[23] Danzer, K., *Z. Chem.*, **14**, 73 (1974).
[24] Danzer, K., *Z. Chem.*, **15**, 158 (1975).
[25] Danzer, K., *Z. Chem.*, **15**, 326 (1975).
[26] Huber, J. F. K. and Smit, H. C., *Z. Anal. Chem.*, **245**, 84 (1969).
[27] Grotch, S. L., *Anal. Chem.*, **42**, 1214 (1970).
[28] Palm, E., *Z. Anal. Chem.*, **256**, 25 (1975).
[29] Massart, D. L., *J. Chromatog.*, **79**, 157 (1973).
[30] Clerc, J. T., Kaiser, R., Rendl, J., Spitzy, H., Zettler, H., Gottschalk, G., Malissa, H., Schwarz-Bergkampf, E. and Werder, R. D., *Z. Anal. Chem.*, **272**, 1 (1974).
[31] Massart, D. L. and Smits, R., *Anal. Chem.*, **46**, 283 (1974).
[32] Dupuis, F. and Dijkstra, A., *Anal. Chem.*, **47**, 379 (1975).
[33] Grys, St., *Z. Anal. Chem.*, **273**, 177 (1975).
[34] Eskes, A., Dupuis, F., Dijkstra, A., De Clercq, H. and Massart, D. L., *Anal. Chem.*, **47**, 2168 (1975).
[35] Eckschlager, K., *Collection Czech. Chem. Commun.*, **41**, 2527 (1976).
[36] Eckschlager, K., *Anal. Chem.*, **49**, 1265 (1977).
[37] Eckschlager, K., *Z. Anal. Chem.*, **277**, 1 (1975).
[38] Eckschlager, K., *Collection Czech. Chem. Commun.*, **42**, 225 (1977).
[39] Eckschlager, K., *Collection Czech. Chem. Commun.*, **43**, 231 (1978).
[40] Ehrlich, G., *Chem. Anal. (Warsaw)*, **21**, 303 (1976).
[41] Michalowski, T., Parczewski, A. and Rokosz, A., *Chem. Anal. (Warsaw)*, **21**, 979 (1976).
[42] Danzer, K., *Z. Chem.*, **18**, 104 (1978).
[43] van Marlen, G. and Dijkstra, A., *Anal. Chem.*, **48**, 595 (1976).
[44] Souto, J. and Gonzales de Valesi, A. J., *J. Chromatog. Sci.*, **64**, 135 (1970).
[45] Dupuis, F. and Dijkstra, A., *Z. Anal. Chem.*, **290**, 357 (1978).
[46] Dupuis, F., Dijkstra, A. and van der Maas, J. H., *Z. Anal. Chem.*, **291**, 27 (1978).
[47] Heite, F. H., Dupuis, F., Van't Klooster, H. A. and Dijkstra A., *Anal. Chim. Acta,* **103**, 313 (1978).
[48] Danzer, K. and Eckschlager, K., *Talanta,* **25**, 725 (1978).
[49] Eckschlager, K., *Collection Czech. Chem. Commun.*, **44**, 2373 (1979).
[50] Liteanu, C. and Rîcă, I., *Anal. Chem.*, **51**, 1986 (1979).

[51] Brillouin, L., *La science et la théorie de l'information,* p. 196, Masson, Paris (1959).
[52] Novitskii, P. V., *Osnovi teorii informatsii izomerezelnykh ustroistv,* Izd. Energia, Leningrad (1968).
[53] Fitzgerald, J. J. and Winefordner, J. D., *Rev. Anal. Chem.*, 2, 299 (1975).

Chapter 4

Stability of Analytical Systems

4.1 INTRODUCTION

The performance of an analytical system can be assessed in terms of quality of results, and also the length of time for which it will function continuously without breakdown or deterioration of performance. The system becomes defective when the errors consistently exceed certain preselected limits.

Analytical systems are characterized by a particularly complex functional structure, because they can include chemical, mechanical, electrical and optical elements. If the quality parameters of these elements change irreversibly in time, corresponding changes occur in the quality characteristics of the analytical system. An examination of quality parameters in general gives an insight into those which affect the analytical results.

In the most general terms, the variation in a quality parameter with time can be represented as the sum of a random reversible process associated with the normal fluctuations of the parameter under working conditions, and a semi-random irreversible process due to wear and aging of the element. In other words, the analytical errors can often be resolved into a stationary random component and a systematic time-dependent component. Thus, for example, in weighing on a given balance, the weighing error alters with time, and has a random component due to the many minute fluctuations in the operation of the working parts during the weighing, and a long-term systematic and time-dependent component due to wear.

Optical elements (mirrors, filters, lenses, prisms, lattices, etc.) change with time because of irreversible wear phenomena, so the quality parameters which correspond to them (the coefficients of reflection, transmission, dispersion, selectivity, etc.) also change in time.

The physico-chemical properties of the materials used in the analysis process also change with time. Solutions may lose solvent by evaporation, or solute by chemical interaction with the environment (the solvent, the atmosphere or the container) or may become contaminated. Contamination of the sample – an extremely important source of error in trace analysis – is often time-dependent.

Chromatographic columns are also affected by irreversible phenomena, so the packing must be periodically reactivated or renewed.

The transducers and sensors used in the measuring elements of analytical systems are also affected by irreversible phenomena and they must be periodically calibrated, renewed or reconditioned from time to time.

Generally, for qualitative characterization of technical systems or elements, we use the concept of 'stability'. By this we mean the quality of performance as a function of time. The most commonly used stability parameters are:

(a) the reliability factor, measured by the probability that an element, or a system of elements, is in proper working order at a certain level of quality during the time interval $(0,t)$;

(b) the probability that the system is working at any given moment;

(c) the utilization ratio, measured by the fraction of the total time for which the system is working at the required quality level.

The statistical theory of the stability of analytical systems is an application of the general theory of stability, and includes the statistical methods of evaluating and controlling the quality of the systems as a function of time.

These methods can be separated into three classes:

(a) methods for evaluating, analysing and increasing the stability;

(b) methods for control of the quality;

(c) methods for regulation of the systems.

The increasing complexity of analytical systems, which include a large number of varied elements, demands statistical evaluation and analysis of the stability, so that the design and operation of the system can be optimized.

To reduce the effects of wear and aging, two preventive measures can be taken: (a) regular checks on the quality state of the system, and (b) regular maintenance and servicing of working parts, replacement of defective or substandard parts, and standardization of the materials used, etc.

The organization of these measures on statistical principles is the most reliable way of controlling the quality of analytical systems, especially if an electronic computer is used for the purpose. The best method of directing and controlling an analytical system is with an on-line process computer, which gives feedback control of the analysis process and removes the need for an operator in certain steps of the process.

X-Ray emission and fluorescence spectrometers, electron-beam micro-analysers and other modern chemical analysers are now generally equipped with microprocessors giving electronic command and control of the analyser; feedback for correcting errors caused by detuning or other misadjustments; calculation and processing of data and conveyance of results to the user, *etc.*

The need to increase the stability of the analytical system is of current interest because of the demand for effective quality control of chemical composition etc., in the chemical and metallurgical industries, environmental control, etc. The high signal-to-noise ratio required by trace analysis can be achieved only by analytical systems which have a high degree of stability.

4.2 STABILITY OF ANALYTICAL SYSTEMS

Consider an analytical system for which the input is a homogeneous material of constant composition and the output is the set of measurement results (analytical signals) obtained by repeating the analysis (Fig. 4.1).

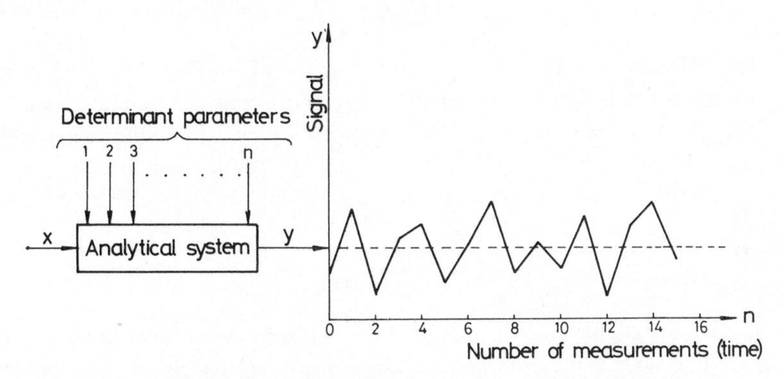

Fig. 4.1 Fluctuation, with time, of the signal given by an analytical system; x – the sum of the input variables; y – the sum of the output variables.

Under these conditions, the fluctuation of the results is caused only by the variation with time of the determining parameters of the analytical system. Hence information about the stability of the analytical system is obtained by examining such a set of data.

Examination of the stability must sometimes consider the total time, and sometimes the effective time (the time during which the system is operating). When the effective time is considered, the time per analytical cycle can be taken as a unit, and the stability examined in relation to the number of cycles. This can also be done when the analysis process is repeated at equal time intervals, the time unit in this case being the interval between the starts of two successive cycles.

The stability of a system must be either stationary or non-stationary with respect to time. It is stationary if the distribution functions of the signals, analytical results, and analytical errors remain unchanged irrespective of the time at which the process was started. In that case the results fluctuate between two constant limits.

The stability is non-stationary when the distribution functions of the results and errors change with time.

Detailed examination of the parameters of the process results in further division of these two classes into several limiting types, some of which are discussed below.

4.2.1 Stationary stability without systematic deviation

Writing the error of an analytical determination as $(\Delta c)_t$, where the index t signifies the constant appearance of the error on repetition of the analysis process, we obtain the following two conditions defining stationary stability without systematic deviation:

$$M(\Delta c)_t = 0 \tag{4.1}$$

$$\mathrm{var}(\Delta c)_t = \sigma^2 \tag{4.2}$$

This means that the distribution function of the errors of analysis is independent of time, has a mean equal to zero and a constant standard deviation.

The limiting case is the strictly stationary stability without systematic deviation. In this case it is required that, besides meeting conditions (4.1) and (4.2), the errors of analysis at different times should be mutually independent, i.e.:

$$\mathrm{cov}[(\Delta c)_{t_k}, (\Delta c)_{t_{k+a}}] = 0 \tag{4.3}$$

In such a case the measurement results conform to a model similar to that given in Fig. 4.2, representing the measurement results obtained in emission

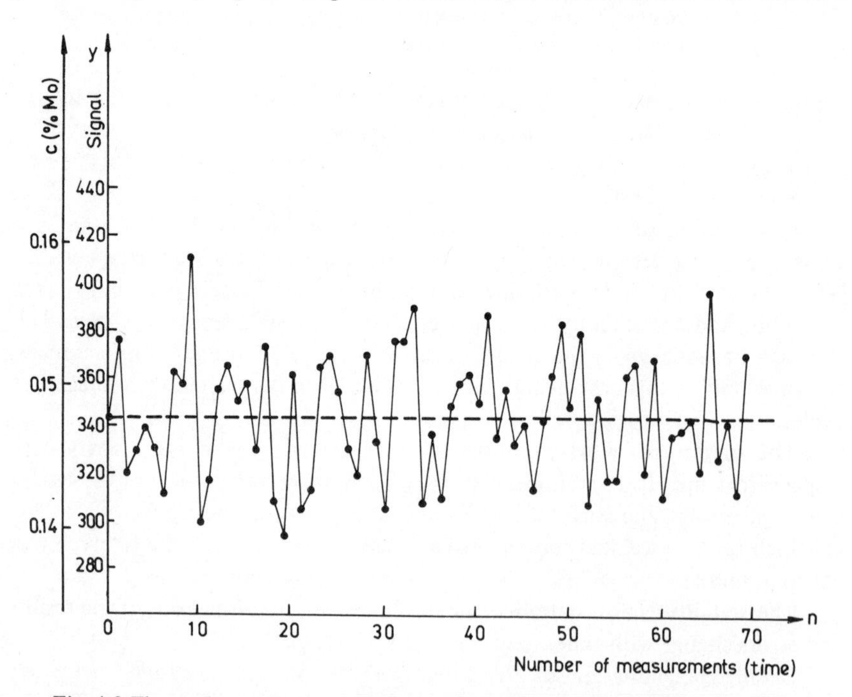

Fig. 4.2 The random variation, with time, of the measurement results in a strictly stationary system.

spectrometry determination of molybdenum in a steel sample with a molybdenum concentration $c_{Mo} = 0.148\%$.

In many cases condition (4.3) is not met by the analytical system and there is a time correlation between the errors at different, but close, moments of time. A model of this type is given in Fig. 4.3, where the random process includes an oscillatory component, the period and amplitude of which are relatively large. Consequently, there is a positive correlation between the fluctuations of results taken at close times; it is manifested by the difference between consecutive results being generally smaller than the difference between two results well separated in time.

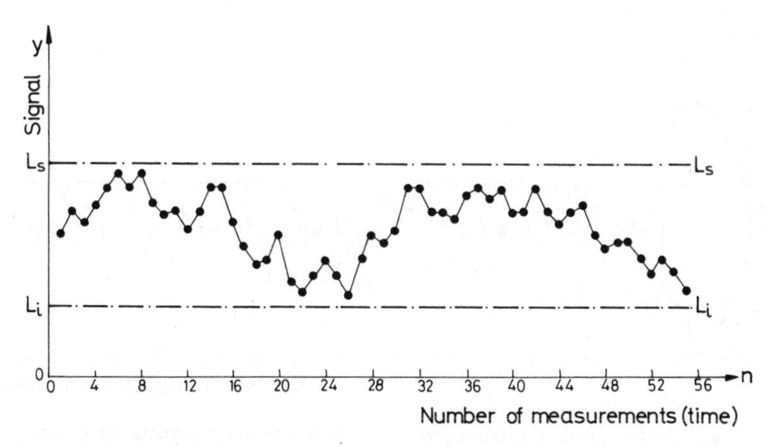

Fig. 4.3 The model of a stationary covariant system with a positive correlation. L_s – the upper limit; L_i – the lower limit.

For a strictly stationary system [Eq. (4.3)], the sample mean of n results $y_1, y_2, \ldots, y_n$

$$\bar{y}_n = \sum_{i=1}^{n} y_i/n \tag{4.4}$$

estimates the general mean, and the corresponding variance is:

$$\mathrm{var}(\bar{y}_n) = \frac{1}{n}\mathrm{var}(y) \tag{4.5}$$

where var(y) is the general variance of the individual results.

In a covariant stationary system, i.e. when there is a time correlation between the results, the variance of the mean of n successive results depends upon the degree of correlation of the results:

$$\mathrm{var}(\bar{y}_n) = \frac{1}{n}\mathrm{var}(y) + \frac{2}{n}\sum_{k=1}^{n}\left(1 - \frac{k}{n}\right)\mathrm{cov}(y_1, y_k) \tag{4.6}$$

It follows that analytical systems with covariant stationary stability with positive correlation behave, over short time intervals, as if they were non-stationary. Equation (4.6) shows that, in order to estimate correctly the general variance in the case of a covariant stationary system, the number of data should be large enough to cover the time interval within which there are correlations, or the data should be taken at such intervals that the correlation is absent.

4.2.2 Stationary stability with constant systematic deviation

A model of this system is given in Fig. 4.4.

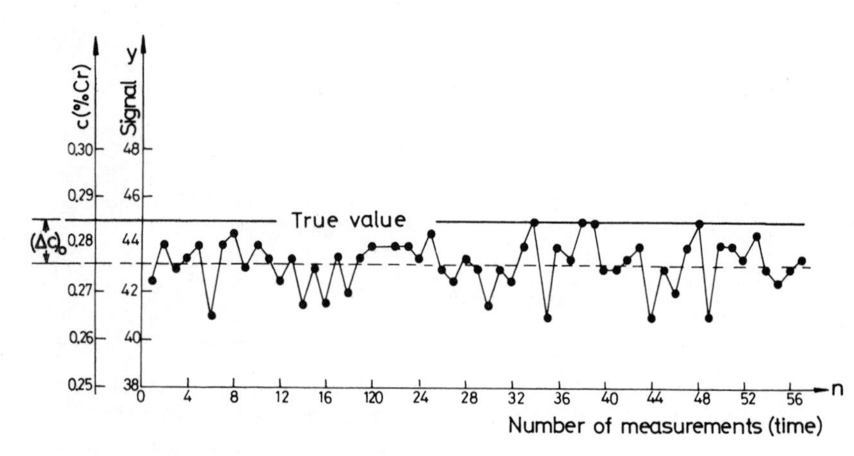

Fig. 4.4 The model of a stationary system with constant systematic deviation.

The analysis error $(\Delta c)_t$ comprises two terms at any given moment: a constant term $(\Delta c)_0$, and a random stationary term $(\Delta c')_t$

$$(\Delta c)_t = (\Delta c)_0 + (\Delta c')_t \tag{4.7}$$

In this case, the component $(\Delta c')_t$ can also have either a strictly stationary or a covariant stationary behaviour.

Evaluation of the systematic component $(\Delta c)_0$ and correction for it will convert such a system into one of stationary stability without systematic deviation.

4.2.3 Stationary stability with random systematic deviation

A model for the time variation of the analysis error in this system is shown in Fig. 4.5. In this case the error comprises a stationary random component and a systematic component which changes in random steps.

A stability system of this kind appears when a determining parameter of the analytical system varies in steps. This can happen, for example, in trace analysis when the purity of the determination reagent varies from batch to batch. So

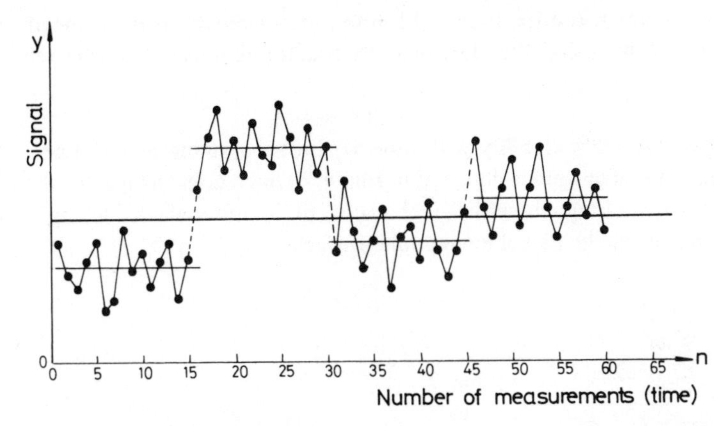

Fig. 4.5 The model of a stationary system with a random systematic deviation.

long as the same batch is used (provided there is no steady deterioration from day to day), the results will all be affected by the same systematic error, but this error can change when a new batch of reagent is used.

4.2.4 Stationary stability with intermittent gross error

This system is shown schematically in Fig. 4.6.

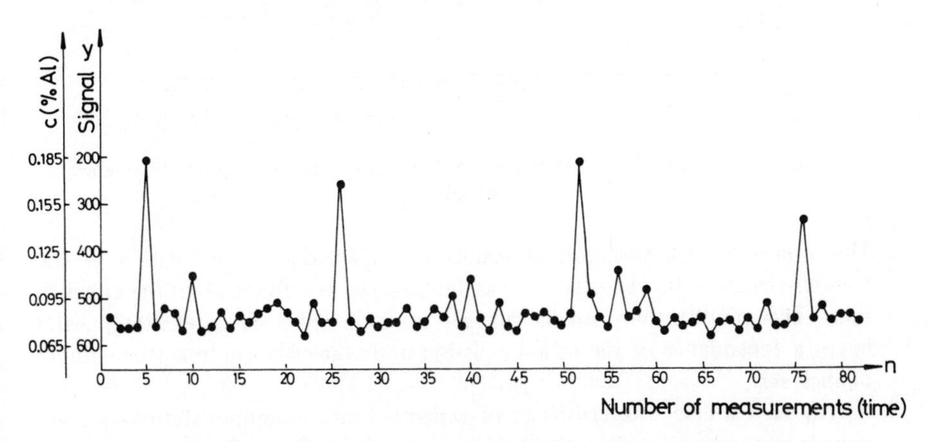

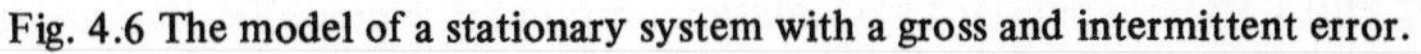

Fig. 4.6 The model of a stationary system with a gross and intermittent error.

The error includes a small stationary random component, and also a gross component which has a variable amplitude and frequency. The source of the gross error must be sought in the appearance of faulty states of the system, with one or several of the determining parameters taking values which are far outside the normal working range. A high frequency of gross intermittent errors indicates a low level of knowledge, organization and exploitation of the analytical system.

Given enough results, those with intermittent gross errors can be identified and removed by using the statistical criteria for removing extreme values (see p. 43).

4.2.5 Non-stationary stability with time-dependent systematic deviation

This type of system is depicted in Fig. 4.7, which gives the results of repeated measurements obtained, without changing the counter-electrode, by emission spectrometry for the nickel in a sample of steel.

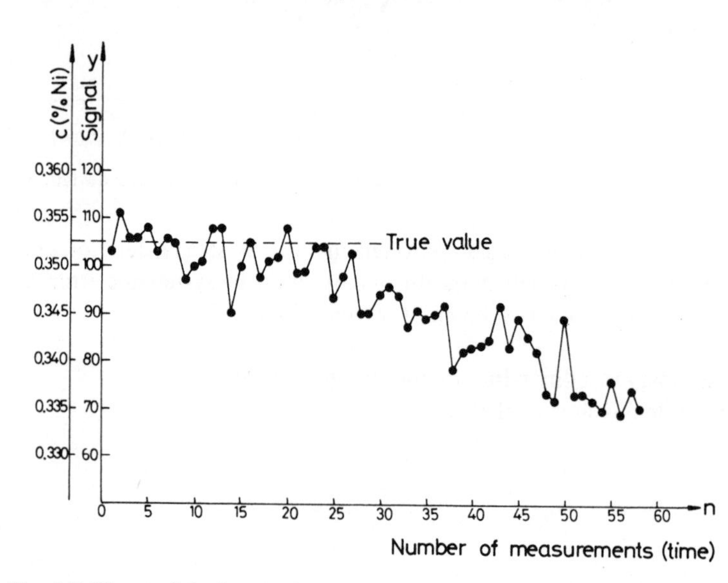

Fig. 4.7 The model of a stationary system with a time-dependent systematic deviation.

This non-stationary variation of results is explained by modification of the counter-electrode by the repeated excitations, i.e. as a function of the effective time. This modification entails changes in the conditions of excitation, and hence a dependence of the nickel emission upon the effective time (the number of analyses).

In such a case, the error can be separated into a random stationary component and a time-dependent systematic component. Generally, the determining parameters of an analytical system in a stability state of this kind are subject to reversible changes, and also to irreversible time-dependent changes which can be caused by various processes, such as wear, aging, misadjustment, etc.

4.2.6 Non-stationary stability with time-dependent variance

An example of a non-stationary system with a variance monotonically increasing with time is given in Fig. 4.8.

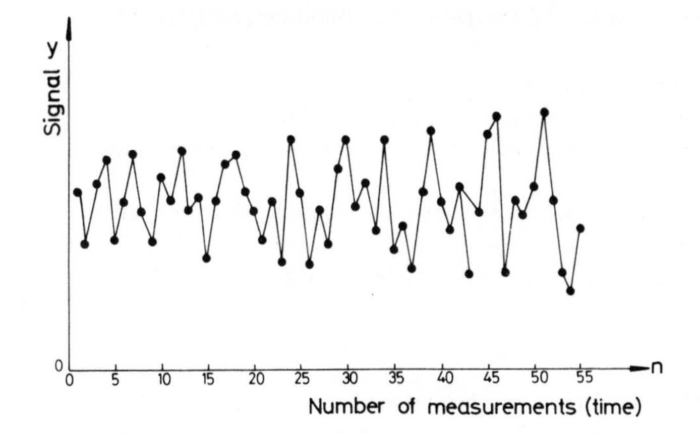

Fig. 4.8 The model of a non-stationary system with a monotonically increasing variance.

Stability states of this type are due to an increase (with time) of the fluctuations of the determining parameters of the system. Such an increasing variance may appear in any analytical system if the proper preventive measures are not taken for removing the effects of wear, aging, or misadjustments.

We have given just a few limiting types of stability states for analytical system. Much more complex situations can arise in practice. Thus, for example, time-dependent systematic deviation and a variance increase may be found simultaneously, or a process leading to gross errors may be superimposed on a time-dependent systematic deviation, and so on.

The following important observations result from this section:

(a) In any analysis process the determination parameters are subject to fluctuations around the prescribed working conditions, and also to time-dependent irreversible changes. Because of these two different types of effect, the analysis errors will behave as a function of time that may have both stationary random components and time-dependent systematic components. Consequently, all analytical systems will have a tendency towards non-stationary behaviour. Stationary behaviour is accomplished and maintained only by planning proper preventive measures for preserving the quality of the system as a function of time.

(b) The study and evaluation of the stability of analytical systems must be based on the methods of mathematical statistics, by considering as large a number of measurement results as possible, obtained over as long a time interval as possible; the conclusions can be extrapolated only a little way in time.

(c) An analytical system must be so conceived, organized, and conducted, that the behaviour of the analysis errors with time is close to a Gaussian stationary model without systematic deviation, with as small a variance as possible.

4.3 STATISTICAL METHODS FOR ASSESSING STABILITY

4.3.1 Regression analysis

Consider a series $y_1, y_2, \ldots, y_n$ of analytical results obtained by repeatedly analysing a homogeneous material of constant composition, at equal time intervals.

The function which estimates the dependence of the mean on the number of determinations n (i.e. on the time, expressed in units equal to the time interval between the starts of two successive analytical cycles) can be written as

$$\bar{y} = \bar{y}'_0 + b'n \tag{4.8}$$

Obviously, if the experimental data are obtained with a stationary system, b' will be either zero or fluctuate around it, and $\bar{y}'_0$ will be an estimate of the mean of the results. If b' is significantly different from zero, this indicates the existence of a non-stationary system with a time-dependent systematic deviation.

We can also obtain information about the stability of the analytical system by evaluating the correlation coefficient between the result and time. The correlation coefficient r must always lie in the interval $-1 \leqslant r \leqslant 1$. For a stationary system, the correlation coefficient between results and time will be zero, but not in the case of a non-stationary regime with time-dependent stationary deviation.

In a covariant stationary system with a relatively large period of oscillation, when only a small time interval is covered by the series of data $y_1, y_2, \ldots, y_n$, the values taken by the parameters b' and r can be significantly different from zero, as if there were a non-stationary system with time-dependent systematic deviation. If, however, the series of data is extended over a longer period. the parameters b' and r will tend towards zero if the system is covariant and stationary.

Example 4.1. For the data given in Fig. 4.7, the linear dependence between the mean signal and time (expressed as the number of determinations) is

$$\bar{y} = 110.4 - 0.61n$$

The value $b' = -0.61$ is significantly different from zero, so the system is in a non-stationary state with time-dependent systematic deviation (Fig. 4.9). Consequently each signal measurement error includes a systematic component of $-0.61n$ digits.

4.3.2 Statistical hypothesis tests for stability

To verify hypotheses about the random behaviour of a set of data there are certain tests for randomness.

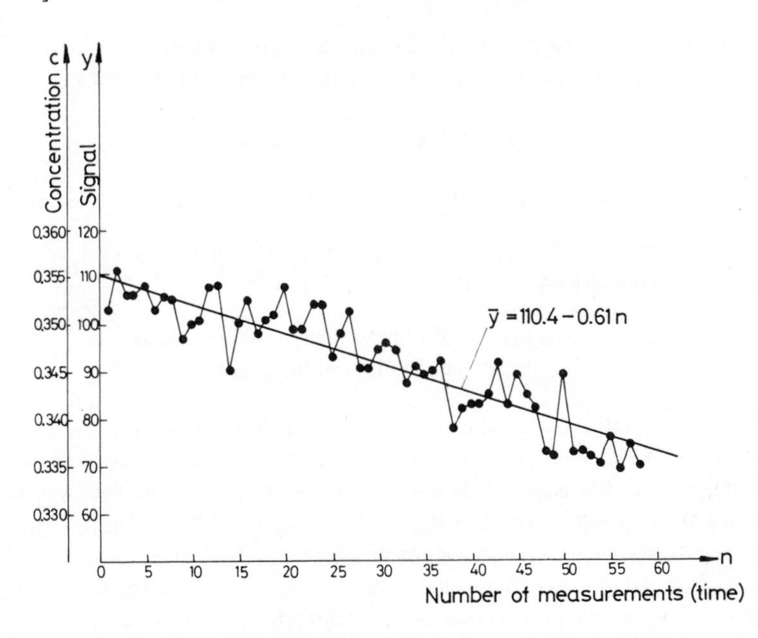

Fig. 4.9 The dependence between the mean result and the signal.

(a) *Number of consecutive similar derivatives and length of iterations.* Consider a series of data $y_1, y_2, \ldots, y_n$ obtained in examining the stability of an analytical system. We write the results in the order in which they were obtained, and label them according to whether they are smaller or greater than the means (say 'a' and 'b' respectively). Thus we obtain a series such as:

$$\text{aabaaabbbbab} \tag{4.9}$$

The number of consecutive iterations supplies information about the stability of the system.

It can be proved that, for a large set of independent and random data belonging to a normal distribution (i.e. a stationary system), the number u of consecutive like deviations in such an experimental series is normally distributed, with mean $M(u) = (n+2)/2$ and variance $\sigma_u^2 = (n-1)/4$.

Consequently, the experimental quantity $z_{\text{exp}} = [u+\frac{1}{2}-(n+2)/2]/\sqrt{(n-1)/4}$ has a distribution $N(0, 1)$, so that, for a significance level α, the critical regions for rejecting the hypothesis of a strictly stationary regime are:

$$z_{\text{exp}} < -|z_{\alpha/2}|; z_{\text{exp}} > |z_{\alpha/2}| \tag{4.10}$$

For a small number of data ($n < 30$), from the statistics of the variable u we can determine two critical values $u'_{\alpha/2}$ and $u''_{\alpha/2}$ for a significance level α, such that, if the experimental value u_{exp} is contained in the critical regions

$$u_{\mathrm{exp}} < u'_{\alpha/2} \,; u_{\mathrm{exp}} > u''_{\alpha/2} \tag{4.11}$$

the hypothesis of a strictly stationary system must be rejected.

Example 4.2. Marking the results in the way just indicated, we obtain the following series for the data given in Fig. 4.2:

abaaaaabbbaabbbbabaabaabbbaabaabbbaaa
bbbbbabaaaabbbbabaabbabaaaaabaaab

For the hypothesis that the system is both strictly stationary and Gaussian, $M(u) = 36$, and $\sigma_u^2 = 17.25$ for $n = 70$. The number of consecutive like deviations in the series of data is $u = 32$, so $z_{\mathrm{exp}} = -0.9$. The table of the normal distribution, for $\alpha = 0.10$, gives $z_{0.05} = 1.65$, so the hypothesis of a strictly stationary system cannot be rejected.

In a similar way, the hypothesis of a strictly stationary system can be verified by a test which takes into consideration the statistics of the length of an iteration of like deviations.

If k_{exp} is the experimentally observed length of an iteration, for a significance level α we can define the critical region

$$k_{\mathrm{exp}} > k_\alpha \tag{4.12}$$

for which the hypothesis of a strictly stationary system is rejected.

The critical value k_α, for a given significance level α, is obtained as follows:

$$k_\alpha = \frac{\log \dfrac{-0.4343\,n}{\log(1-\alpha)}}{\log 2} - 1 \tag{4.13}$$

For the data in example 4.2, i.e. $n = 70$ and $\alpha = 0.05$, Eq. (4.13) yields $k_\alpha = 9.4$. Since the longest iteration in the experimental series takes the value $k_{\mathrm{exp}} = 5 < k_\alpha = 9.4$, the hypothesis of a stationary system cannot be rejected.

(b) *Mean of squares of successive differences.* Again consider a series of experimental data $y_1, y_2, \ldots, y_n$. If the data belong to a strictly stationary Gaussian system with variance σ^2, the quantity:

$$\delta^2 = \sum_{i=1}^{n-1} (y_{i+1} - y_i)^2/(n-1) \tag{4.14}$$

will estimate the general variance by the relation:

$$\delta^2 = 2\sigma^2 \tag{4.14'}$$

Hence the mean of the quantity $\eta = \delta^2/s^2$ should be 2 (s^2 is the sample variance for the set of data).

It can be proved that, for a large set of experimental data ($n>25$), the quantity $\epsilon = 1 - (\eta/2)$ has a normal distribution with mean equal to zero, and variance:

$$\sigma_\epsilon^2 = \frac{n-2}{(n+1)(n-1)} \tag{4.15}$$

Consequently, the experimental variable $z_{exp} = \epsilon/\sigma_\epsilon$ has a distribution $N(0, 1)$. Hence, for a given significance level α, the critical regions for rejecting the hypothesis of a strictly stationary system are

$$z_{exp} < -|z_{\alpha/2}|; z_{exp} > |z_{\alpha/2}| \tag{4.16}$$

If the experimental value z_{exp} is in either critical region, the hypothesis of a strictly stationary Gaussian system will be rejected.

An experimental value $z_{exp} > z_{\alpha/2}$ is due either to a stationary covariant regime with positive correlation, or to a non-stationary regime with time-dependent systematic deviation.

For a small number of data ($n<25$), from the statistics of η, the critical quantities $\eta'_{\alpha/2}$ and $\eta''_{\alpha/2}$ can be defined so that the following critical regions can be established for rejecting the hypothesis of a strictly stationary Gaussian system:

$$\eta_{exp} < \eta'_{\alpha/2}; \eta_{exp} > \eta''_{\alpha/2} \tag{4.17}$$

Example 4.3. Table 4.1 gives two series of measurements obtained with an infrared spectrometer, for the purpose of examining the stability of the system under different analytical conditions. From the data, verify the hypothesis of a strictly stationary system.

For condition I, we find:

$$\delta^2 = 3.387, s^2 = 2.629, \eta = 1.479$$

$$\epsilon = 0.261, \sigma_\epsilon^2 = 0.0321, z_{exp} = 1.45$$

so the hypothesis of a strictly stationary system cannot be rejected ($z_{0.05} = 1.65$).

For condition II,

$$\delta^2 = 5.838, s^2 = 5.258, \eta = 1.11$$

$$\epsilon = 0.45, \sigma_\epsilon^2 = 0.0321, z_{\text{exp}} = 2.5$$

Consequently, for analytical condition II, the hypothesis of a strictly stationary system must be rejected.

Table 4.1 Values of the analytical signal obtained with an infrared spectrometer (UR-10) by integration (12 sec)

Measurement number	Condition I	Condition II
1	2144	2185
2	2234	2236
3	2244	2144
4	2196	2140
5	2285	2233
6	2255	2187
7	2166	2159
8	2068	2307
9	2183	2151
10	2166	2237
11	2119	2349
12	2121	2311
13	2251	2218
14	2217	2347
15	2220	2194
16	2146	2273
17	2109	2308
18	2208	2329
19	2188	2363
20	2164	2326
21	2226	2304
22	2198	2230
23	2196	2202
24	2179	2291
25	2179	2272
26	2286	2315
27	2225	2343
28	2216	2379
29	2175	2328

The correctness of the interpretation follows from an examination of Fig. 4.10, where the two series of data are presented graphically. For condition II, the deviation from a strictly stationary system is evidenced by the appearance of a time-dependent systematic deviation.

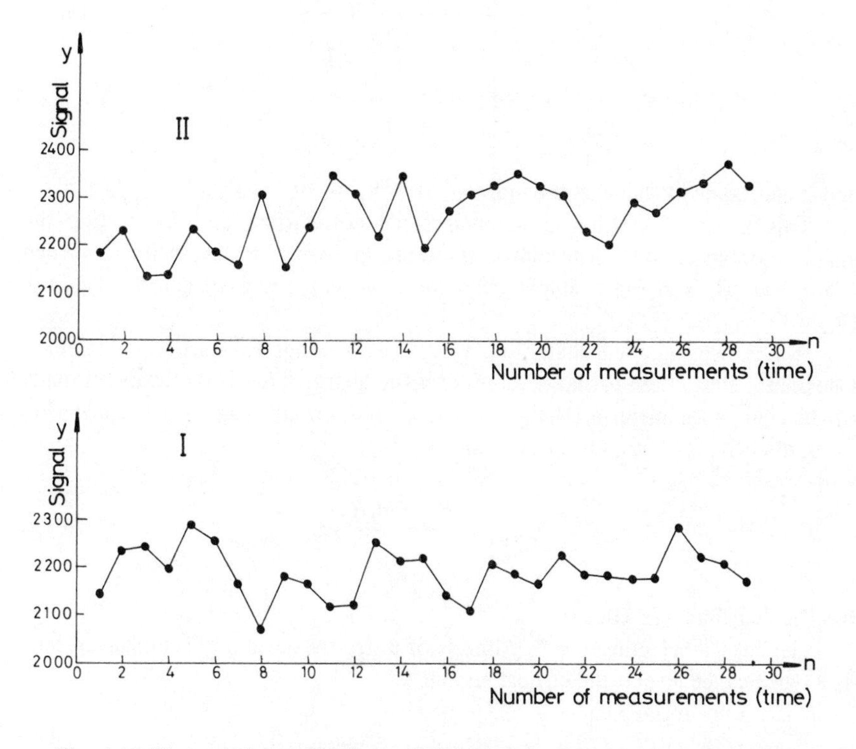

Fig. 4.10 The variation, with time, of the signal from an infrared spectrometer (type U-10) under two sets of analytical conditions.

(c) *The serial correlation coefficient.* By serial correlation we mean the correlation between pairs of equidistant data in an experimental series. That is, for the series of experimental data $y_1, y_2, \ldots, y_n$, the correlation between the data pairs y_i and y_{i+h}, where h is the difference in position of the data in the series.

For example, for a series of $n = 12$ data, the pairs which are to be correlated for $h = 3$ are:

$$(y_1,y_4), (y_2,y_5), (y_3,y_6), (y_4,y_7), (y_5,y_8), (y_6,y_9),$$

$$(y_7,y_{10}), (y_8,y_{11}), (y_9,y_{12}), (y_{10},y_1), (y_{11},y_2), (y_{12},y_3)$$

The correlation coefficient is given by:

$$R_h = \frac{\sum_{i=1}^{n} y_i y_{i+h} - \frac{\left(\sum_{i=1}^{n} y_i\right)^2}{n}}{\sum_{i=1}^{n} y_i^2 - \frac{\left(\sum_{i=1}^{n} y_i\right)^2}{n}} \tag{4.18}$$

and it can take any value in the interval $-1 \leqslant R_h \leqslant +1$.

This is the definition of *circular* serial correlation, and for testing the stability of an analytical system it is better to omit the data pairs for which $i > n - h$. Thus in the example given we must omit the pairs (y_{10},y_1), (y_{11},y_2), (y_{12},y_3).

It has been proved that if $y_1, y_2, \ldots, y_n$ belongs to a strictly stationary Gaussian system, then for large values of n the quantity R_1 is practically normally distributed, with mean $M(R_1) = -1/(n-1)$ and variance $\sigma^2_{R_1} = (n-2)/(n-1)^2$. Consequently, for large values of n, the variable:

$$z_{\text{exp}} = \frac{R_1 - M(R_1)}{\sigma_{R_1}} \tag{4.19}$$

has the distribution $N(0, 1)$.

Hence, for rejecting the hypothesis of a strictly stationary Gaussian system, at a significance level α the critical regions are:

$$z_{\text{exp}} < -|z_{\alpha/2}| \,;\, z_{\text{exp}} > |z_{\alpha/2}| \tag{4.19'}$$

If the experimental quantity z_{exp} is in either critical region, the hypothesis that the system is strictly stationary and Gaussian will be rejected.

In the case of a small number of data, from the statistics of the variable R_1, the critical values $R'_{\alpha/2}$ and $R''_{\alpha/2}$ for rejecting the hypothesis are:

$$R_{\text{exp}} < R'_{\alpha/2} \,;\, R_{\text{exp}} > R''_{\alpha/2} \tag{4.20}$$

It is a simple matter to apply the serial correlation for $h = 1$. If h and n have no common factor, then R_h and R_1 have identical distributions. Particularly, if n is a prime number, all the $n-1$ serial coefficients will be identically distributed. Thus, in a practical application, in order to simplify the calculations we can remove data from the experimental series so as to satisfy these conditions. The values of the variable R_1 are given in Table XII in the Appendix.

For a stationary covariant system, the higher the positional difference h, the less significant the serial correlation coefficient R_h. On the other hand, for a non-stationary system with time-dependent systematic deviation, the R_h correlation coefficient remains significant even at higher values of h.

Example 4.4. We consider again the data of example 4.3. For the data for analytical condition I, and $h = 1$, the value of the serial correlation coefficient is $R_1 = 0.24$. For $\alpha = 0.05$ this value is not significant ($R'_{0.025} = -0.44$, and $R''_{0.025} = 0.26$).

For analytical condition II, for $h = 1$ the serial correlation coefficient is $R_1 = 0.406$, which is a significant value for $\alpha = 0.05$.

Hence, as before (Example 4.3) the hypothesis of a strictly stationary system must be rejected for analytical condition II.

4.3.3 Variance analysis

Among various applications of variance analysis to stability problems we mention:

(a) the verification of the hypothesis of a stationary system, by use of single-factor variance analysis;

(b) the application of multi-factor variance analysis for separating the general variance into components associated with the various influencing factors; by this means we can establish the factors which have the greatest effect on the instability of the system, and hence devise methods for improving the performance.

Here we shall give only the test based on single-factor variance analysis.

Consider the experiment outlined in Table 4.2. The data collected at times

Table 4.2 Data obtained for examination of the stability of an analytical system by single-factor variance analysis

Series number	Time 1	2	...	$t-1$	t
1	y_{11}	y_{21}	...	$y_{t-1\,1}$	y_{t1}
2	y_{12}	y_{22}	...	$y_{t-1\,2}$	y_{t2}
⋮			⋮		
$n-1$	$y_{1\,n-1}$	$y_{2\,n-1}$	...	$y_{t-1\,n-1}$	$y_{t\,n-1}$
n	y_{1n}	y_{2n}	...	$y_{t-1\,n}$	y_{tn}

1, 2, . . . , t are listed in the columns of the table. Obviously, the times are chosen so that the intervals between them are much larger than the time required for obtaining each series of measurements.

The series of data are not required to be of the same size, but to simplify the presentation we shall assume that they are.

The general variance of the data listed in Table 4.2 is given by:

$$s_{gen}^2 = \frac{\sum_{i=1}^{t}\sum_{j=1}^{n}(y_{ij}-\bar{y})^2}{nt-1} \tag{4.21}$$

The variance within a table column is given by:

$$s_{col}^2 = \frac{\sum_{i=1}^{t}\sum_{j=1}^{n}(y_{ij}-\bar{y}_{i.})^2}{nt-t} \tag{4.22}$$

where $\bar{y}$ = the general mean of all the measurement results,
$\bar{y}_{i.}$ = the mean of the results within one column,
y_{ij} = an individual measurement result,
n = the number of results within a column,
t = the number of columns.

We calculate the ratio $F_{exp} = s_{gen}^2/s_{col}^2$, and if $F_{exp} > F_{(\alpha;nt-1,\ nt-t)}$, it will be decided that s_{gen}^2 is significantly greater than s_{col}^2, and the hypothesis of a strictly stationary system will be rejected.

The variance s_{gen}^2 can be significantly greater than s_{col}^2 either because of a stationary covariant system, or because of a non-stationary system. To decide between them we first pair the columns of Table 4.2 (1 with 2, 3 with 4, 5 with 6, . . . , $t-1$ with t), obtaining a new table with $t/2$ new columns (if t is even), then we calculate the variance within the $t/2$ columns, as follows:

$$(s_{col}^2)_2 = \frac{\sum_{i=1}^{t/2}\sum_{j=1}^{2n}(y_{ij}-\bar{y}_{i.})^2}{(t/2)\,2n-t/2} \tag{4.23}$$

If $(s^2_{col})_2$ is not significantly different from s^2_{gen} we decide that there is a stationary covariant system. If $(s^2_{col})_2$ is significantly different from s^2_{gen} we continue by pairing the new columns to obtain a table with $t/4$ columns. The variance $(s^2_{col})_4$ within the columns of the new table is again calculated as shown above, and compared with the general variance s^2_{gen}. If $(s^2_{col})_4$ is not significantly different from s^2_{gen}, we decide in favour of the hypothesis of a stationary covariant system.

Usually, a stationary covariant system can be identified after the first or second grouping of columns.

In the case of a non-stationary regime with time-dependent systematic deviation, s^2_{gen} continues to be significantly greater than the variance within the columns even after the columns have been grouped.

Example 4.5. Table 4.3 lists the results obtained by emission spectrometry for a steel sample, at 14 different times.

From the data we have: $s^2_{gen} = 0.492$ and $s^2_{col} = 0.143$. The ratio $F_{exp} = s^2_{gen}/s^2_{col} = 3.58$. Because $F_{(0.01;55,42)} = 2.04$, we decide that the system is not in a strictly stationary regime.

By grouping columns 1 and 2, 3 and 4, . . . , 13 and 14, a new table with 7 columns is obtained, each with 8 data. The variance within the columns of this table is $(s^2_{col})_2 = 0.376$, which is not significantly different from the general variance $s^2_{gen} = 0.492$.

Hence, after this step of data processing, we decide that the system is in a stationary covariant state.

Table 4.3 The values of the analytical signal obtained, as a function of time, for Ti in a sample of steel by emission spectrometry.

	Time													
Measurement number	1	2	3	4	5	6	7	8	9	10	11	12	13	14
1	9.5	9.5	10.0	11.0	10.5	9.5	10.0	10.0	8.5	10.0	9.5	12.0	10.5	10.5
2	9.0	10.0	10.0	10.5	11.0	9.5	10.0	10.0	9.5	10.0	9.5	11.0	10.0	10.0
3	9.5	10.0	10.0	11.0	11.0	9.5	10.5	11.0	9.0	10.0	9.5	11.0	10.5	10.5
4	9.5	10.0	10.0	11.0	11.0	10.0	11.0	11.5	9.0	10.0	10.5	11.0	10.5	10.5

4.4 RELIABILITY OF ANALYTICAL SYSTEMS

We say that an analytical system is inadequate for solving a particular problem if the errors of analysis are greater than certain preselected values. The concept of reliability is introduced in order to characterize an analytical system with respect to the two extreme values (limits) for the analysis errors. It is defined as the probability of obtaining results with errors within these limits.

The definition is illustrated in Fig. 4.11, which gives a representation of the probability density function associated with the results obtained for a material containing a known amount of the component to be determined. In the figure $f(R)$ is the probability density function of the results and $\bar{R}$ is its mean, R_0 is the true value and E_i and E_s are the limits for the analysis errors.

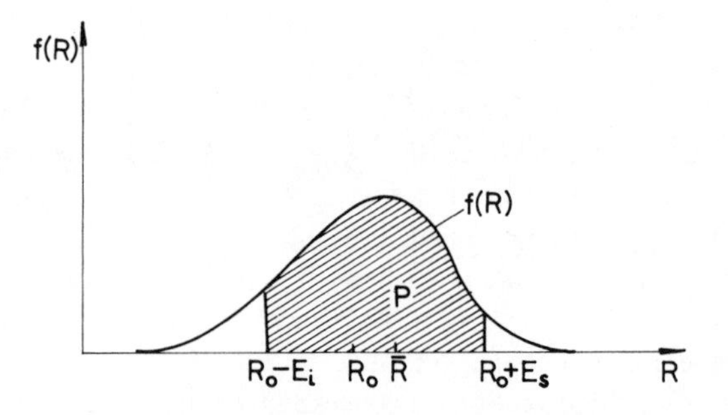

Fig. 4.11 The definition of the reliability of an analytical system with respect to two limits for the analysis errors.

The reliability of the system with respect to E_i and E_s is given by the integral

$$P = \int_{R_0-E_i}^{R_0+E_s} f(R)\mathrm{d}R \tag{4.24}$$

This integral measures the hatched areas in Fig. 4.11.

The reliability of a stationary analytical system remains the same irrespective of the time at which the analytical process is begun. For a non-stationary system, however, reliability is time-dependent. If, after some particular time, the reliability falls below a certain preselected critical value, it is natural to say that the analytical system becomes deficient at that time.

As with any other technical system, an analytical system can go into a deficient state either gradually or suddenly. For example, if a titrant undergoes deterioration on storage there will be a systematic component of the analysis errors, dependent on the age of the solution (the total time). In such a case, the error of analysis, $E(t)$, is the sum of two components: $\bar{e}(t)$, the systematic component and e, a stationary random component:

$$E(t) = \bar{e}(t) + e \tag{4.25}$$

This model is illustrated graphically in Fig. 4.12, where E_i and E_s are the analysis error limits, and $f_1(E), f_2(E), \ldots, f_n(E)$ are the analysis error probability densities at times $t_1, t_2, \ldots, t_n$.

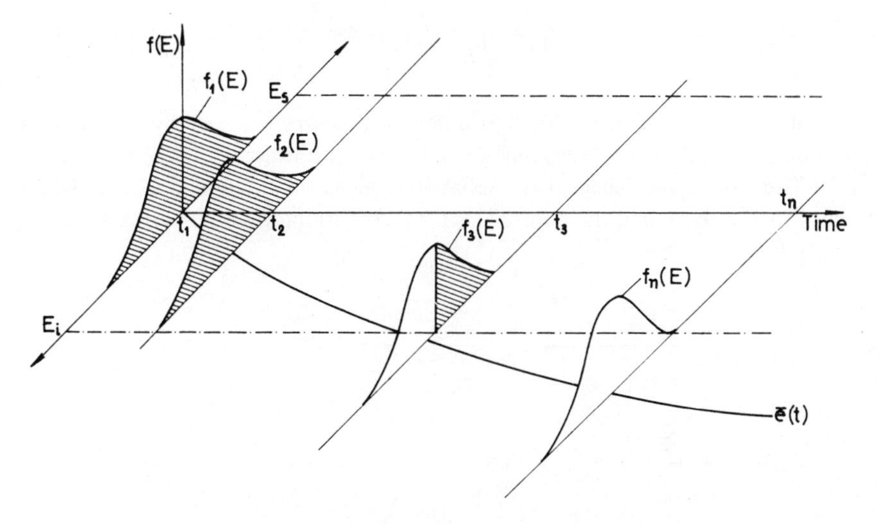

Fig. 4.12 The model of increase of analysis errors with time, for an analytical system with a time-dependent parameter.

For an analytical system of this kind the reliability P decreases monotonically with time, because the probability density for the errors of analysis is drifting towards one of the limits for the analysis errors.

At time t_1, the probability density of the analysis errors is almost completely contained between the error limits so the reliability will be practically equal to unity.

$$P_1 = \int_{E_i}^{E_s} f_1(E)\mathrm{d}E = 1 \tag{4.26}$$

At time t_2, the probability density of the analysis errors is displaced a bit towards E_i, but the analysis errors still do not go beyond this limit. At time t_3, however, the shift of the probability density is so large that many of the results will have errors that are outside the limit E_i. Hence the reliability at time t_3 will be less than unity:

$$P_3 = \int_{E_i}^{E_s} f_3(E)dE < 1 \tag{4.27}$$

Finally, at time t_n and afterwards, the probability density is completely outside the limit E_i. Hence, the reliability will be practically zero:

$$P_n = \int_{E_i}^{E_s} f_n(E)dE = 0 \tag{4.28}$$

A stability regime which leads to a model of this type is caused by the variation of one or several of the determining parameters with time.

The time-dependence of the reliability values for the model in Fig. 4.12 is given in Fig. 4.13, and the point at which reconditioning (restandardization etc.) is needed is indicated.

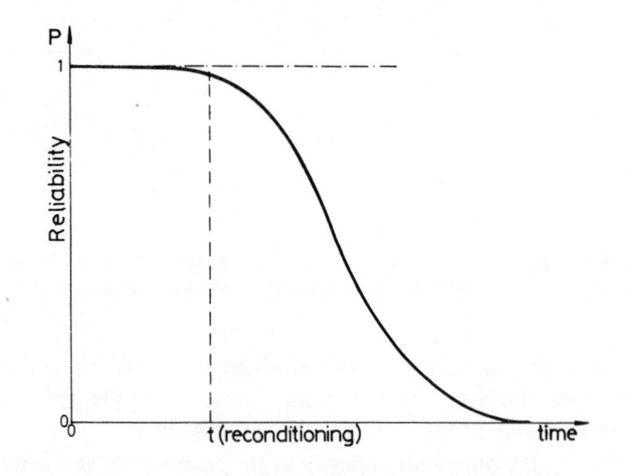

Fig. 4.13 The dependence between the reliability coefficient and time, for an analytical system in a non-stationary stability state.

Analysis systems which correspond to models of this kind should be avoided if possible. If they *must* be used, the time-dependent parameter(s) leading to decreasing reliability should be identified, and the effect(s) eliminated or the onset of deterioration forecast so that reconditioning can be initiated in time to keep the reliability P as close to unity as possible.

The procedure of estimating the reliability of an analytical system is given in detail in Section 2.5.3.

4.5 DETERIORATION AND REGULATION OF ANALYTICAL SYSTEMS

The parameters of an analytical system are subject to irreversible phenomena or, in other words, any analytical system is subject to deterioration. Two preventive measures can be taken to maintain the quality of performance.

(a) Control analyses (i.e. checks with standard substances) to obtain information about the quality of the system, and foreseeable changes in quality.

(b) Overhaul or servicing of the system to re-establish or improve the quality of performance.

Measures of type (b) include adjustment, cleaning, replacement, etc. It is a complex problem to decide when to take such steps. Usually this is done on the basis of knowledge of the physico-chemical processes etc. causing the deterioration, and experience accumulated in use of the analytical system. Frequency of replacement of reagents etc., is usually established from the knowledge of their chemistry (e.g., the need to use freshly prepared solutions of easily oxidized reagents). Checking of mechanical, electrical, or electronic equipment is based on experience.

A graphical model of deterioration and the methods of dealing with it is given in Fig. 4.14, where U_0 is the reference (desired) value, prescribed for the determining parameter, and U_i and U_s are the corresponding tolerance limits.

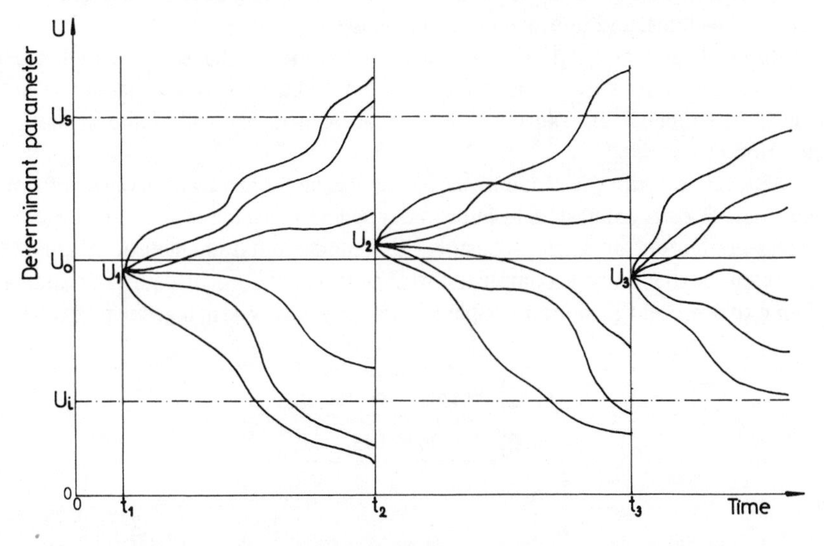

Fig. 4.14 A graphical model for deterioration and regulation of a parameter used for determination.

The routine maintenance sets the reference value of the determining parameter at U_n, which is randomly distributed but as close as possible to the value U_0 prescribed by the working instructions. During subsequent use the determining parameter changes along a trajectory like those shown schematically in the figure.

Mathematically, the process of deterioration can be expressed by the equation:

$$U(t) = U_0 + u_j + At \tag{4.29}$$

where $u_j = U_0 - U_j$, U_0 being the prescribed value and U_j the value after regulation, and A is the rate of deterioration.

That is, the reference value $U(t)$ at any time t can be resolved into a constant term U_0, expressing the prescribed (standard) value, a random term u_j accounted for by inability to reset U exactly to U_0, and a term which is a product of the time and the rate of the deteriorative processes, which is a random quantity. Thus $U(t)$ is a complex and divergent function of time, associated with a stochastic process.

Maintenance must be planned so the tolerance limits are never exceeded during use of the system.

4.6 IMPROVEMENT OF STABILITY

The problem of improving the stability and quality of analytical systems is very complex, and is dealt with in two ways: research on and design of new analytical systems, and improvement of existing systems.

Research and design is usually devoted to finding the simplest and safest operational and constructional principles, development of the most stable and reliable instruments and automation and automatic control of the system in operation.

Automatic control is accomplished by coupling the apparatus to an electronic computer. Special computers have been devised and developed for this purpose—process computers or microprocessors. The fundamental characteric of process computers is that they accomplish a real-time processing of data. The general scheme of a process computer coupled to an analytical system is given in Fig. 4.15.

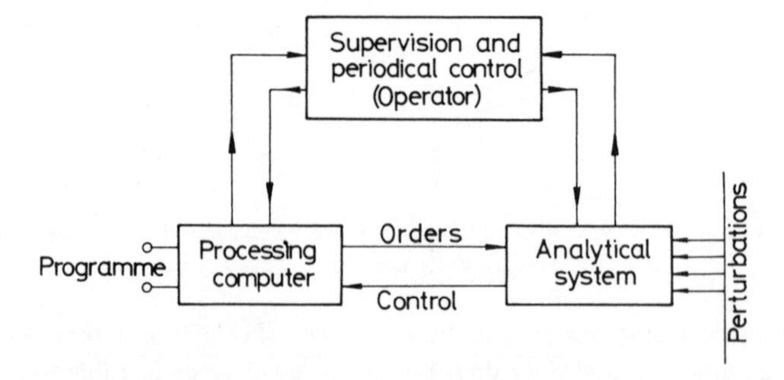

Fig. 4.15 Process computer on-line with an analytical system.

Remarkable achievements have been made in analytical chemistry by utilization of process computers, both in the exploitation of analytical systems, and in the maintenance of stability.

The conditions in which analyses are performed will influence, to a large degree, the magnitude of the errors. Thus, the stability of analytical instruments is affected by the ambient temperature, vibration, accidental knocks, chemical reagents, etc. The laboratory should be designed so that sources of instrument noise are eliminated as far as possible.

To utilize to the full the quality of performance of an analytical system, the equipment must be properly conditioned and maintained, and working schedules should be optimized.

4.7 STATISTICAL CONTROL OF THE STABILITY OF A WORKING ANALYTICAL SYSTEM

At any moment, the quality of an operating analytical system depends on the distribution function of the analytical errors. For a stationary Gaussian system, this distribution function is characterized by the mean of the errors, $\overline{\Delta c}$, and the error variance $\sigma^2_{\Delta c}$, so for statistical control of the quality of a working system, this mean and variance have to be evaluated.

For quality control, good results can be obtained by using control charts. These are diagrams showing the limits of variation statistically allowed for the mean and sample variance of the analysis errors in a stationary state. For a stationary Gaussian system without systematic error, the statistical limits for the error mean are:

$$E_{\mathrm{s}} = +z_P \frac{\sigma_{\Delta c}}{\sqrt{n}}$$

$$E_{\mathrm{i}} = -z_P \frac{\sigma_{\Delta c}}{\sqrt{n}} \qquad (4.30)$$

where n is the number of control determinations, and z_P the normalized deviate. The value of z_P depends upon the confidence probability P, which is preselected.

The control limits for the sample variance s^2 obtained from n determinations are calculated by using the χ^2 distribution (see 2.5.2), and are:

$$0 < s^2 < \frac{\chi^2_P \sigma^2_{\Delta c}}{n-1} \qquad (4.31)$$

The statistical control of an operating analytical system consists in introducing a known sample (or samples) in the process of analysis, at certain time intervals and several times in each case; the sample (or samples) must have a known concentration of the component (or components) to be determined. The sample mean and variance for the analysis errors are then compared with the statistical control limits, which have to be prescribed beforehand. As long as the mean and the sample variance remain within the corresponding control statistical limits, the analytical system is used in the normal way. When they fall outside the limits, work is stopped until the source of error has been found and rectified.

Example 4.6. In a stationary state without systematic deviation ($\overline{\Delta c} = 0$), an analytical system has an error variance of $\sigma^2_{\Delta c} = 0.000016$. For 5 control determinations and a confidence probability $P = 0.997$ ($z_{0.997} = 2.97$), the statistical control limits for the means of the errors are: $E_i = -0.0041$ and $E_s = +0.0041$. These limits, as well as the results obtained after several cycles of statistical control, are given in Fig. 4.16.

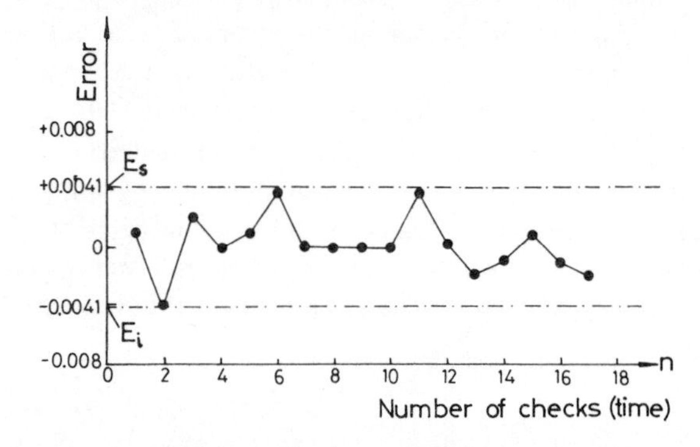

Fig. 4.16 Statistical control chart for the mean of analysis errors.

For a sample size of 5 determinations and a confidence probability $P = 0.975$, the upper limit of the preselected sample variance is $\sigma_s^2 = 0.00005$. This limit and several control results are shown in Fig. 4.17.

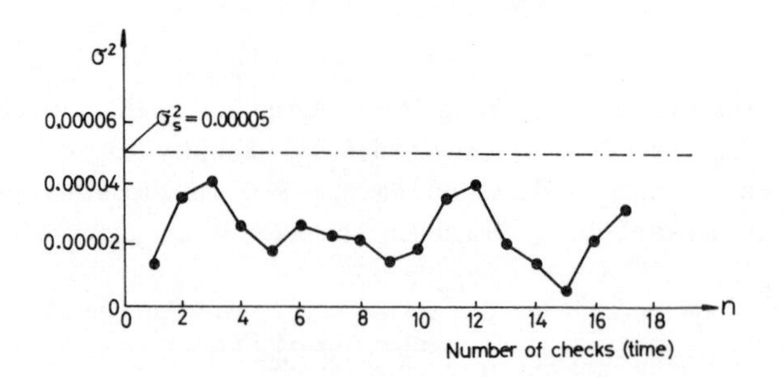

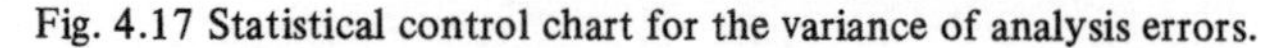

Fig. 4.17 Statistical control chart for the variance of analysis errors.

Chapter 5

Relation between Signal and Concentration

5.1 INTRODUCTION

This chapter is concerned with several problems to be solved when calibrating analytical systems in general, and which are particularly important in trace analysis. For simplicity we shall deal with an analytical system with one channel of measurement, shown in generalized form in Fig. 5.1.

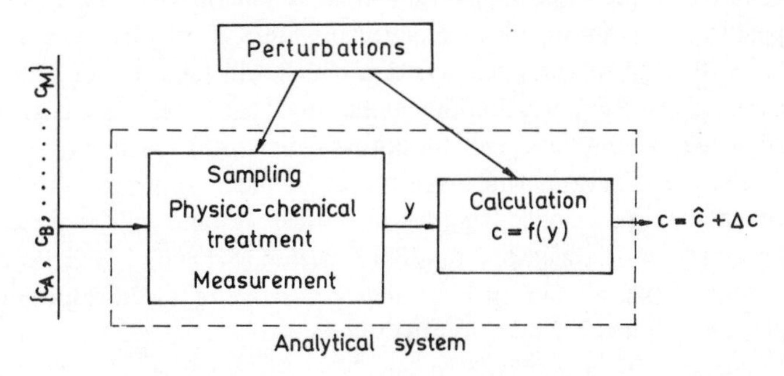

Fig. 5.1 The general structure of an analytical system with one measurement channel.

The input of the analytical system is the set of materials to be investigated, symbolically designated as:

$$c_A, c_B, \ldots, c_M \quad .$$

More specifically, the sample to be analysed contains several components (A, B, . . . , M), and the amounts of these, c_A, c_B, . . . , c_M, can vary randomly (within certain limits) from one sample to another. Every sample undergoes a well-defined treatment before the analytical measurement is made. This measurement gives the result y (the analytical signal); in other words, y is the response of the analytical system. From the signal the result c of the analytical determination is obtained by using the function $c = f(\bar{y})$ where $\bar{y}$ is an estimate of the expected response. The significance of this function will be treated in detail in this chapter.

From Fig. 5.1 it is clear that an analytical system always operates in real conditions, that is, it is always subject to various perturbations. Consequently, there is always a certain degree of uncertainty in the results.

To make the treatment completely general, we shall assume that the response of the analytical system depends on the amounts of all the components of the sample:

$$y = f(c_A, c_B, \ldots, c_M) \tag{5.1}$$

This relation between the response y and the composition of the sample will be called the **response function of the analytical system.** An analytical system with several measurement channels will have several response functions – one for each channel.

Under real conditions, i.e. when perturbations are present, the response function has a random character. In other words, an analytical system operating in real conditions is characterized by a response function with aleatory coefficients. The simplest way to demonstrate this is to make a series of analyses on samples from a material with constant composition; except by chance, a different value for the response of the system will be obtained each time because of the perturbations acting on the analytical system. Thus, there is a set of states associated with a given analytical system, which will take one or other of these states according to the perturbations acting on it, and each state corresponds to a certain value of the response function. When the experiment is rigorously designed, the perturbations will be restricted as much as possible, and the set of states which can be reached by the analytical system will be confined to a restricted group, so the aleatory random function (5.1) will have a limited range of fluctuation. Thus, we can speak of an expected value for the response function and shall use the following notation for this value:

$$\hat{y} = f_0(c_A, c_B, \ldots, c_M) \tag{5.2}$$

The expected value also has another significance; it is the most probable value of the response function. Consequently, at a given time the response of the analytical system can be resolved into two terms:

$$y = \hat{y} + \Delta y \tag{5.3}$$

The first term $\hat{y}$ is the expected value of the response function, and the second term Δy is the deviation from it.

In analytical practice we try to prescribe the conditions so that y depends only on the concentration c of the component to be determined. The response function will then be

$$y = f(c) \tag{5.4}$$

Under such conditions, Eq. (5.3) for the system response becomes:

$$y = f_0(c) + \Delta y \tag{5.5}$$

Further, the analytical conditions are chosen so as to minimize the fluctuations of the response function, i.e. to minimize Δy.

There are two types of mathematical model for the response function. One is linear, for example in gravimetric, titrimetric or polarographic methods where there is a linear relationship between the amount of the component to be determined and the precipitate mass, titration volume or diffusion current. The other is exponential, for example in potentiometric and spectral methods. In potentiometry there is a logarithmic dependence (the Nernst equation) between the electrode potential (the response) and the amount of component to be determined (the determinand). In spectral emission methods, the intensity of the emitted radiation is an exponential function of the amount of determinand, and a similar relation holds for the absorption methods (the Lambert–Beer law).

Linear response functions are generally preferred, and when the response function is non-linear, a linearization procedure is commonly used, with a suitable change of variables.

When the analytical experiment is rigorously organized so that the conditions required by the central limit theorem are satisfied (see 2.2), the random term Δy will have a mean equal to zero and a normal (or close to normal) distribution, that is:

$$\Delta y \Rightarrow N(0, \sigma_{\Delta y}) \tag{5.6}$$

Usually, the standard deviation $\sigma_{\Delta y}$ depends on the amount of determinand as shown in Fig. 5.2, which is valid in most cases [1–11].

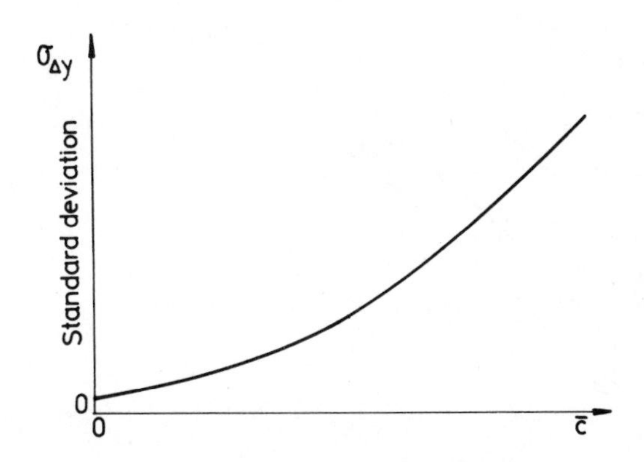

Fig. 5.2 The model for dependence of the standard deviation of the response on the amount to be determined.

5.2 INTERFERENCE

When the response depends not only upon the amount of determinand, but also upon the amount of the other components in the sample, we say that there is interference. Other terms are also in use for the same effect, namely inter-influence or the matrix effect.

Let c be the amount of the component to be determined, and $c_A, c_B, \ldots, c_M$ the amounts of the components producing the interference. Then, the general equation (5.3) for the response of the system becomes

$$y = f_0(c, c_A, c_B, \ldots, c_M) + \Delta y \tag{5.7}$$

Consequently, the expected value of the response will be:

$$\hat{y} = f_0(c, c_A, c_B, \ldots, c_M) \tag{5.8}$$

Depending on the nature of the function (5.8) the interference can be additive or multiplicative. In **additive interference** the expected value can be split into two terms:

$$\hat{y} = f_0(c) + f(c_A, c_B, \ldots, c_M) \tag{5.9}$$

The effect of additive interference by a single component is shown in Fig. 5.3 for three values of the amount of interferent. Any change in the amount of the interfering component will cause a parallel shift of the response function in the co-ordinate plane.

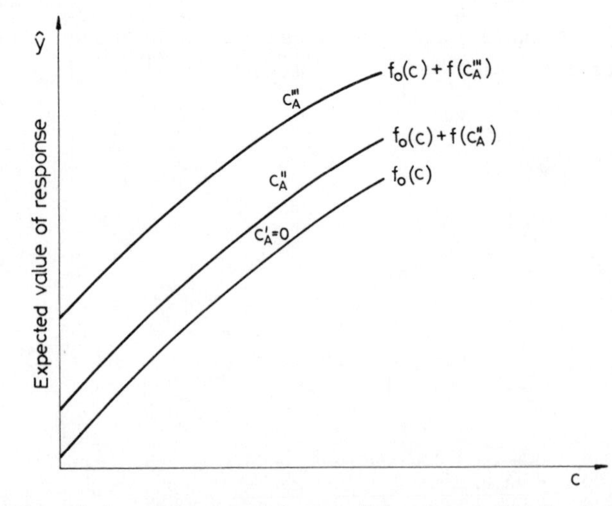

Fig. 5.3 The variation of the expected value of the response function in the case of an additive type of interference.

Common examples are found in chromatography (overlap of the peaks) or spectroscopy (spectral line or band overlap). This type of interference may appear in any analytical system.

In **multiplicative interference**, the interference depends both on the amounts of interferents and on the amount of the component to be determined. In this case the response function splits into two terms of the form:

$$\hat{y} = f_0(c) + f(c, c_A, c_B, \ldots, c_M) \tag{5.10}$$

where the second term is a function of both the determinand and interferent concentrations.

The effect of multiplicative interference is illustrated in Fig. 5.4, where $f(c, c''_A)$ and $f(c, c'''_A)$ represent the interference term for two non-zero levels of the amount of interferent A.

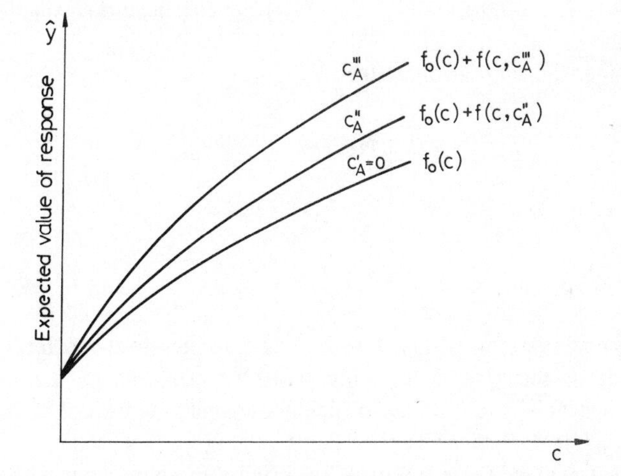

Fig. 5.4 The variation of the expected value of the response function in the case of a multiplicative-type of interference.

Multiplicative interference is found in spectroscopy, for example, when the emission line intensity for the determinand is altered by a constant fraction by the presence of a fixed amount of other components.

Several concepts have been introduced and defined in relation to interference. Thus, the concepts of specificity and selectivity (long used in classical analytical chemistry) have been redefined in accordance with recent developments in chemical analysis.

Kaiser [12, 13], discussing the analytical methods used for simultaneous determination of several components, defines an analytical procedure as completely selective if it allows the components of a sample to be determined independently.

With Kaiser's notation and his linear model, the following system of equations for the response functions is obtained in the case of a material with n components:

$$
\begin{aligned}
y_1 &= \gamma_{11}c_1 + \gamma_{12}c_2 + \ldots + \gamma_{1n}c_n \\
y_2 &= \gamma_{21}c_1 + \gamma_{22}c_2 + \ldots + \gamma_{2n}c_n \\
&\ldots\ldots\ldots\ldots\ldots\ldots\ldots\ldots\ldots\ldots \\
y_n &= \gamma_{n1}c_1 + \gamma_{n2}c_2 + \ldots + \gamma_{nn}c_n
\end{aligned}
\tag{5.11}
$$

where $y_1, y_2, \ldots, y_n$ are the responses given by the n channels of measurements, $c_1, c_2, \ldots, c_n$ the amounts of the n components present in the sample, and γ_{ik} (with $i = 1, 2, \ldots, n$; $k = 1, 2, \ldots, n$) the coefficients of the linear response functions.

Consider the **calibration matrix**:

$$
\begin{bmatrix}
\gamma_{11}, \gamma_{12}, \ldots, \gamma_{1n} \\
\gamma_{21}, \gamma_{22}, \ldots, \gamma_{2n} \\
\gamma_{n1}, \gamma_{n2}, \ldots, \gamma_{nn}
\end{bmatrix}
\tag{5.12}
$$

An analytical system which can be used to determine several components is completely selective if the only non-zero elements of the matrix of the response coefficients are of the principal diagonal, i.e. the coefficients γ_{ii}, where $i = 1, 2, \ldots, n$.

A treatment of the concept of selectivity, in connection with interference, has been given by Gottschalk [14]. To characterize the selectivity, he has introduced and defined the concept of the **interference limit** for a component as the largest amount of that component for which interference is not observable. When the determination of a component is prevented by the presence of a small amount of another component, the determination is said to be **blocked**. Further, for quantitative characterization of interference, Gottschalk has introduced the concept of the **interference ratio**.

Closely connected with the phenomenon of interference is the concept of **resolution**, largely used in spectroscopy and chromatography, in connection with the degree of separation of two spectral lines or two chromatographic peaks. The direct measurement of two neighbouring spectral lines or chromatographic peaks (or any two characteristics which tend to overlap) is possible only if the distance between them is greater than a certain critical value.

Given a Gaussian profile for the distribution of the two neighbouring characteristics, the resolution R_s is defined as shown in Fig. 5.5.

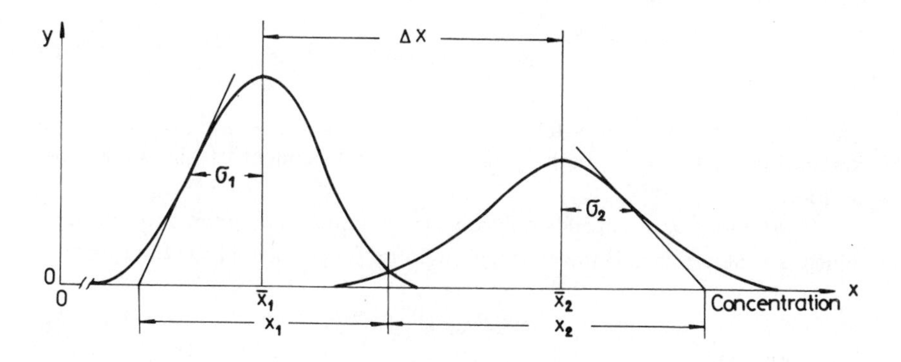

Fig. 5.5 The definition of resolution.

The degree of separation is evaluated by the quantity:

$$R_s = 2 \frac{\Delta x}{x_1 + x_2} \tag{5.13}$$

The condition for 98% separation is:

$$\Delta x = 2\sigma_1 + 2\sigma_2 \tag{5.14}$$

i.e. $R_s = 1$. If this condition is fulfilled, the two characteristics can be individually evaluated, i.e. in such cases it is considered that there is no interference at all.

The concept of resolution can be extended to other cases. Thus, two components can be separated by extraction only when the difference between their extraction coefficients is larger than a certain critical value; similar considerations apply to the selective precipitation of one component in the presence of another component or the separation of two components by electrolytic deposition, and so on.

The concepts of interference, selectivity and resolution are closely interdependent.

5.3 BACKGROUND

The background signal, denoted here by y_0, is the set of values which characterize the response of the analytical system when the sample at the input does not contain the component to be determined.

When there is interference, the background is complex, because its values are correlated with the composition of the sample to be analysed. If there is no interference, the background signal takes the form

$$y_0 = \hat{y}_0 + \Delta y_0 \tag{5.15}$$

where $\hat{y}_0$ is the expected value for the background signal, and Δy_0 its random fluctuation; in this case, both $\hat{y}_0$ and Δy_0 are independent of the sample composition.

If the response y depends only on the amount c of component to be determined (the determinand) the expected response function will have the form:

$$\hat{y} = \hat{y}_0 + f_0(c) \tag{5.16}$$

Under rigorous working conditions, i.e. when the perturbations are minimal, the random component Δy_0 will have a mean equal to zero and a normal (or close to normal) distribution, that is $\Delta y \Rightarrow N(0;\sigma_{\Delta y_0})$.

It is convenient and useful to express $\hat{y}_0$ in terms of the equivalent amount of determinand, c_b. The significance of the expected value $\hat{y}_0$ of the background signals and its expression as an equivalent amount of determinand (c_b) are given in Fig. 5.6.

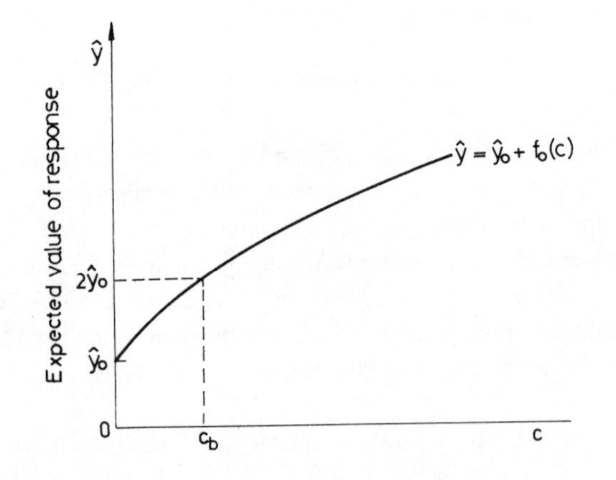

Fig. 5.6 The significance of the expected value of the background signal and its expression as an equivalent amount of component to be determined.

Obviously, in trace analysis, the analytical conditions must be chosen so as to minimize both the expected value $\hat{y}_0$ of the background signal, and its fluctuations Δy_0.

Usually, the correction for the background signal is achieved by subtracting the signal for a blank from the signal given by the sample to be analysed. However, extra caution is required for background correction in the case of a non-linear calibration function [15].

In modern analytical chemistry, reduction of the background signal and its fluctuation by using adequate treatment and principles of measurement is an efficient method for increasing the signal-to-noise ratio, i.e. for increasing the quality of performance of the analytical system (this problem will be discussed in detail in the last chapter).

5.4 SENSITIVITY

The response function expected in the absence of interference is $\hat{y} = f_0(c)$, and the sensitivity is defined as the partial derivative $\partial\hat{y}/\partial c$. The higher the value of the derivative, the more sensitive the analytical system. In the case of an analytical system with several channels, each channel will be characterized by its own sensitivity.

When the response function is linear, the sensitivity will be constant and equal to the slope, that is:

$$\partial\hat{y}/\partial c = b, \tag{5.17}$$

whereas in the case of a non-linear response function it will depend on the amount of determinand.

From the system of equations (5.11) for the linear response functions in the case of n components, the partial sensitivities with respect to the various components of the sample are, by definition:

$$\gamma_{ik} = \partial y_i/\partial c_k \tag{5.18}$$

According to Kaiser [12, 13] and Junker and Bergmann [16], the sensitivity $H_{(n)}$ of an analytical system which can determine n components is measured by the determinant of (5.12), formed by the coefficients of the response functions. The determinant is symbolically written as:

$$H_{(n)} = \text{det.}\,(\gamma_{ik}) \tag{5.19}$$

where $i, k = 1, 2, \ldots, n$.

In chemical analysis generally, and in trace analysis especially, the analytical system and conditions must be carefully selected in order to maximize the elements of the principal diagonal of the determinant of the response coefficients (5.12), and at the same time to minimize the absolute values of the other elements. These conditions ensure a high sensitivity and selectivity for the system.

5.5 CALIBRATION FUNCTION AND ANALYTICAL FUNCTION

As shown before, the response function of an analytical system is random in character, and the expected value for it is considered.

We use the term calibration function [12, 17, 18] of an analytical system for an estimation value for the expected value of the response function.

In the absence of interference, the calibration function can be written as:

$$\bar{y} = f_0(c) \tag{5.20}$$

where $\bar{y}$ is an estimate of the expected value of the response, i.e. the mean response.

Hence, the expected value of the calibration function will be the function $\hat{y} = f_0(c)$, that is, it will be the same as the expected value of the response function.

The analytical function, or (more completely) the analytical calculation function is by definition the inverse of the calibration function. In the absence of interference, it is

$$c = f(\bar{y}) \tag{5.21}$$

If interference occurs, however, it must be taken into account in the determination of the calibration function and calculation of the results. For example, when the interference is caused by the components A, B, . . . , M, the calculation function will have the form:

$$c = f(\bar{y}, c_A, c_B, \ldots, c_M) \tag{5.22}$$

and all the appropriate quantities need to be known.

For multi-channel analytical systems, the calculation function (5.22) can be expressed in terms of the responses corresponding to the interfering components, that is:

$$c = f(\bar{y}, \bar{y}_A, \bar{y}_B, \ldots, \bar{y}_M) \tag{5.23}$$

Obviously, when interference is present, the correct evaluation of the results is greatly facilitated by using a multi-channel system coupled to an electronic computer. In this case, concomitantly with the response y from the channel of the component to be determined, the responses y_A, y_B, . . . , y_M from the channels of the interfering components are also evaluated, and the result of interest calculated by using the function (5.23) (established beforehand).

In the calculation functions (5.21)-(5.23), the mean response values are used because these functions are obtained from a set of analytical data, so that the responses are effectively sample means.

Both the calibration function and the calculation function are evaluated from a set of experimental data, the calibration data. In the absence of interference, the data consist of a set of pairs (c, y) relating amount of component to system response, and obtained by applying the analytical system to standard samples of accurately known composition. If interference occurs, however, the calibration data must also include the amounts of the interferents or the corresponding analytical signals. During the entire calibration, the working conditions must be kept constant and preferably optimal, and all subsequent analyses must be done under precisely the same conditions.

The calibration function and/or calculation function can be obtained from the calibration data either by graphical methods or by the least-squares method (see e.g. [19–22]).

The calibration error ϵ is measured by the deviation of the calibration function from the expected value of the response function, that is:

$$|\bar{y} - \hat{y}| = \epsilon \tag{5.24}$$

The smaller ϵ, the more accurate the calibration.

The fundamental conditions for a correct calibration are as follows.

(a) There must be good agreement between the real model of the response function and the model used in the calculation or assumed for calculating the parameters.

(b) The composition of the standard samples must be rigorously known. These samples must be representative of the materials to be analysed.

(c) The method of calculation must be correct.

If these conditions are satisfied, the probability that $\bar{y}$ coincides with $\hat{y}$ will converge towards 1 as the number of calibration data increases:

$$\lim_{n \to \infty} [P|\bar{y} - \hat{y}| = 0] = 1 \tag{5.25}$$

5.6 THE LINEAR CALIBRATION FUNCTION

Linear calibration functions are generally preferred in chemical analysis, particularly in trace analysis. When the system response is non-linear it is customary to change the variables to obtain linearization.

When interference is absent and the response is linear, the general expression for the response function is:

$$y = \hat{y} + \Delta y = a + bc + \Delta y \tag{5.26}$$

where $\hat{y} = a + bc$ is the expected value, and Δy the fluctuation.

The calibration function, which estimates $\hat{y}$, will be written as

$$\bar{y} = a' + b'c \tag{5.27}$$

where a' and b' are estimation values for the intercept and the slope respectively, i.e. for the background signal and the sensitivity.

However, no matter how rigorously done the evaluation of the amount of component in the standard samples, there will always remain a certain degree of uncertainty in c. Hence the value of c for a standard sample composition is best written as

$$c = \hat{c} + \Delta c, \tag{5.28}$$

where $\hat{c}$ is the true value (which cannot be *precisely* known), and Δc the deviation from the true value.

As an analytical system is working under real conditions, the system response y will also have an uncertainty:

$$y = \hat{y} + \Delta y \tag{5.29}$$

where $\hat{y}$ is the expected value and Δy the random error, with a normal (or close to normal) distribution.

The method of calibration always assumes that Δc is small enough to be neglected. Hence, when there is no interference,

$$y \Rightarrow N(a + bc; \sigma_{\Delta y}) \tag{5.30}$$

that is, analytical responses are assumed to be normally distributed, with the mean linearly dependent on the amount of determinand but the standard deviation $\sigma_{\Delta y}$ independent of it.

The graphical method is the simplest for evaluating the calibration function, the line best fitting the points being judged by eye.

The alternative mathematical procedure is the least-squares method (Section 2.7.11).

The sum of the squares of the absolute deviations from the general function

$$\hat{y} = a + b\hat{c} \tag{5.31}$$

is minimized with respect to the two parameters.

For a set of n pairs of calibration data, $(c_1, y_1), \ldots, (c_n, y_n)$, the estimation parameters are:

$$a' = \bar{y} - b'\bar{c} \tag{5.32}$$

$$b' = \frac{S_{cy}}{S_{cc}} \tag{5.33}$$

where

$$S_{cc} = \sum_{i=1}^{n} c_i^2 - \frac{\left(\sum_{i=1}^{n} c_i\right)^2}{n} \qquad S_{cy} = \sum_{i=1}^{n} c_i y_i - \frac{\sum_{i=1}^{n} c_i \sum_{i=1}^{n} y_i}{n}$$

$$\bar{c} = \frac{\sum_{i=1}^{n} c_i}{n} \quad \text{and} \quad \bar{y} = \frac{\sum_{i=1}^{n} y_i}{n}$$

The variance $\sigma_{\Delta y}^2$ of the signal fluctuations is estimated by

$$s_{\Delta y}^2 = \frac{S_{yy} - b' S_{cy}}{n - 2} = s_y^2 \tag{5.34}$$

where

$$S_{yy} = \sum_{i=1}^{n} y_i^2 - \frac{\left(\sum_{i=1}^{n} y_i\right)^2}{n}$$

The estimation variance corresponding to the parameters a' and b' evaluated in this way is given by:

$$s_{a'}^2 = \frac{s_{\Delta y}^2 \sum_{i=1}^{n} c_i^2}{n \sum_{i=1}^{n} (c_i - \bar{c})^2} \tag{5.35}$$

$$s_{b'}^2 = \frac{s_{\Delta y}^2}{\sum_{i=1}^{n} (c_i - \bar{c})^2} \tag{5.36}$$

The confidence intervals for the true values of the parameters a and b, corresponding to a probability P, are:

$$a' - |t_{\alpha/2;n-2}| s_{a'} < \hat{a} < a' + |t_{\alpha/2;n-2}| s_{a'} \tag{5.37}$$

$$b' - |t_{\alpha/2;n-2}| s_{b'} < \hat{b} < b' + |t_{\alpha/2;n-2}| s_{b'} \tag{5.38}$$

where $\alpha = 1 - P$.

From the statistics of the parameters a' and b', confidence intervals can be evaluated for the expected value of the response function [Eq. (2.105)], and also for the individual values of the response function [Eq. (2.106)]:

$$a' + b'c - |t_{\alpha/2;n-2}|s_{\Delta y}\sqrt{\frac{1}{n} + \frac{(c - \bar{c})^2}{S_{cc}}} < \hat{y} < a' + b'c + |t_{\alpha/2;n-2}|s_{\Delta y}\sqrt{\frac{1}{n} + \frac{(c - \bar{c})^2}{S_{cc}}} \tag{5.39}$$

and

$$a' + b'c - |t_{\alpha/2;n-2}|s_{\Delta y}\sqrt{1 + \frac{1}{n} + \frac{(c - \bar{c})^2}{S_{cc}}} < y < a' + b'c + |t_{\alpha/2;n-2}|s_{\Delta y}\sqrt{1 + \frac{1}{n} + \frac{(c - \bar{c})^2}{S_{cc}}} \tag{5.40}$$

The significance of these two confidence intervals is shown graphically in Fig. 5.7.

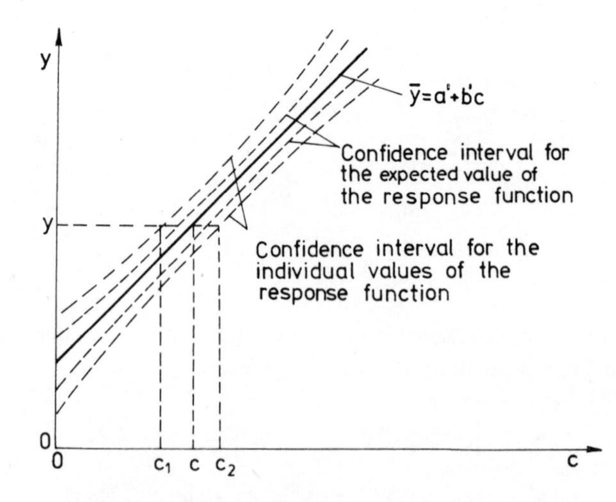

Fig. 5.7 The significance of the confidence intervals for the expected value and the individual values of the response function.

The confidence interval for the individual values is called the dispersion band, and that for the expected value is also called the confidence interval for the calibration function. The alternative names just given accord less well with the concept of the response function.

For estimating a' and b' [Eqs. (5.32) and (5.33)], only the pairs of values (c_i, y_i) within the limits of the confidence band [(5.40)] will be used.

Rigorous evaluation of the calibration functions is a matter of first importance. From (5.35), to obtain the best evaluation of a' a large number of standard samples with a low level of determinand must be used. On the other hand, from (5.36) correct estimation of b' requires not only that the number of standard samples be large, but also that the range of content be as wide as possible, covering all possible applications.

The simplest way to obtain the analytical result is by reading the calibration graph, as shown in Fig. 5.7, which also shows the significance of the confidence interval for $\hat{c}$, obtained by taking the values c_1 and c_2 corresponding to y on the hyperbolic confidence limits of the response function for a certain probability. Then $c_1 < \hat{c} < c_2$.

An alternative for linear response is to solve the calibration function (5.27) for c:

$$c = \frac{\bar{y} - a'}{b'} \tag{5.41}$$

Other statistical procedures for estimating the linear function parameters are known. Aivazian and Bogdanovskii [23] give the 'Wald', 'Bartlett', and 'Gauss-Brenen' procedures for estimation of the slope. Knecht and Stark [24] give two new procedures for the evaluation of the calibration function parameters. One is based on percental deviation from the general function, the other on the logarithms of the calibration data. They show that for low concentrations, the calibration function is more correctly evaluated by this procedure than by the classical least-squares method. When the procedure based on percentual deviations is used, the two parameters (slope and intercept) are given [24–25] by:

$$b' = \frac{\sum_{i=1}^{n} \frac{1}{c_i^2} \sum_{i=1}^{n} \frac{y_i^2}{c_i^2} - \left(\sum_{i=1}^{n} \frac{y_i}{c_i^2} \right)^2}{\sum_{i=1}^{n} \frac{1}{c_i^2} \sum_{i=1}^{n} \frac{y_i}{c_i} - \sum_{i=1}^{n} \frac{1}{c_i} \sum_{i=1}^{n} \frac{y_i}{c_i^2}} \tag{5.42}$$

$$a' = \frac{\sum_{i=1}^{n} \frac{y_i}{c_i^2} - b' \sum_{i=1}^{n} \frac{1}{c_i}}{\sum_{i=1}^{n} \frac{1}{c_i^2}} \tag{5.43}$$

Knecht and Stark have also introduced a **coefficient of quality**, g, instead of the correlation coefficient. It is evaluated by

$$g = \frac{\sum_{i=1}^{n} (\Delta c_i\%)^2}{n - 1} \tag{5.44}$$

where $\Delta c_i\%$ is the relative deviation of the value of the amount of component in the standard sample from the corresponding value obtained from the calculation function.

Plesch [26], on the other hand, considers that the classical least-squares method is *not* adequate for evaluation of the calibration function at small concentrations in X-ray fluorescence work, and for this region recommends the two Knecht and Stark methods.

Boček and Novák [27] have shown that in chromatography the standard deviations for the system response do not usually remain constant for different levels of determinand. They recommend the 'mean slope' method instead of the classical least-squares method.

5.7 STATISTICAL HYPOTHESES AND THE LINEAR RESPONSE FUNCTION

There are several problems and hypotheses that can be formulated with respect to the response function, for example:

(a) establishing the significance of a' and b' in comparison with certain values a_0 and b_0;

(b) establishing the significance of two or several estimation values for the calibration function;

(c) testing the hypothesis that the standard deviation of the response fluctuations is independent of the amount of component;

(d) testing the hypothesis that the residual error is normally distributed;

(e) establishing the significance of the residual error, etc.

The significance of the estimation values a' and b' in comparison with the values a_0 and b_0 can be established by means of the statistical variables $a' - a_0$ and $b' - b_0$. As shown in Section 2.7.1.1, for a significance level α, the hypothesis of a significant difference will be accepted if:

$$\frac{|a' - a_0|}{s_{a'}} \geqslant |t_{\alpha/2;n-2}| \tag{5.45}$$

or

$$\frac{|b' - b_0|}{s_{b'}} \geqslant |t_{\alpha/2;n-2}| \tag{5.46}$$

In the case of two estimation values for the calibration function:

$$\bar{y} = a'_1 + b'_1 c \tag{5.47}$$

$$\bar{y} = a'_2 + b'_2 c \tag{5.48}$$

the hypothesis of a significant difference between them is accepted if either

$$\frac{|a'_1 - a'_2|}{s_{\Delta a'}} \geqslant |t_{\alpha/2;n_1+n_2-4}| \tag{5.49}$$

or

$$\frac{|b'_1 - b'_2|}{s_{\Delta b'}} \geqslant |t_{\alpha/2;n_1+n_2-4}| \tag{5.50}$$

where n_1 and n_2 are the numbers of pairs of calibration data used in evaluation of the functions, and $s_{\Delta a'}$ and $s_{\Delta b'}$ are the standard deviations for the differences $|a' - a|$ and $|b' - b|$; $s_{\Delta a'} = \bar{s}_{a'}\sqrt{\frac{1}{n_1} + \frac{1}{n_2}}$, and $s_{b'} = \bar{s}_{\Delta b'}\sqrt{\frac{1}{n_1} + \frac{1}{n_2}}$, where $\bar{s}_{\Delta a'}$ and $\bar{s}_{\Delta b'}$ are obtained by using Eq. (2.56) for $\nu_1 = n_1 - 2$, and $\nu_2 = n_2 - 2$ degrees of freedom.

The hypothesis that the standard deviation is independent of the amount of determinand can be verified by the Bartlett test (see 2.6.3.3). For this purpose, we evaluate the sample variances $s_1^2, s_2^2, \ldots, s_n^2$ corresponding to different values $c_1, c_2, \ldots, c_n$ of the amount of determinand (the number n of different values c_i should be as large as possible). The hypothesis that the standard deviation is constant will be accepted if, by using the Bartlett test at the chosen significance level, it is found that the set of sample variances $s_1^2, s_2^2, \ldots, s_n^2$ is homogeneous.

The residual term Δy_i is, by definition, the deviation of the response y_i (for a value c_i of the amount of determinand) from the corresponding value of the calibration function:

$$\Delta y_i = y_i - (a' + b'c_i) \tag{5.51}$$

Obviously, a measure of the residual term is the standard deviation $s_{\Delta y}$, which can be evaluated for a set of calibration data according to Eq. (5.35).

The residual term should be normally distributed, and to check this hypothesis several standard samples are used, giving a set of pairs of calibration data (c_i, y_i), from which the calibration function is estimated and the residual term calculated for each pair. The hypothesis is tested by using the χ^2 test or the

λ test (Kolmogorov). If the hypothesis must be rejected, this will indicate that the model assumed for the dependence between response and amount is significantly different from the real situation.

The value and significance of the standard deviation $s_{\Delta y}$ contains information about the degree of agreement between the real (but unknown) model for the response function and the model assumed for parameter evaluation, and also information regarding the quality of the analytical system in general.

Suppose we have a set of n standard samples, in which the amounts of the component of interest are c_i ($i = 1, 2, \ldots, n$), such that the set is representative for all materials to be analysed subsequently, and assume that m analytical measurements are made for each standard. The set of calibration data will then have the form (c_i, y_{ij}), where $i = 1, 2, \ldots, n$, and $j = 1, 2, \ldots, m$, and we can estimate the calibration function, the sample variance $s_{\Delta y}^2$ for the fluctuations of the residual term, and also the variance

$$s_{\Delta y}'^2 = \frac{\sum_{i=1}^{n} \sum_{j=1}^{m} (\bar{y}_i - y_{ij})^2}{mn - n} \tag{5.52}$$

of the response fluctuations around the sample means corresponding to the n levels of c. Here the $\bar{y}_i$ values are the sample means for the corresponding responses from the n levels of c and the y_{ij} values are the individual responses.

The significance of the sample variance $s_{\Delta y}^2$ in comparison with the variance $s_{\Delta y}'^2$ can arise from the following causes:

(a) difference between the real and assumed models of the response function;
(b) interference;
(c) uncertainty in c_i of the standards;
(d) instability of the analytical system.

The evaluation of the variance $s_{\Delta y}^2$ and its significance in comparison with the variance $s_{\Delta y}'^2$ is only the first step in the examination and improvement of the quality of performance for a system.

Let us consider a practical example for the case of a linear calibration function.

Example 5.1. The calibration data for copper in standard steel samples by an X-ray fluorescence method are given in Table 5.1.

With these data, we shall exemplify several of the theoretical problems given above. Note that the numbers in the calculations are rounded off to a realistic number of significant figures.

(a) *Testing the hypothesis of constant standard deviation*

The sample variances calculated for the columns of Table 5.1 are: $s_1^2 = 36.5$; $s_2^2 = 62.5$; $s_3^2 = 24.8$; $s_4^2 = 53.5$; $s_5^2 = 24.0$; $s_6^2 = 55.8$; $s_7^2 = 58.5$; $s_8^2 = 35.8$; $s_9^2 = 35.7$.

Table 5.1 Calibration data

	Standard sample	1	2	3	4	5	6	7	8	9
	Amount of Cu(%)	0.10	0.23	0.17	0.30	0.015	0.43	0.73	0.23	0.47
System response	1	694	1211	953	1503	339	2028	3252	1244	2175
	2	685	1228	951	1515	352	2012	3247	1250	2172
	3	685	1220	960	1510	345	2031	3264	1239	2186
	4	697	1209	968	1498	345	2026	3244	1234	2175
	5	684	1212	950	1499	349	2028	3253	1240	2171

The ratio of the extreme values is:

$$F_{\text{exp}} = s_2^2/s_5^2 = 2.60$$

Because $F_{\text{exp}} < 6.39 = F_{0.05;\nu_1 = \nu_2 = 4}$, we cannot reject the hypothesis that the standard deviation is constant irrespective of the amount of copper present.

(b) *Estimation of the parameters of the calibration function*

It is assumed that there is a linear dependence between response and amount of copper. From (5.32) and (5.33), the estimation values obtained by the least-squares method are: $a' = 283.7$ and $b' = 4057$. Hence, the calibration function and the analytical calculation function are:

$$\bar{y} = 283.7 + 4057\,c$$

and

$$c = \frac{\bar{y} - 283.7}{4057}$$

(c) *Estimation of the variance of the residual term*

From Eq. (5.34), the variance of the residual term is

$$s_{\Delta y}^2 = 176.8$$

(d) *Estimation of the variances of* a′ *and* b′

From (5.35) and (5.36), these variances are found to be:

$$s_{a'}^2 = 13.3 \text{ and } s_{b'}^2 = 93.4$$

(e) *Evaluation of the confidence intervals*

From (5.37) and (5.38), the confidence intervals for the true values of a and b are:

$$283.7 - |t_{\alpha/2;43}| \times 3.65 < \hat{a} < 283.7 + |t_{\alpha/2;43}| \times 3.65$$

and

$$4057 - |t_{\alpha/2;43}| \times 9.7 < \hat{b} < 4057 + |t_{\alpha/2;43}| \times 9.7$$

From (5.39) and (5.40), the confidence intervals for the individual values of the response function and for the expected value are

$$283.7 + 4057\,c - |t_{\alpha/2;43}| \times 13.3 \sqrt{1 + \frac{1}{45} + \frac{(c-0.297)^2}{1.893}} <$$

$$y < 283.7 + 4057\,c + |t_{\alpha/2;43}| \times 13.3 \sqrt{1 + \frac{1}{45} + \frac{(c-0.297)^2}{1.893}}$$

and

$$283.7 + 4057\,c - |t_{\alpha/2;43}| \times 13.3 \sqrt{\frac{1}{45} + \frac{(c-0.297)^2}{1.893}} <$$

$$\hat{y} < 283.7 + 4057\,c + |t_{\alpha/2;43}| \times 13.3 \sqrt{\frac{1}{45} + \frac{(c-0.297)^2}{1.893}}$$

(f) *Testing the hypothesis of a normal distribution of the residual term*

In Table 5.2 we give the results obtained with the χ^2 test for the hypothesis of a normal distribution of the residual term $\Delta y_i = y_i - (283.7 + 4057\, c_i)$ for the calibration data given in Table 5.1 (the method for testing hypotheses concerning the nature of a distribution was given in Section 2.6.5).

Because (from the statistical tables) $\chi^2_{0.95;5} > \chi^2_{\text{exp}} = 3.595$, the hypothesis of a normal distribution of the residual term cannot be rejected.

Table 5.2 Testing the hypothesis of the normal distribution of the residual term by using the χ^2 test

Variation interval	Frequency of occurrence in the interval	z_i	$F(z_i)$	Probability for the interval	$\frac{(n_i - np_i)^2}{np_i}$
$\Delta y_i \leqslant -20$	4	−1.50	0.067	0.067	0.320
$-20 < \Delta y_i \leqslant -10$	5	−0.75	0.227	0.160	0.670
$-10 < \Delta y_i \leqslant 0$	17	0	0.500	0.273	1.800
$0 < \Delta y_i \leqslant +10$	11	+0.75	0.773	0.273	0.130
$+10 < \Delta y_i \leqslant +20$	5	+1.50	0.933	0.160	0.670
$\Delta y_i > +20$	3		1.000	0.067	0.005
Σ	$n = 45$	–	–	–	$\chi^2_{exp} = 3.595$

(g) *Establishing the significance of the residual term*

From (5.34), the sample variance of the total residual term is $s^2_{\Delta y} = 176.8$; the corresponding number of degrees of freedom is $\nu_1 = 43$.

From (5.52), the sample variance for the fluctuations of the individual response values around the mean values is $s'^2_{\Delta y} = 43$ and corresponds to $\nu_2 = 36$ degrees of freedom.

The F-test (see Section 2.6.3) is used to establish the significance of the two variances $s^2_{\Delta y}$ and $s'^2_{\Delta y}$. The ratio of these variances is $F_{exp} = s^2_{\Delta y}/s'^2_{\Delta y} = 4.11$. As $F_{exp} > 1.71 = F_{0.05;43,36}$, it follows that for a significance level $\alpha = 0.05$, the hypothesis of a significant difference between the two sample variances must be accepted. The cause of this significant difference would need further statistical investigation.

5.8 THE LINEAR CALIBRATION FUNCTION $\bar{y} = b'c$

The linear model $\bar{y} = b'c$ for the calibration function is obtained by background correction done by subtracting the system response for a blank sample ($c = 0$) from that for the sample to be analysed.

Linear calibration functions of the type $\bar{y} = b'c$ pass through the origin, and are used in spectral, electrochemical, chromatographic and other methods of analysis.

The results can also be evaluated by simple proportion from the system responses for the standard sample and the sample analysed, and the composition of the standard. The statistical aspects of this procedure have been discussed, for the case of spectrophotometric methods, by Francis and Sobel [28].

If the composition of the standard is accurately known, and the results of measurements on the standard and the sample are independent but have the same variance σ_y^2, the variance of the analytical result is:

$$\sigma_{c_x}^2 = c_s^2 \sigma_y^2 \left(\frac{1}{\hat{y}_s^2} + \frac{\hat{y}_x^2}{\hat{y}_s^4} \right) \tag{5.53}$$

where subscript x indicates the sample and s the standard.

This equation is easily established from the fact that the variance of the ratio of two independent variables is given by:

$$\text{var} \left(\frac{y_x}{y_s} \right) = \left(\frac{1}{\hat{y}_s^2} \right) \text{var} \, (y_x) + \left(\frac{\hat{y}_x^2}{\hat{y}_s^4} \right) \text{var} \, (y_s) \tag{5.54}$$

where $\hat{y}$ is the expected value and y an individual value.

When background correction is necessary, there is another analytical variant, the method of standard addition. Two aliquots of the sample to be analysed are taken and a known amount c_s of determinand is added to one of them. The system response is measured for each sample. The analytical result is calculated from

$$c_x = c_s \frac{y_x}{y_{x+s} - y_x} \tag{5.55}$$

where y_{x+s} is the response from the sample to which c_s was added, and y_x is the response from the other sample.

Assuming again that the measurements are independent, the variance is

$$\sigma_{c_x}^2 = \frac{c_s^2 \sigma_y^2}{(\hat{y}_{x+s} - \hat{y}_x)^4} (\hat{y}_{x+s}^2 + \hat{y}_x^2) \tag{5.56}$$

A statistical discussion of the method of standard addition, applied to X-ray fluorescence analysis, is given by Plesch [29].

5.9 THE METHOD OF MULTIPLE STANDARD ADDITION

In this method, as utilized in trace analysis, different small known amounts of determinand are added to a series of aliquots from the sample. Then each of these samples is analysed.

A graphical plot of the results (or responses) versus the amounts of standard added usually gives a straight line. If no background correction is necessary, the amount of determinand is given by the intercept of the straight line with the abscissa.

If the calibration function is non-linear, it is linearized before the graph is drawn.

When background corrections are required, a supplementary set of samples which do not contain the component to be determined is prepared. In all other respects these samples should have similar composition to the material to be analysed. The multiple standard addition method is applied to these samples. The relation between the system response and the added standard amounts gives the **background correction function**, the correction itself being given by the intercept on the abscissa.

The principle of multiple standard additions is illustrated graphically in Fig. 5.8.

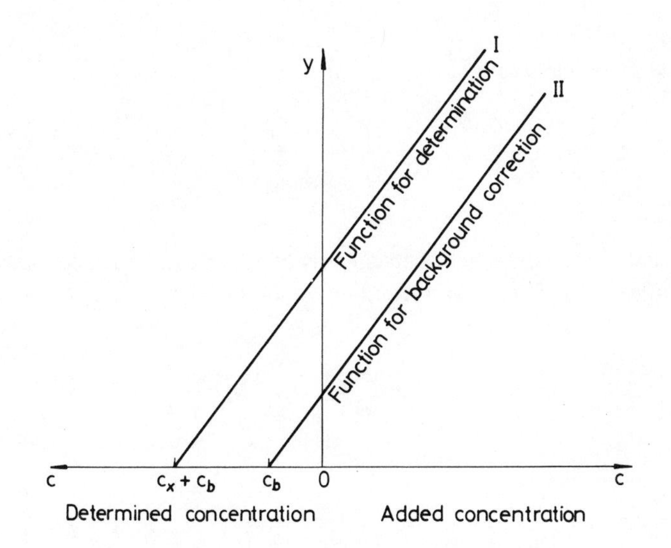

Fig. 5.8 The principle of the method with several standard additions.

When background correction is required, the amount c_x of the component to be determined is given by:

$$c_x = c_{x+b} - c_b \qquad (5.57)$$

where c_b is the background value, and c_{x+b} is the sum of the amount of the component to be determined and the background value.

The two straight lines in Fig. 5.8 can be evaluated either graphically or by the least-squares method. If the lines are represented by

$$\bar{y} = a'_1 + b'_1 c \tag{5.58}$$

$$\bar{y} = a'_2 + b'_2 c \tag{5.59}$$

the intercepts on the abscissa will be:

$$c_{\mathrm{x+b}} = -\frac{a'_1}{b'_1} \tag{5.60}$$

and

$$c_{\mathrm{b}} = -\frac{a'_2}{b'_2} \tag{5.61}$$

Hence, the analytical result is:

$$c_{\mathrm{x}} = \frac{a'_1}{b'_1} - \frac{a'_2}{b'_2} \tag{5.62}$$

If a'_1 and b'_1 are estimated independently of a'_2 and b'_2, the variances associated with the intercepts of lines I and II in Fig. 5.8 on the abscissa are

$$\sigma^2_{\mathrm{x+b}} = \frac{1}{b_1^2}\sigma^2_{a'_1} + \frac{a_1^2}{b_1^4}\sigma^2_{b'_1} \tag{5.63}$$

and

$$\sigma^2_{\mathrm{b}} = \frac{1}{b_2^2}\sigma^2_{a'_2} + \frac{a_2^2}{b_2^4}\sigma^2_{b'_2} \tag{5.64}$$

so when background correction is necessary the variance of the analytical results will be:

$$\sigma^2_{\mathrm{c}} = \sigma^2_{\mathrm{x+b}} + \sigma^2_{\mathrm{b}} \tag{5.65}$$

Let n_1 and n_2 be the numbers of calibration data used for lines I and II respectively. From (5.34) we can estimate the sample variances of the residual terms $(s^2_{\Delta y})_{\mathrm{I}}$ and $(s^2_{\Delta y})_{\mathrm{II}}$. From these we obtain $\bar{s}^2_{\Delta y} = [\nu_1(s^2_{\Delta y})_{\mathrm{I}} + \nu_2(s^2_{\Delta y})_{\mathrm{II}}]/(\nu_1+\nu_2)$. This value is used, together with (5.35) and (5.36), to calculate the parameter sample variances $s^2_{a'_1}$, $s^2_{a'_2}$, $s^2_{b'_1}$, and $s^2_{b'_2}$, each with $n_1 + n_2 - 4$ degrees of freedom. These values are used in (5.63), (5.64), and (5.65) to evaluate the sample variances $s^2_{\mathrm{x+b}}$, s^2_{b}, and s^2_{c}.

For a chosen probability $P = 1 - \alpha$, the confidence interval for the true amount of the component determined will be obtained by using the t distribution, and the result is:

$$c_{\mathrm{x}} - |t_{\alpha/2;n_1+n_2-4}|s_{\mathrm{c}} < \hat{c} < c_{\mathrm{x}} + |t_{\alpha/2;n_1+n_2-4}|s_{\mathrm{c}} \tag{5.66}$$

The method of multiple standard addition has also been used in trace analysis, e.g. in potentiometric methods [30–36]. A statistical investigation of the performance of this method, as applied to atomic-absorption spectrometry, has been given by Larsen *et al.* [37].

We shall now exemplify the application of this procedure to the evaluation of analysis results and their confidence intervals.

Example 5.2. The results of an analytical experiment for the spectrophotometric determination of the amount of phenol in industrial waters by the method of multiple standard addition are given in Table 5.3. The reagent was *p*-nitroaniline. Each measurement was corrected for background.

Table 5.3 Analytical data for the determination of phenol in industrial waters

Number of the water sample	1	2	3	4	5	6	7
Amount (c) of added phenol (mg)	0	0.10	0.15	0.20	0.30	0.40	0.50
Absorbance (y)	0.140	0.310	0.385	0.460	0.640	0.820	0.980

By the least-squares methods, we find that the equation relating absorbance y to amount c is:

$$\bar{y} = 0.135 + 1.691c$$

The intercept, which is the analytical result, is:

$$c = 0.0797 \text{ mg}$$

By using (5.34) and (5.35), the sample variances for the parameters a'_1 and b'_1 are found to be:

$$s^2_{a'_1} = 0.00003092 \text{ and } s^2_{b'_1} = 0.000381$$

By using (5.65), the value of the sample variance of the result is

$$s^2_{\text{c}} = 0.0000162,$$

and the corresponding number of degrees of freedom is $\nu = 5$.

From the t distribution, the confidence interval for the true amount of the component determined (the true value of the intercept on the abscissa) at a confidence probability P is given by:

$$0.0797 - |t_{\alpha/2;5}|\, 0.004 < \hat{c} < 0.0797 + |t_{\alpha/2;5}|\, 0.004$$

where $\alpha = 1 - P$.

For example, taking the confidence probability $P = 0.90$, we have $t_{0.05;5} = 2.015$, so the confidence interval is:

$$0.0716 \text{ mg} < \hat{c} < 0.0875 \text{ mg}$$

5.10 THE INTERSECTION OF TWO LINEAR REGRESSIONS

This problem appears in the case of linear titrations, for example conductometric, amperometric, thermometric, photometric, oscillometric titrations. From the expressions for the two regression lines, $y = a'_1 + b'_1x$ and $y = a'_2 + b'_2x$, which are both obtained by the least-squares method, the estimation value of the point of intersection will be [38-40]:

$$\bar{x}_i = \frac{a'_2 - a'_1}{b'_1 - b'_2} = \frac{\Delta a'}{\Delta b'} \tag{5.67}$$

Because a confidence band, bounded by two hyperbolic arcs [Eq. (5.39)], is associated with each regression line, it is obvious that the point of intersection, x_i, is only a mean value, with which a certain confidence interval is associated.

If the signal values both before and after the point of intersection, are normally distributed around the line with a constant standard deviation, the point of intersection and its statistical confidence interval will have the significance shown in Fig. 5.9 [41].

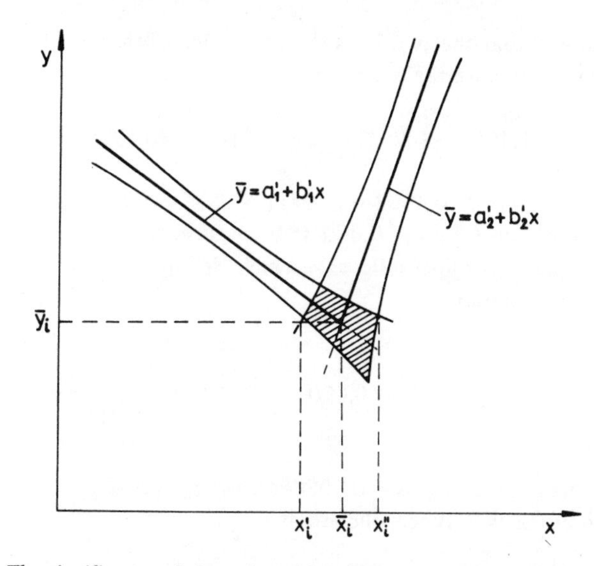

Fig. 5.9 The significance of the point of intersection and the confidence interval resulting from the crossing of two linear regressions [41]. (By permission of Pergamon Press).

The intersection point $\bar{x}_i$ is estimated by the projection of the intersection onto the abscissa. The confidence interval (x_i', x_i'') for the true value of the equivalence point is given by the projection on the abscissa of the common surface delimited by the four hyperbolic arcs.

Example 5.3. The results of a conductometric titration are given in Table 5.4, where $y = (1000 - a)/a$, in which a is the bridge reading.

Table 5.4 Data from a conductometric titration of $0.1M$ HCl with $0.1M$ NaOH [40]

V (ml of $0.1M$ NaOH) (x)	2	4	6	8	10	12	14
$(1000 - a)/a$ (y)	1.265	1.141	1.028	0.906	0.777	0.641	0.510
V (ml of $0.1M$ NaOH) (x)	16	17	18	20	22	24	
$(1000 - a)/a$ (y)	0.372	0.388	0.441	0.544	0.644	0.752	

If we assume that the two lines should be linear, then according to (5.32) and (5.33), they have the equations:

$$(\bar{y})_1 = 1.403 - 0.0637x \quad \text{and} \quad (\bar{y})_2 = -0.4908 + 0.0517x,$$

so from (5.67), $\bar{x}_i = (-0.4908 - 1.403)/(-0.0637 - 0.0517) = 16.41$ ml. From (5.34), we obtain $(s^2_{\Delta y})_1 = 0.0001069$; because from (5.39) $t_{(\alpha/2=0.05;\nu_1=6)} = 1.943$, the two hyperbolic arcs which delimit the confidence band are given by the equation

$$y_i = \bar{y}_i \pm 0.0201 \sqrt{0.125 + \frac{(x_i - 9)^2}{168}}$$

For the second line, $(s^2_{\Delta y})_2 = 0.0000042$ and $t_{(\alpha/2=0.05;\nu_2=3)} = 2.353$, so the equation of the two hyperbolic arcs is

$$y_i = \bar{y}_i \pm 0.0048 \sqrt{0.2 + \frac{(x_i - 20.2)^2}{32.8}}$$

Hence, the lower value x'_i of the confidence interval is obtained by solving the equation:

$$1.403 - 0.0637\,x'_i - 0.0201 \sqrt{0.125 + \frac{(x'_i - 9)^2}{168}}$$

$$= -0.4908 + 0.0517\,x'_i + 0.0048 \sqrt{0.2 + \frac{(x'_i - 20.2)^2}{32.8}}$$

which gives $x'_i = 16.27$ ml, and the higher value x''_i is obtained from the equation

$$1.403 - 0.0637\,x''_i + 0.0201 \sqrt{0.125 + \frac{(x''_i - 9)^2}{168}}$$

$$= -0.4908 + 0.0517\,x''_i - 0.0048 \sqrt{0.2 + \frac{(x''_i - 20.2)^2}{32.8}}$$

which gives $x''_i = 16.57$ ml. Hence $16.27 \text{ ml} < \hat{x}_i < 16.57$ ml.

Jander and co-workers [41] approximate by two straight lines the hyperbolae which delimit the confidence interval for the two lines and find that the length of the confidence interval of the intersection point, $\Delta x_i = x_i'' - x_i'$, is formed by two equal segments and has the value

$$\Delta x_i = 2\frac{s_{b_1'}t_1(\bar{x}_1 - \bar{x}_i) + s_{b_2'}t_2(\bar{x}_2 - \bar{x}_i)}{b_2' - b_1'} \tag{5.68}$$

where $s_{b_1'}$ and $s_{b_2'}$ are the sample standard deviations for the parameters b_1' and b_2' of the two experimental straight lines, $\bar{x}_1$ and $\bar{x}_2$ are the mean values for x (concentration, titration volume) before and after the point of intersection, which have been used in the calculation, t_1 and t_2 are the values of the t distribution corresponding to $\nu_1 = n_1 - 2$, and $\nu_2 = n_2 - 2$ degrees of freedom.

With the data given in Table 5.4, from (5.36) we obtain $s_{b_1'} = 7.98 \times 10^{-4}$, and from (5.36) we obtain $s_{b_2'} = 3.58 \times 10^{-3}$. Further, $t_1 = t_{(\alpha/2=0.05;\nu_2=6)} = 1.943$, and $t_2 = t_{(\alpha/2=0.05;\nu_2=3)} = 2.353$, so from (5.68) $\Delta x_i = 0.354$. Hence, $x_i' = \bar{x}_i - \Delta x_i/2 = 16.23$ ml, and $x_i'' = \bar{x}_i + \Delta x_i/2 = 16.59$ ml. Thus $16.23 \text{ ml} < \hat{x}_i < 16.59$ ml.

Since the point of intersection $\bar{x}_i$ belongs to one of the response functions, then as shown in Fig. 5.7, a certain confidence interval is associated with it. Similarly, if it is regarded as belonging to the other response function, there is another confidence interval associated with it, as shown in Fig. 5.10.

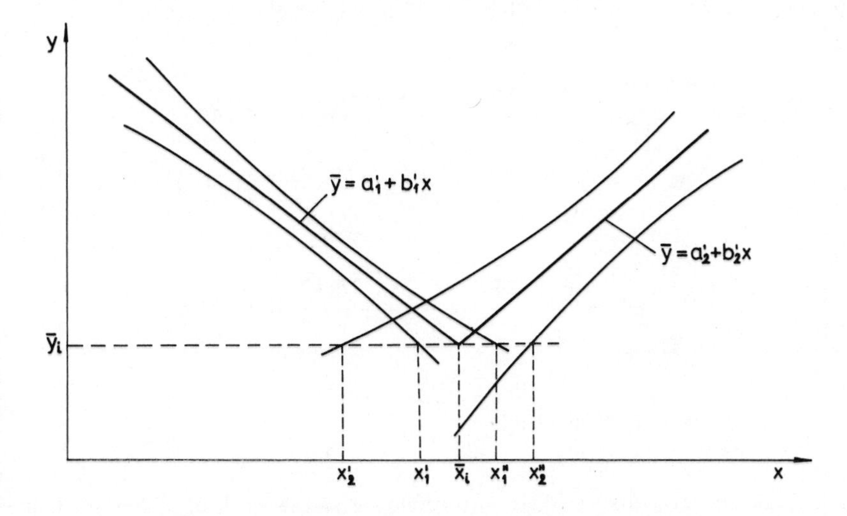

Fig. 5.10 The significance of the point of intersection and its separate confidence intervals belonging to two regression lines [42]. (By permission of Pergamon Press).

Because the intersection point belongs concomitantly to the two response functions, the two segments which together compose the confidence interval will be obtained by averaging the segments of the two separate confidence intervals [42]. Hence:

$$\frac{(n_1-2)x'_1 + (n_2-2)x'_2}{n_1+n_2-4} < \hat{x}_i < \frac{(n_1-2)x''_1 + (n_2-2)x''_2}{n_1+n_2-4} \tag{5.69}$$

The two values of the limits of the confidence interval will be the two solutions of the second degree equations

$$\bar{y}_i = a'_1 + b'_1 x_i \pm |t_{(\alpha/2;n_1-2)}| (s_{\Delta y})_1 \sqrt{\frac{1}{n_1} + \frac{(x_i-\bar{x}_1)^2}{S_{xx}}} \tag{5.70}$$

and

$$\bar{y}_i = a'_2 + b'_2 x_i \pm |t_{(\alpha/2;n_2-2)}| (s_{\Delta y})_2 \sqrt{\frac{1}{n_2} + \frac{(x_i-\bar{x}_2)^2}{S_{xx}}} \tag{5.71}$$

With $\alpha = 0.10$, for the data in Table 5.4 we obtain $x'_1 = 16.02; x''_1 = 16.81$; $x'_2 = 16.34; x''_2 = 16.49$. Hence the weighted means are

$$x'_i = \frac{6 \times 16.02 + 3 \times 16.34}{9} = 16.13 \text{ ml}$$

and

$$x''_i = \frac{6 \times 16.81 + 3 \times 16.49}{9} = 16.71 \text{ ml}$$

so 16.13 ml $< \hat{x}_i <$ 16.71 ml.

Lark and co-workers [43], with reference to the work of Fisher [44] and Kastenbaum [45] on the statistical problem of the intersection of two experimental lines, have shown that the limits of the confidence interval for the true value of the point of intersection of two straight lines are the roots of the equation:

$$\bar{x}_i^2[(\Delta b')^2 - t^2 s_{\Delta b'}^2] - 2\bar{x}_i[\Delta a'\Delta b' - t^2 s_{\Delta a'\Delta b'}] + [(\Delta a')^2 - t^2 s_{\Delta a'}^2] = 0 \tag{5.72}$$

where t is Student's parameter at a confidence probability corresponding to the number of degrees of freedom $\nu = \nu_1 + \nu_2 = n_1 + n_2 - 4$, $s_{\Delta a'}^2$ is the sample variance for the variable $\Delta a'$:

$$s_{\Delta a'}^2 = s_{a_1'}^2 + s_{a_2'}^2 \tag{5.73}$$

$s_{\Delta b'}^2$ is the sample variance for the variable $\Delta b'$:

$$s_{\Delta b'}^2 = s_{b_1'}^2 + s_{b_2'}^2 \tag{5.74}$$

and $s_{\Delta a'\Delta b'}^2$ is the sample variance for the variable $\Delta a'\Delta b'$:

$$s_{\Delta a'\Delta b'} = \bar{s}_{\Delta y}^2(\bar{x}_1/S_{xx} + \bar{x}_2/S_{xx}) \tag{5.75}$$

where $\bar{s}_{\Delta y}^2 = \dfrac{\nu_1(s_{\Delta y}^2)_1 + \nu_2(s_{\Delta y}^2)_2}{\nu_1 + \nu_2}$, and $S_{xx} = \sum_{i=1}^{n} x_i^2 - \dfrac{\left(\sum_{i=1}^{n} x_i\right)^2}{n}$.

With $\alpha = 0.10$, so that $P = 1 - \alpha = 0.90$, we obtain, for the data given in Table 5.4: $s_{\Delta a'}^2 = 0.0009631$, $s_{\Delta b'}^2 = 0.0000026$, $\bar{s}_{\Delta y}^2 = 0.0000727$, $s_{\Delta a'\Delta b'} = 0.00004867$; with $\nu = 9$, $t = 1.833$, so from (5.72) $x_i' = 16.29$ ml and $x_i'' = 16.54$ ml. Hence: 16.29 ml $< \hat{x}_i <$ 16.54 ml.

The precision of the point of intersection and the corresponding statistical confidence interval can be found in the simplest way by considering the law of addition of variances.

From (5.54) we obtain the following equation for the variance associated with the point of intersection:

$$\sigma_{\bar{x}_i}^2 = \frac{1}{(\Delta b')^2}\sigma_{\Delta a'}^2 + \frac{(\Delta a')^2}{(\Delta b')^4}\sigma_{\Delta b'}^2 \tag{5.76}$$

Further, if the variables a'_1 and a'_2, and b'_1 and b'_2, are independent, the variance of the variables $\Delta a'$ and $\Delta b'$ will be:

$$\sigma^2_{\Delta a'} = \sigma^2_{a'_1} + \sigma^2_{a'_2} \tag{5.77}$$

$$\sigma^2_{\Delta b'} = \sigma^2_{b'_1} + \sigma^2_{b'_2} \tag{5.78}$$

When the sample values are considered, the intersection point variance (5.76) becomes

$$s^2_{\bar{x}_i} = \frac{1}{(\Delta b')^2} s^2_{\Delta a'} + \frac{(\Delta a')^2}{(\Delta b')^4} s^2_{\Delta b'} \tag{5.79}$$

Taking into account the t distribution for the intersection point sample values, i.e. $t = (\bar{x}_i - \hat{x}_i)/s_{\bar{x}_i}$, the confidence interval for the true value $\hat{x}_i$ of the intersection point is [42]:

$$\bar{x}_i - |t_{\alpha/2;n_1+n_2-4}| s_{\bar{x}_i} < \hat{x}_i < \bar{x}_i + |t_{\alpha/2;n_1+n_2-4}| s_{\bar{x}_i} \tag{5.80}$$

If there are n_1 experimental data before the intersection point and n_2 experimental data after it, the variance $s_{\bar{x}_i}$, evaluated by taking into account all the $(n_1 + n_2)$ experimental data, will correspond to $\nu = n_1 + n_2 - 4$ degrees of freedom.

For the data of Table 5.4 $s^2_{\bar{x}_i} = 0.1256$, and with $\alpha = 0.10$, $P = 0.90$, $t = 1.833$ ($\nu = 9$), it results that $t_{(\alpha/2;\nu)} s_{\bar{x}_i} = 0.648$. So finally we have 15.77 ml $< \hat{x}_i <$ 17.06 ml.

5.11 NON-LINEAR CALIBRATION AND CALCULATION FUNCTIONS

In chemical analysis, there are many cases when non-linear calibration functions must be used. Examples can be given from emission spectroscopy, absorption spectroscopy, X-ray fluorescence, etc.

When interference is absent, the calibration function can be estimated, as in the linear case, by either the graphical or the least-squares method, from a set of calibration data (y_i, c_i).

For a parabolic model of the response function

$$y = b_0 + b_1 c + b_2 c^2 + \Delta y \tag{5.81}$$

the parameters b_0, b_1 and b_2 are estimated by the least-squares method. The corresponding system of equations (2.109) was given in Section 2.7.1.2.

In the case of a non-linear model, the analytical calculation function cannot be obtained by relating the calibration function to the amount to be determined, as is done in the linear case.

Instead, the mathematical expression of the analytical calculation function for a non-linear model can be obtained by using the least-squares method, but reversing the role of the two variables; the system response y is regarded as the independent variable, and the amount in the standard sample as the dependent variable.

In this case, the correctness of the calibration and the agreement between the real response function and the model assumed for parameter evaluation will again be measured by the value of the residual term (i.e. the deviation Δy from the calibration function).

Example 5.4. The analytical data obtained with an emission spectrometer for a set of standard samples in order to establish the calibration function and the analytical calculation function for determination of copper in steel are given in Table 5.5.

Table 5.5 Calibration data for the determination of copper in steel

Standard samples	1	2	3	4	5	6	7	8	9	10	11
c_{Cu} (%)	0.019	0.033	0.067	0.26	0.47	0.16	0.36	0.047	0.084	0.072	0.70
System response (digits)	26	38	63	182	301	113	213	50	81	68	366
	28	33	72	166	284	119	216	48	86	68	380
	27	34	60	178	281	119	223	47	79	72	400
	27	41	68	160	270	118	247	50	74	75	375
	30	38	64	161	278	115	238	50	70	62	387

Assuming initially a linear model for the response function, the least-squares method gives the calibration function as

$$\bar{y} = 28.51 + 523.9\,c$$

The variance of the residual term in this case is $(s^2_{\Delta y})_1 = 115.9$.

With a parabolic model (5.81) for the response function, however, the least-squares method gives the following equation for the calibration function:

$$\bar{y} = 19.93 + 638.4c - 173.8c^2,$$

and the variance of the residual term will be $(s^2_{\Delta y})_2 = 66.7$, a smaller value than in the first case.

Because $(s^2_{\Delta y})_1$ is significantly larger than $(s^2_{\Delta y})_2$, it follows that the model (5.81) for the system response function better approximates the real model.

The same conclusion also follows from Fig. 5.11, where the linear and the parabolic calibration functions are both represented graphically, as well as the mean values given in Table 5.5 for the system responses for the calibration samples.

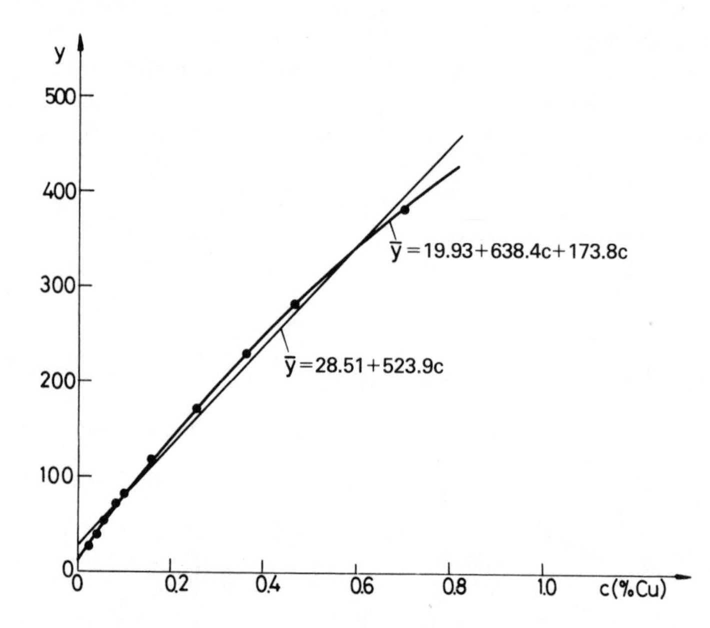

Fig. 5.11 The linear and parabolic calibration function for the data of example 5.4.

For estimating the analytical calculation function parameters, we accept the parabolic model. With y as the independent variable, and c as the dependent variable, the least-squares method gives

$$c = -0.02656 + 0.001439y - 0.00000118y^2$$

for the analytical calculation function. The variance of the residual term is found to be $s^2_{\Delta c} = 0.00025907$. As this is a very small value, it indicates that the analytical calculation function has been correctly evaluated.

5.12 CALIBRATION AND ANALYTICAL CALCULATION FUNCTIONS DEPENDING ON SEVERAL VARIABLES

It was pointed out in Section 5.2 that the response of an analytical system may be dependent not only on the amount of determinand, but also on the amounts of other components in the sample; this is called interference. It is obvious that, for a given working procedure, the nature and composition of the samples to be analysed cannot be arbitrarily varied without entering the range within which interference occurs. Interference will be absent for only a limited range of composition.

To widen the range of samples that can be analysed we must use calibration and analytical calculation functions of several variables.

Such functions are always used when corrections for interference phenomena are required. Their use has been made much easier by the on-line use of electronic computers with analytical systems. Thus, modern emission and X-ray fluorescence spectral systems are equipped with microprocessors which make possible the use of the complex analytical functions which can correct for the matrix effects.

To illustrate the method of using these calibration and analytical calculation functions, we shall refer to X-ray fluorescence methods [15, 46–51]. One of the models used in the X-ray fluorescence calculations is [48]

$$c_i = \sum_{i=1}^{n} \beta_i I_i + \sum_{i=1}^{n-1} \beta_{ij} I_i I_j + e_i \qquad (5.82)$$

$$i < j \leqslant n$$

where c_i is the amount of the ith component in the sample, I_i is the relative intensity of the X-radiation for this component (ratio of the intensity from the sample to the intensity from the pure component) and e_i is the residual (error) term. Such a function must be evaluated for each component, and each takes into account all the components present in the sample. Obviously, the analytical system must have at least one measurement channel for each component.

The coefficients, β_i and β_{ij}, of function (5.82) are evaluated by the least-squares method from a large set of calibration data, obtained from as large as possible a number of standard samples of rigorously known composition, which are representative of the material to be analysed.

We shall now consider an example of interference correction done with an analytical function of two variables.

Example 5.5. The X-ray fluorescence measurements obtained for arsenic and tungsten in a set of standard steel samples are given in Table 5.6. The analytical calculation function, with correction for tungsten interference in arsenic determination, is calculated by the least-squares method.

Table 5.6 Analytical measurements for arsenic and tungsten determination

Standard sample	Content c_{As} (%)	Content c_W (%)		System response for As and W (digits)				
1	0.003	0.003	I_{As}:	916	916	916	902	911
			I_W:	133	129	132	134	128
2	0.032	0.000	I_{As}:	1136	1131	1129	1130	1149
			I_W:	114	131	118	120	117
3	0.016	0.008	I_{As}:	993	1022	1005	995	1007
			I_W:	171	186	176	178	171
4	0.023	0.020	I_{As}:	1108	1080	1082	1094	1091
			I_W:	227	242	233	223	238
5	0.050	0.050	I_{As}:	1291	1311	1311	1305	1308
			I_W:	385	377	375	381	383
6	0.045	0.145	I_{As}:	1311	1324	1319	1308	1309
			I_W:	745	744	750	748	772
7	0.015	0.580	I_{As}:	1248	1253	1251	1251	1270
			I_W:	2525	2530	2586	2552	2562
8	0.025	0.050	I_{As}:	1123	1119	1101	1119	1103
			I_W:	346	357	378	361	356
9	0.030	0.260	I_{As}:	1274	1289	1257	1275	1270
			I_W:	1286	1294	1293	1309	1326
10	0.015	0.015	I_{As}:	997	1020	984	999	1004
			I_W:	192	181	180	176	181

If the effect of tungsten on the arsenic signal is ignored, the linear calibration function for arsenic calculated from the data pairs (c_{As}, I_{As}) in Table 5.6 by the least-squares method is:

$$I_{As} = 938.7 + 7940c_{As}$$

The corresponding residual-term variance is $(s^2_{\Delta I})_1 = 6707$. This large variance of the residual term indicates that the linear model is not adequate. A possible reason for this large variance may be the presence of interference.

Now let us still use a linear model, but assume that the arsenic signal depends not only on the amount of arsenic but also on the amount of tungsten. The least-squares method applied to the data of Table 5.6 now gives the following equation for the calibration function:

$$I_{As} = 880.4 + 8214c_{As} + 453.3c_W,$$

and the corresponding variance of the residual term $(s^2_{\Delta I})_2$ is 243.

The fact that $(s^2_{\Delta I})_2 \ll (s^2_{\Delta I})_1$ shows that the second model, which takes into account the tungsten interference, is much closer to reality. Hence, the analytical calculation function must also take into account the tungsten interference.

From the analytical data in Table 5.6 we can now evaluate, by the least-squares method, the analytical calculation function for arsenic. This function, expressed as a linear dependence of the amount of arsenic (c_{As}) on the arsenic and tungsten channel responses (I_{As} and I_W), is given by:

$$c_{As} = -0.1050 + 0.0001215\, I_{As} - 0.0000131\, I_W$$

The small value of the residual-term variance $s^2_{\Delta c} = 0.0000030$ indicates that this function is well-corrected for the effect of interference due to tungsten, and can be utilized in determination of arsenic.

REFERENCES

[1] Doerffel, K., *Z. Anal. Chem.*, 185, 91 (1962).
[2] Püschel, R., *Mikrochim. Acta*, 783 (1968).
[3] Koch, W. and Sauer, K. H., *Arch. Eisenhüttenwiss.*, 35, 861 (1964).
[4] Lassner, E., *Z. Anal. Chem.*, 222, 170 (1966).
[5] Nalimov, V. V., Nedler, V. V. and Menshova, N. P., *Zavodsk. Lab.*, 27, 861 (1961).
[6] Schwarz-Bergkampf, E., *Z. Anal. Chem.*, 221, 143 (1966).
[7] Miskary'ants, V. G., Kaplan, B. Ya. and Nedler, V. V., *Zavodsk. Lab.*, 37, 170 (1971).
[8] Zilbershtein, Kh. I. and Legeza, S. S., *Zh. Prikl. Spektrosk.*, 8, 6 (1968).
[9] Zilbershtein, Kh. I. and Legeza, S. S., *Zh. Priklad. Spektrosk.*, 8, 531 (1968).
[10] Zilbershtein, Kh. I., Polivanova, N. G. and Fratkin, Z. G., *Zh. Priklad. Spektrosk.*, 11, 204 (1969).
[11] Zilbershtein, Kh. I., *Zh. Prikl. Spektrosk.*, 14, 12 (1971).
[12] Kaiser, H., *Z. Anal. Chem.*, 260, 252 (1972).
[13] Kaiser, H., *Pure Appl. Chem.*, 34, 35 (1973).
[14] Gottschalk, G., *Z. Anal. Chem.*, 276, 257 (1975).

[15] van der Linden, W. E., *Z. Anal. Chem.*, **269**, 26 (1974).
[16] Junker, A. and Bergmann, G., *Z. Anal. Chem.*, **272**, 267 (1974).
[17] Gottschalk, G., *Z. Anal. Chem.*, **275**, 1 (1975).
[18] Eckschlager, K., *Chem. Listy*, **68**, 750 (1974).
[19] Acton, F. S., *Analysis of Straight-Line Data*, Wiley, New York (1959).
[20] Draper, N. R. and Smith, H., *Applied Regression Analysis*, Wiley, New York (1966).
[21] Brownlee, K. A., *Statistical Theory and Methodology in Science and Engineering*, p. 308. Wiley, New York (1960).
[22] Lark, P. D., Craven B. R. and Bosworth, R. C. L., *The Handling of Chemical Data*, p. 136. Pergamon, London (1968).
[23] Aivazian, S. A. and Bogdanovskii, I. M., *Zavodsk. Lab.*, **40**, 285 (1974).
[24] Knecht, J. and Stark, G., *Z. Anal. Chem.*, **270**, 97 (1974).
[25] Smith, E. D. and Mathews, D. M., *J. Chem. Educ.*, **44**, 757 (1967).
[26] Plesch, R., *Z. Anal. Chem.*, **275**, 269 (1975).
[27] Boček, P. and Novák, J., *J. Chromatog.*, **51**, 375 (1970).
[28] Francis, M. and Sobel, E., *Anal. Chem.*, **42**, 314 (1970).
[29] Plesch, R., *Z. Anal. Chem.*, **261**, 97 (1972).
[30] Gran, G., *Analyst*, **77**, 661 (1952).
[31] Liberti, A. and Mascini, M., *Anal. Chem.*, **41**, 679 (1969).
[32] Jagner, D. and Pavlova, V., *Anal. Chim. Acta*, **60**, 153 (1972).
[33] Frant, M. S., Ross, J. W. Jr. and Riseman, J. H., *Anal. Chem.*, **44**, 2227 (1972).
[34] Buffle, J., Parthasarathy, N. and Monnier, D., *Anal. Chim. Acta*, **59**, 427 (1972).
[35] Buffle, J., *Anal. Chim. Acta*, **59**, 439 (1972).
[36] Parthasarathy, N., Buffle, J. and Monnier, D., *Anal. Chim. Acta*, **59**, 447 (1972).
[37] Larsen, I. L., Hartmann, N. A. and Wagner, J. J., *Anal. Chem.*, **45**, 1511 (1973).
[38] Mika, J., *Z. Anal. Chem.*, **106**, 248 (1936).
[39] Liteanu, C. and Cörmöş, D., *Talanta*, **7**, 18 (1960).
[40] Liteanu, C. and Hopîrtean, E., *Studia Univ. Babeş-Bolyai, Ser. Chem.* (1), **11**, 135 (1966).
[41] Jandera, P., Kolda, S. and Kotrlý, S., *Talanta*, **17**, 443 (1970).
[42] Liteanu, C., Rîcă, I. and Liteanu, V., *Talanta*, **25**, 593 (1978).
[43] Lark, P. D., Craven, B. R. and Bosworth, R. C. L., *The Handling of Chemical Data*, p. 177. Pergamon, London (1968).
[44] Fisher, R. A., *Statistical Methods for Research Workers*, 13th Ed. Oliver & Boyd, Edinburgh (1963).
[45] Kastenbaum, M. A., *Biometrics*, **15**, 323 (1959).
[46] Lucas-Tooth, H. J. and Price, B. J., *Metallurgia*, **64**, 149 (1961).
[47] Alley, B. J. and Myers, R. H., *Anal. Chem.*, **37**, 1685 (1965).

[48] Stephenson, D. A., *Anal. Chem.*, **43**, 310 (1971).
[49] Blokhin, M. A., Belov, V. T., Duimakaev, Sh. I. and Topova, L. N., *Zavodsk. Lab.*, **39**, 1081 (1973).
[50] Belov, V. T. and Duimakaev, Sh. I., *Zavodsk. Lab.*, **40**, 958 (1974).
[51] Plesch, R., *Z. Anal. Chem.*, **272**, 262 (1974).

Chapter 6

Analytical Signal Detection

6.1 INTRODUCTION

We shall now discuss several fundamental problems of analytical signal detection, in terms of detection theory, which itself is a part of information theory.

Detection theory has been developed on the basis of the theory of probability and the statistical theory of decisions. The fundamental ideas in detection theory were first used about 35 years ago, in the field of signal transmission and reception. It has since undergone great development, parallel to its extension in the already familiar field of radar and telecommunications and has been applied in several other fields (astronomy, meteorology, seismology, management, etc.). It will be shown below that it can also be used to formulate a rigorous theory of analytical detection. Detection theory is now applied in the design and achievement of new analytical systems of measurement, and in their utilization.

The result of an analytical measurement is conveniently called an analytical signal. In general it is a function of several parameters, $x_1, x_2, \ldots, x_n$, one of which is usually time. Thus:

$$y = f(x_1, x_2, \ldots, x_n) \tag{6.1}$$

For extraction of the desired information, the analytical system usually allows the evaluation of some characteristic of the signal (generally the amplitude) as a function of a single parameter x which is modified, according to a given programme, within a certain range. For example, in spectroscopy we evaluate the change in amplitude of the emitted or absorbed signal $I(\lambda)$, when the wavelength λ is varies over a particular range (λ_a, λ_b). In this way we obtain a spectrogram similar to that shown in Fig. 6.1.

In polarography, the variation in amplitude of the limiting diffusion current is recorded as a function of the applied voltage, giving a polarogram. Similarly, a chromatogram is obtained by evaluating the change in amplitude of a signal (conductivity, absorption, refraction, etc.) as a function of the time taken for material to migrate through the chromatographic medium.

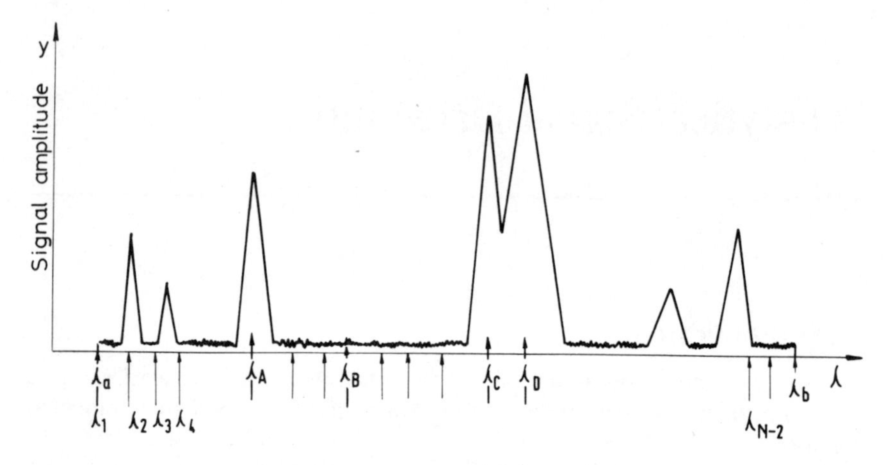

Fig. 6.1 Typical spectrogram of a material.

In many cases, the analytical measurement consists only in evaluating the amplitude of the signal at a given value of the parameter. In what follows, we shall use the term analytical signal in both the senses just illustrated.

Usually, analytical measurements are made on chemical systems which are in a non-stationary state (i.e. their properties vary with time). Further, the speed of response of the measuring instruments is always finite. Because of this, the amplitude of the analytical signal is also usually non-stationary. A model for the variation of the signal amplitude with time is given schematically in Fig. 6.2.

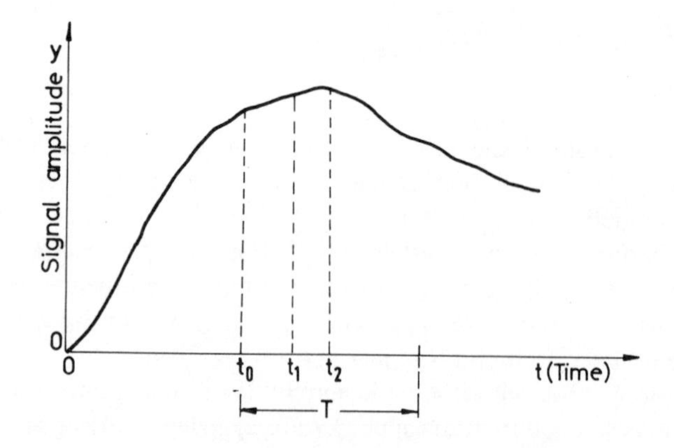

Fig. 6.2 A model for the time variation of the amplitude of a signal.

In such cases, to extract the maximum amount of information, the signal must be measured at a given time, e.g. $t_0, t_1, t_2, \ldots$, in Fig. 6.2. For example, in colorimetric methods, to calibrate the system and ensure accurate results for a reaction product which is unstable or results from a slow reaction, the colour intensity of the product must be measured at a certain time after the start of the reaction. In arc or spark emission spectroscopy, the physical and chemical processes from the plasma resulting from the sample are usually non-stationary; the amplitude of the emitted lines varies with time, so a strict time schedule must be kept. Indeed, some sort of time schedule is necessary for almost any analytical system.

It is important to remember that a signal, no matter how complex or rigorously defined, can always be represented in multi-dimensional space. Referring to Fig. 6.1, within the interval (λ_a, λ_b) we can set N points at the wavelengths $\lambda_1, \lambda_2, \ldots, \lambda_N$. Let $I_1, I_2, \ldots, I_N$ be the measured amplitudes at these points. In an N-dimensional space where the N axes correspond to the N quantities $\lambda_1, \lambda_2, \ldots, \lambda_N$, the signal can be represented by a point with coordinates $I_1, I_2, \ldots, I_N$, or by a vector $\vec{I}$ (Fig. 6.3). Because N can be as large as we wish, a signal can be represented to any degree of rigour.

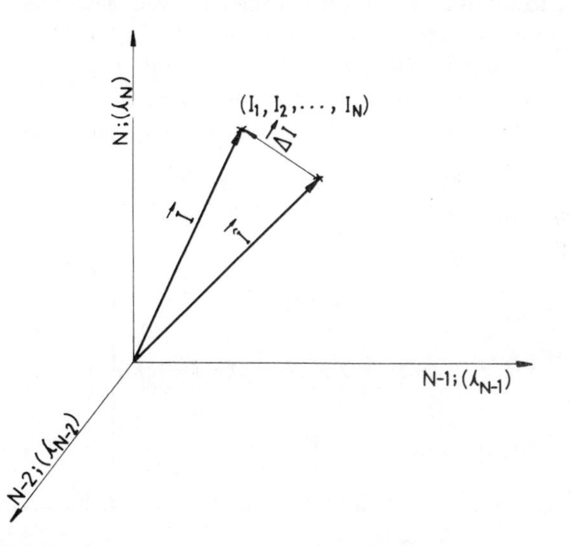

Fig. 6.3 The representation of a signal in multi-dimensional space.

Under real conditions, the signals will always include noise. We shall write $\vec{I} = \vec{\tilde{I}} + \vec{\Delta I}$, where $\vec{\tilde{I}}$ is the true but unknown value of the signal, and $\vec{\Delta I}$ is the perturbation effect (the noise).

The problems of signal interpretation posed by modern science and technology to analytical chemistry can be divided into three classes, as follows.

6.1.1 Analytical detection

This is an analytical investigation for the purpose of establishing the presence or absence of one or several of the components in a material.

In the case of establishing the presence or absence of a component on the basis of a single measurement, we have to decide between the two possible hypotheses: H_0 – the component is not present; H_1 – the component is present.

The amount of component to be detected is a variable quantity which takes the value $c_0 = 0$ if the component is absent, or a value $c_i > 0$ if it is present. The signal amplitude is a function of the amount of component to be detected, so the detection problem for one component, in the case when there is no interference, belongs to the general problem of detection of a known signal, having a variable amplitude, in the presence of noise. For the detection of one component, there are only two possible states for the investigated material: the state '0', without the component to be detected (i.e. the amount of the component to be detected is $c_0 = 0$) or the state '1', with the component to be detected (at a given moment, the amount of the component to be detected has a value $c_i > 0$ out of a set of possible values).

When a hypothesis corresponds to a multitude of values (points) of a parameter, it is said to be complex. Hence, in the analytical detection of one component, the hypothesis H_0 is simple, and the hypothesis H_1 is complex. This is shown schematically in Fig. 6.4, where only one point in the domain (space) of the parameters is associated with the hypothesis H_0, while a multitude of points ($c_i > 0$), represented by the surface C, is attached to the hypothesis H_1.

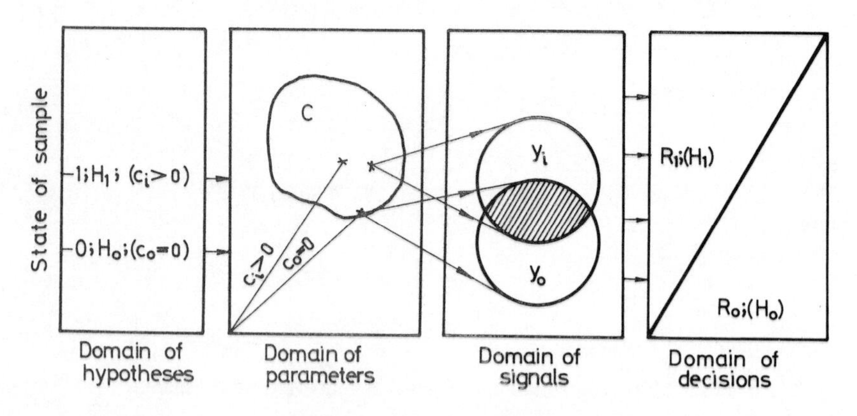

Fig. 6.4 The structure of a detection process.

Owing to perturbations, there is not a one-to-one correspondence between the domain of the parameters and the domain of the analytical signals; for one point in the parameter domain there is a set of corresponding points in the signal domain. Consequently, for two distinct points in the parameter domains, the corresponding set of points in the signal domain may have common elements. Because the correspondence between parameter and signal is governed by the laws of probability, statistical criteria must be used for basing a decision on a signal. For detection of a single component, the decision criterion divides the signal domain into two regions R_0 and R_1 such that the hypothesis chosen is H_0 when the signal pertains to the region R_0 and H_1 when it pertains to the region R_1.

Thus a detection process involves (1) the connection of the domain of hypotheses to the domain of signals and (2) the connection of the domain of signals to the domain of decisions (Fig. 6.5). Because (1) has statistical features, (2) must be done by means of rules based on statistical criteria.

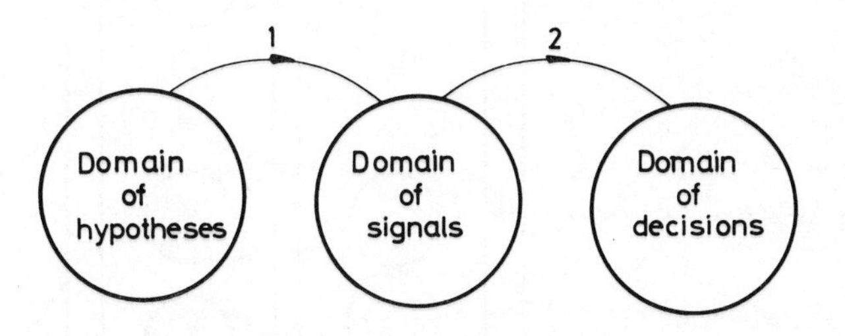

Fig. 6.5 The transformations required in a detection process.

6.1.2 Analytical determination

This is the process of evaluating the amount of one or several of the components of a material. Again, under real conditions, there will be noise and the results will always have a degree of uncertainty, i.e. they will be estimative in character. Two steps are generally involved in an analytical determination: (1) estimation of the amplitude of the signal; (2) conversion of the measurement into an analytical result by use of the calculation function.

Analytical determination can be interpreted by analogy with a detection process for the case of several hypotheses. The amount of component to be determined has, at a given moment, a value c_i out of a set of possible values contained within the domain ($c_0 = 0, c_M$), which is the range of variation of the amount of component. If a suitable quantization unit Δc is selected, this range can be divided into a discrete set of values $c_0 = 0, c_1, c_2, \ldots, c_M$. The process of analytical determination can be viewed as a detection process for a value $c_i (i = 0, 1, \ldots, M)$, out of the set of possible values $c_0, c_1, \ldots, c_M$. This is illustrated in Fig. 6.6.

To each of the values $c_0, c_1, c_2, \ldots, c_M$ we attach a corresponding hypothesis $H_0, H_1, H_2, \ldots, H_M$. Because of the quantization, each hypothesis corresponds to a subset of values of the amount of component. This is shown in Fig. 6.6 by the surfaces $C_0, C_1, C_2, \ldots, C_M$; hence, each hypothesis is complex. Further, because of noise, for a given amount of component there is a set of corresponding observational values. This is shown in Fig. 6.6 by attaching a surface in the signal domain to each value from the parameter domain. Because of the noise, there is overlap of the probability fields for the signals belonging to two distinct values of the amount of component. The decision rule divides the domain of analytical signals into a set of regions $R_0, R_1, R_2, \ldots, R_M$, each associated with one hypothesis. This is, in fact, the calibration process.

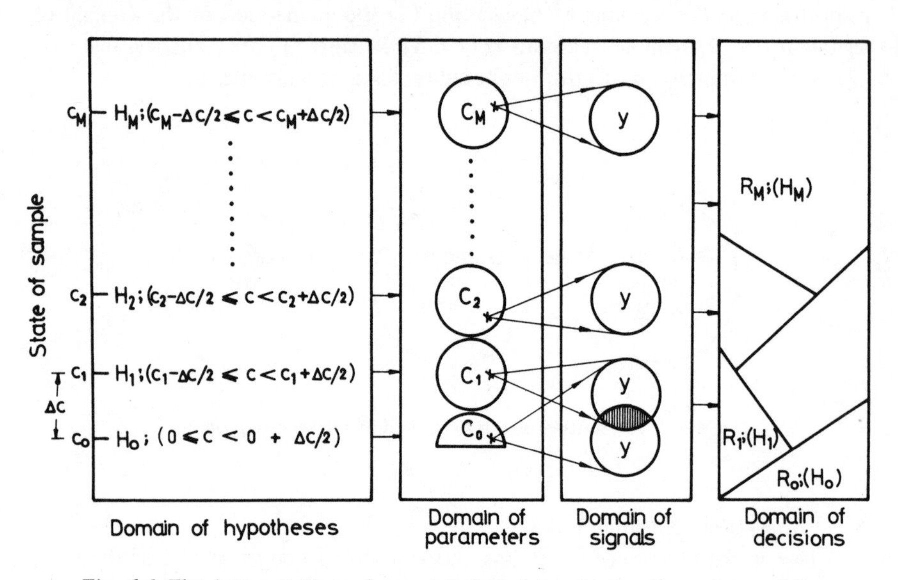

Fig. 6.6 The interpretation of an analytical determination by analogy with a detection process.

Thus an analytical determination is really an analytical detection in the case of multiple hypotheses, so a process of determination has to remove a greater uncertainty than a simple process of analytical detection for one component, and hence gives more information.

6.1.3 Analytical identification

By analytical identification we mean the process of establishing, either completely or partially, the physical or chemical nature of a material.

A complete answer to a problem of analytical identification must state precisely the nature of the material investigated in contradistinction to a problem of analytical detection, where answers such as 'component A is present', or 'components A, B, and C are absent' may be sufficient. The identification of complex materials requires a large number of processes of analytical detection and determination.

It is evident, however, that there are no clear-cut boundaries between analytical detection, determination, and identification; in practice they interpenetrate and affect each other.

6.2 DETECTION OF AN ANALYTICAL SIGNAL IN PRESENCE OF GAUSSIAN NOISE

Detection of a single component is based on measurement of the amplitude of a signal specific for that component. Under real conditions, as shown in Chapter 5, the measurement y will correspond to the sum of the true value $\hat{y}$ and the uncertainty in measurement, Δy:

$$y = \hat{y} + \Delta y \tag{6.2}$$

where Δy has a normal (or close to normal) distribution, with a mean of zero and a standard deviation σ_y, that is:

$$\Delta y \Rightarrow N(0; \sigma_y) \tag{6.3}$$

and if y is a linear function of the amount of component, c, the mathematical model for the results will be:

$$y \Rightarrow N(\hat{y} = \hat{y}_0 + b\hat{c}; \sigma_y) \tag{6.4}$$

Thus, y will be normally distributed, with mean $\hat{y} = \hat{y}_0 + b\hat{c}$ (the true value) which is linearly dependent on the amount $\hat{c}$, and the standard deviation will be σ_y. The term $\hat{y}_0$ is the mean (or true) value for the background signals.

As shown before (Section 5.1) σ_y itself depends on $\hat{c}$, but for narrow ranges of amount of component (especially for small amounts), σ_y can be taken as constant.

The mathematical model (6.4) is illustrated in Fig. 6.7. This model is generally adequate, especially in trace analysis, so it will be taken as the basic model for discussing problems of analytical detection.

From Fig. 6.7 it is obvious that the greater c_i, the smaller the number of common elements of the probability density $p_i(y)$ for the signals corresponding to it and the probability density $p_0(y)$ for the background signals; $p_i(y)$ and $p_0(y)$ are clearly separated only if c_i is above a certain level. Consequently, the situations possible in analytical detection are those shown in Fig. 6.8. The ideal case, when noise is absent, and there is one-to-one correspondence between c and y, is shown in (a). Under real conditions, however, noise causes a set of values of y to correspond to a single value of c, and there are two possibilities. One is shown in (b), where there is no overlap of the probability fields associated with the absence ($c_0 = 0$) and presence ($c_i > 0$) of the component to be detected,

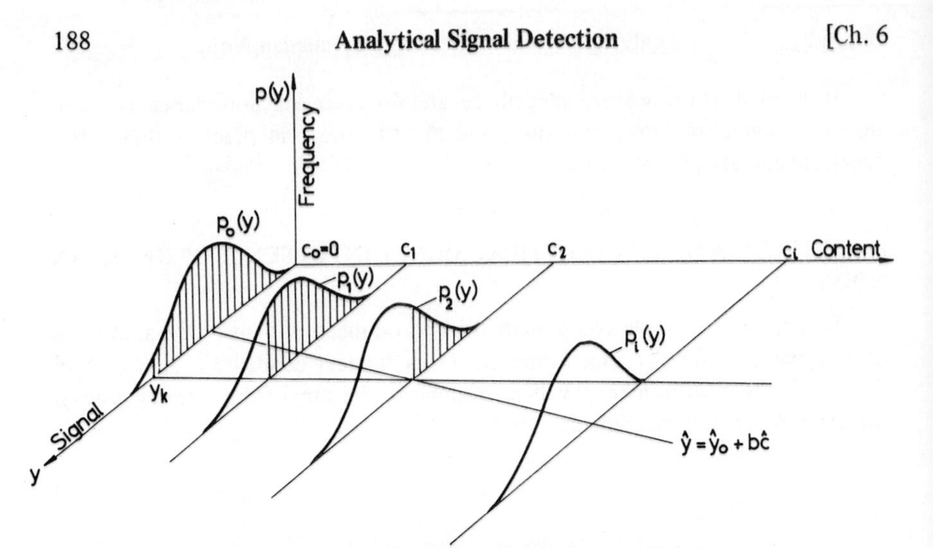

Fig. 6.7 Graphical model for the linear dependence between a signal and the amount of component, in the presence of a Gaussian noise.

and detection presents no problem. In contrast, (c) presents the case when the probability fields do overlap; in that case, detection must necessarily be based on statistical criteria.

A fourth case, when the probability fields are practically completely superimposed, thus making impossible the detection of the component, is shown in (d). This last case occurs when the amount of component to be detected is very small, the noise is large and the sensitivity and selectivity of measurement are poor.

The situations shown in (b), (c), and (d) will appear successively in the analytical system as the amount of component to be detected decreases towards zero, because of the corresponding decrease in the signal-to-noise ratio.

If we vary the number of measurements on which the detection decision is based, the following cases can be distinguished.

(1) We can use a single measurement y; the decision is then based on the probability densities for results of individual measurements, $p_0(y)$ and $p_i(y)$, corresponding to the two hypotheses H_0 and H_1.

(2) We can use N simple measurements, $y_1, y_2, \ldots, y_N$ obtained by repeating the analysis N times, or by taking measurements at N different points of a single signal or by combining these two methods. The results from these N independent measurements can be represented by a point in an N-dimensional space, and the detection decision is based on the probability density functions $p_0(y_1, y_2, \ldots, y_N)$ and $p_i(y_1, y_2, \ldots, y_N)$ associated with the two hypotheses.

(3) The number of measurements can be varied, and after each measurement it is decided whether one of the hypotheses can be accepted or further measurement should be made.

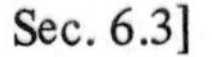

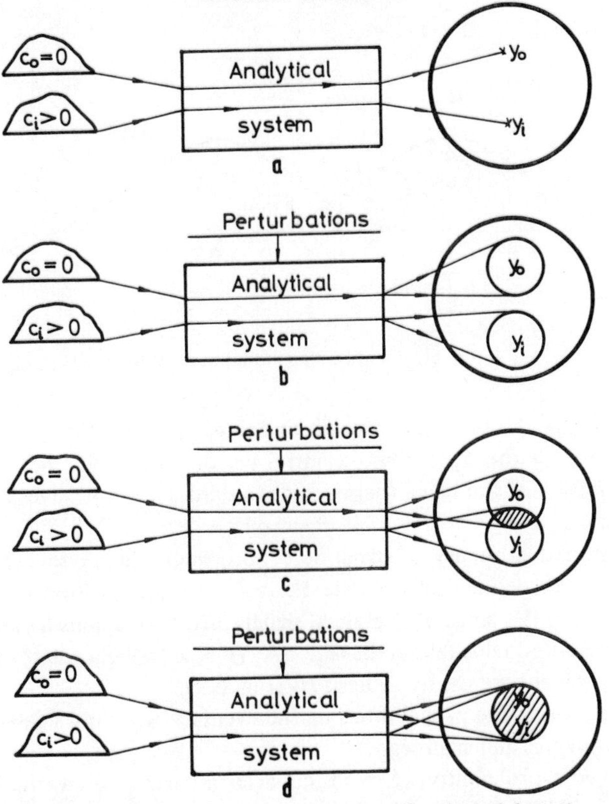

Fig. 6.8 Situations which may occur in a process of analytical detection; c_i – the amount of component to be detected; y_0 – the background signal; y_i – the analytical signal in the presence of the component to be detected. (From [7] by permission of Springer-Verlag)

6.3 DECISION THRESHOLD

For analytical detection, an essential problem is definition of a detection threshold. The component is said to be present when the amplitude of the signal is above a certain critical value (the threshold).

Definition of the threshold implies a complete mathematical model, and this must cover the following fundamental elements.

(a) Errors of interpretation, i.e. taking the noise as the signal, when actually the signal is absent (error of the first kind), or taking the signal as noise, when actually the signal is present (error of the second kind).

(b) Known *a priori* probabilities for the two hypotheses H_0 and H_1.

(c) The cost for each type of decision.

To discuss the concept of detection threshold, we shall consider the simple case of the detection of one component, as illustrated in Fig. 6.7, obtaining the two-step detection model (shown in Fig. 6.9).

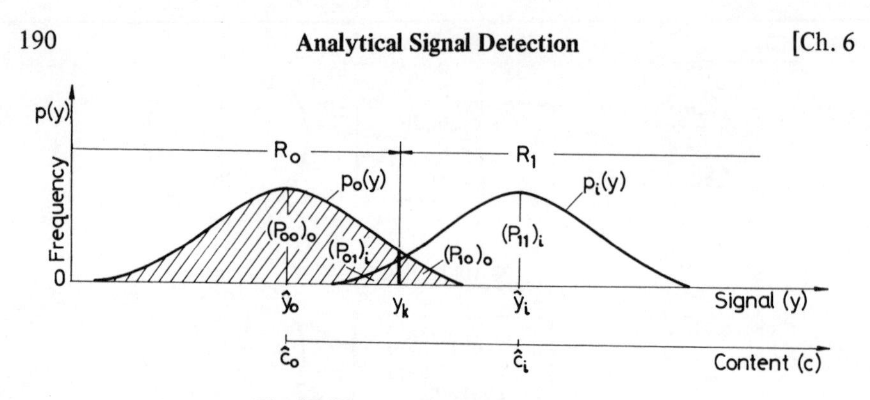

Fig. 6.9 The two-step model of detection.

In this figure the symbols have the meanings given for Eqs. (6.5)-(6.8) and $r = (\hat{y}_i - \hat{y}_0)/\sigma_y$ is the signal-to-noise ratio, i.e. the ratio of the net signal to the standard deviation of the signal (hence this standard deviation serves as a measure for the noise).

In the statistical theory of signal detection, there are several criteria [1-4] by which we can define and evaluate the value of the decision threshold, y_k. This value divides the range of analytical signals into two regions R_0 and R_1, such that, if the measured value falls in the region R_0 ($y \leqslant y_k$) we choose the hypothesis H_0; if $y > y_k$, we choose the hypothesis H_1 (Fig. 6.9).

For this model, the probabilities of the events $y \leqslant y_k$ and $y > y_k$, will have the following values and significance:

$(P_{00})_0$ – the probability of choosing, after a measurement, the hypothesis H_0 when H_0 is indeed true:

$$(P_{00})_0 = \int_{-\infty}^{y_k} p_0(y)\mathrm{d}y \tag{6.5}$$

$(P_{10})_0$ – the probability of choosing, after a measurement, the hypothesis H_1 when actually H_0 is true (the probability of false detection, i.e. the probability of error of the first kind):

$$(P_{10})_0 = \int_{y_k}^{+\infty} p_0(y)\mathrm{d}y \tag{6.6}$$

$(P_{01})_i$ – the probability of choosing, after a measurement, the hypothesis H_0 when H_1 is actually true (the probability of error of the second kind):

$$(P_{01})_i = \int_{-\infty}^{y_k} p_i(y)\mathrm{d}y \tag{6.7}$$

$(P_{11})_i$ – the probability of choosing, after a measurement, the hypothesis H_1 when H_1 is indeed true:

$$(P_{11})_i = \int_{y_k}^{+\infty} p_i(y)dy \tag{6.8}$$

The following equations are obviously satisfied by these probabilities:

$$\begin{aligned}(P_{00})_0 + (P_{10})_0 &= 1\\ (P_{11})_i + (P_{01})_i &= 1\end{aligned} \tag{6.9}$$

The probabilities (6.5)-(6.8), defined for the case of a single measurement, can also be used for an arbitrary number of measurements. For N measurements $(y_1, y_2, \ldots, y_N)$, the probability density functions for H_0 and H_1 are $p_0(Y) = p_0(y_1, y_2, \ldots, y_N)$ and $p_i(Y) = p_i(y_1, y_2, \ldots, y_N)$ respectively.

In order to understand the decision strategy in this case, it is convenient to represent the N measurements as the point with co-ordinates $Y(y_1, y_2, \ldots, y_N)$ in N-dimensional space, as in Fig. 6.10. The strategy of decision in this case consists in dividing the space, by means of a decision surface S_k, into two regions R_0 and R_1 such that if the observation Y is contained in the region R_0 hypothesis H_0 is chosen; if it is in region R_1 hypothesis H_1 is selected (Fig. 6.10). $\hat{Y}_0$ and $\hat{Y}_i$ stand for the true signals (in the absence of noise) corresponding to the N observations, for the hypotheses H_0 and H_1, respectively.

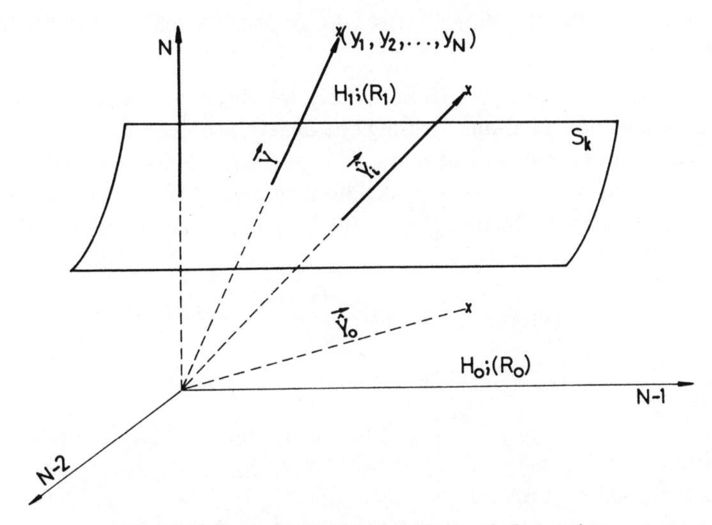

Fig. 6.10 Detection in the case of several observations.

In this case, the probabilities (6.5)-(6.8) will be defined as follows:

$$(P_{00})_0 = \int_{R_0} p_0(Y)\, \mathrm{d}^N Y \tag{6.10}$$

$$(P_{10})_0 = \int_{R_1} p_0(Y)\, \mathrm{d}^N Y \tag{6.11}$$

$$(P_{01})_i = \int_{R_0} p_i(Y)\, \mathrm{d}^N Y \tag{6.12}$$

$$(P_{11})_i = \int_{R_1} p_i(Y)\, \mathrm{d}^N Y \tag{6.13}$$

where $\mathrm{d}^N Y = \mathrm{d}y_1\, \mathrm{d}y_2 \ldots \mathrm{d}y_N$ is the volume element for observations in the N-dimensional space. Thus, for example, the integral (6.10) is the probability of choosing the H_0 hypothesis when it is true, that is, the result Y is represented by a point in the region R_0 when the hypothesis H_0 is true. The significance of the other integrals should be obvious.

6.4 FORMULATION OF DETECTION DECISIONS BY USING LIKELIHOOD CRITERIA

By decision criterion (or decision test) we mean a strategy adopted in formulating detection decisions by using one or several observations.

Several criteria are described in the theory of signal detection. Here we shall describe only those based on the likelihood principle. For this purpose we consider the two-step model of Fig. 6.9. The likelihood ratio is defined by

$$k(y_1, y_2, \ldots, y_N) = \frac{p_i(y_1, y_2, \ldots, y_N)}{p_0(y_1, y_2, \ldots, y_N)} \tag{6.14}$$

where $p_0(y_1, y_2, \ldots, y_N) = p_0(Y)$ and $p_i(y_1, y_2, \ldots, y_N) = p_i(Y)$ are the probabilities that the series of measurements $y_1, y_2, \ldots, y_N$ pertains to the hypotheses H_0 and H_1, respectively.

To formulate detection decisions, the experimental value of $k(y_1, y_2, \ldots, y_N)$

is compared with a threshold value k_0, and the hypothesis is chosen according to the decision rule:

$$k(y_1, y_2, \ldots, y_N) \underset{H_1}{\overset{H_0}{\lessgtr}} k_0 \tag{6.15}$$

The likelihood criteria are also known as likelihood ratio tests.

6.4.1 The Bayes criterion

This is also known as the Bayes decision rule, the minimum risk criterion, or the Bayes test. In order to describe it we shall again consider the two-step model of Fig. 6.9. The threshold value k_0 is evaluated by taking into account the *a priori* probabilities q and $1-q$ for the two hypotheses H_0 and H_1, and also the costs associated with each of the possible decisions, which are:

(a) choosing H_0 when it is valid;
(b) choosing H_1 when H_0 is valid;
(c) choosing H_1, when it is valid;
(d) choosing H_0, when H_1 is valid.

The costs can be represented by a cost matrix written as:

$$C = \begin{bmatrix} C_{00} & C_{01} \\ C_{10} & C_{11} \end{bmatrix} \tag{6.16}$$

where C_{ij} $(i, j = 0;1)$ is the cost when H_i is the hypothesis chosen and H_j is the valid hypothesis.

The mean risk for a decision is:

$$\bar{C} = q \left[C_{00} \int_{-\infty}^{y_k} p_0(y)\mathrm{d}y + C_{10} \int_{y_k}^{+\infty} p_0(y)\mathrm{d}y \right]$$
$$+ (1-q) \left[C_{01} \int_{-\infty}^{y_k} p_i(y)\mathrm{d}y + C_{11} \int_{y_k}^{+\infty} p_i(y)\mathrm{d}y \right] \tag{6.17}$$

It can be shown that the mean risk is minimal in the case when the region R_0 consists of values y which satisfy the condition $k(y) \leqslant k_0$, and the region R_1 of values y for which $k(y) > k_0$. Here $k(y)$ is the likelihood ratio, which, in the case of a single measurement, is given by:

$$k(y) = \frac{p_1(y)}{p_0(y)} \tag{6.18}$$

The threshold value k_0 which corresponds to the minimum risk for the likelihood ratio (the Bayes threshold test) is given by:

$$k_0 = \frac{q(C_{10} - C_{00})}{(1-q)(C_{01} - C_{11})} \tag{6.19}$$

Hence, in a general formulation, for a single measurement, the Bayes decision rule is given by:

$$k(y) \underset{H_1}{\overset{H_0}{\lesseqgtr}} k_0 \tag{6.20}$$

For the two-step model of Fig. 6.9, in the case of analytical detection perturbed by Gaussian noise, we can derive a simple expression for the decision threshold and decision rule. Thus, in the case of Gaussian noise, the probability density functions corresponding to the hypotheses H_0 and H_1 are:

$$p_0(y) = \frac{1}{\sigma_y\sqrt{2\pi}} \exp\left[-\frac{(y-\hat{y}_0)^2}{2\sigma_y^2}\right] \tag{6.21}$$

and

$$p_i(y) = \frac{1}{\sigma_y\sqrt{2\pi}} \exp\left[-\frac{(y-\hat{y}_i)^2}{2\sigma_y^2}\right] \tag{6.22}$$

For the Bayes criterion, the decision threshold y_{k_0} on the signal axis is obtained from the condition that the likelihood ratio corresponding to the signal y_{k_0} should be equal to the threshold value k_0; that is:

$$\frac{p_i(y_{k_0})}{p_0(y_{k_0})} = k_0 \tag{6.23}$$

By substituting the values $p_i(y_{k_0})$ and $p_0(y_{k_0})$, obtained from (6.21) and (6.22), into (6.23) and taking logarithms, we obtain for the threshold value:

$$y_{k_0} = \frac{2\sigma_y^2 \ln k_0 + \hat{y}_i^2 - \hat{y}_0^2}{2(\hat{y}_i - \hat{y}_0)} \tag{6.24}$$

so the decision rule is:

$$y \underset{H_1}{\overset{H_0}{\lesseqgtr}} \frac{\hat{y}_0 + \hat{y}_i}{2} + \frac{\sigma_y^2 \ln k_0}{\hat{y}_i - \hat{y}_0} \tag{6.25}$$

By analogy with (6.17), the mean risk of the decision in the case of several measurements is given by:

$$\bar{C} = q\left[C_{00}\int_{R_0} p_0(Y)\mathrm{d}^N Y + C_{10}\int_{R_1} p_0(Y)\mathrm{d}^N Y\right] + (1-q)\left[C_{01}\int_{R_0} p_i(Y)\mathrm{d}^N Y + C_{11}\int_{R_1} p_i(Y)\mathrm{d}^N Y\right] \quad (6.26)$$

The decision surface S_{k_0} is obtained from the condition that the mean risk be a minimum. If Y represents the point with co-ordinates $(y_1, y_2, \ldots, y_N)$ in N-dimensional space, the region R_0 will be the ensemble of points Y for which the likelihood ratio satisfies $k(Y) \leqslant k_0$, and R_1 the ensemble of points for which $k(Y) > k_0$, where $k(Y) = p_1(Y)/p_0(Y)$, and k_0 is the threshold value given by (6.19).

In a general form, the decision rule for a set of measurements is:

$$k(Y) \underset{H_1}{\overset{H_0}{\lesseqgtr}} k_0 \quad (6.27)$$

Considering again the two-step model of Fig. 6.9, we shall write $y_1, y_2, \ldots, y_N$ for the results of N independent measurements corresponding to one of the two possible hypotheses. The probabilities of these measurements, for the hypotheses H_0 and H_1, are:

$$p_0(Y) = p_0(y_1, y_2, \ldots, y_N) = \left(\frac{1}{\sigma_y^2\sqrt{2\pi}}\right)^N \exp\left[-\frac{\sum_{j=1}^{N}(y_j - \hat{y}_0)^2}{2\sigma_y^2}\right] \quad (6.28)$$

and

$$p_i(Y) = p_i(y_1, y_2, \ldots, y_N) = \left(\frac{1}{\sigma_y^2\sqrt{2\pi}}\right)^N \exp\left[-\frac{\sum_{j=1}^{N}(y_j - \hat{y}_i)^2}{2\sigma_y^2}\right] \quad (6.29)$$

The likelihood ratio and the decision rule are:

$$k(Y) = k(y_1, y_2, \ldots, y_N) = \frac{p_i(y_1, y_2, \ldots, y_N)}{p_0(y_1, y_2, \ldots, y_N)} \quad (6.30)$$

and

$$k(y_1, y_2, \ldots, y_N) \underset{H_1}{\overset{H_0}{\lessgtr}} k_0 \tag{6.31}$$

By substitution of (6.28) and (6.29) into (6.31), the following decision rule is obtained:

$$(\bar{y})_N \underset{H_1}{\overset{H_0}{\lessgtr}} \frac{\hat{y}_0 + \hat{y}_i}{2} + \frac{\sigma_y^2 \ln k_0}{N(\hat{y}_i - \hat{y}_0)} \tag{6.32}$$

where $(\bar{y})_N = \sum_{j=1}^{N} y_j/N$ is the mean of the N measurements.

The Bayes criterion is optimal, in the sense that there is no other strategy that would lead to a smaller risk.

6.4.2 The minimax criterion

When the *a priori* probabilities q and $1-q$ for the two hypotheses H_0 and H_1 are unknown, the Bayes criterion is replaced by the minimax criterion. In this case we use the Bayes strategy with the *a priori* probability value $q = q_0$ corresponding to the maximal Bayes risk.

The significance of the minimax criterion can be seen from Fig. 6.11. The dependence of the Bayes risk function $\bar{c}_{\min}$ on the *a priori* probability q has a maximum. If, instead of the real value of q, we use another value q_1, the risk function would be given by the straight line $\bar{c}(q)$ which is the tangent to the Bayes risk curve at the point $q = q_1$.

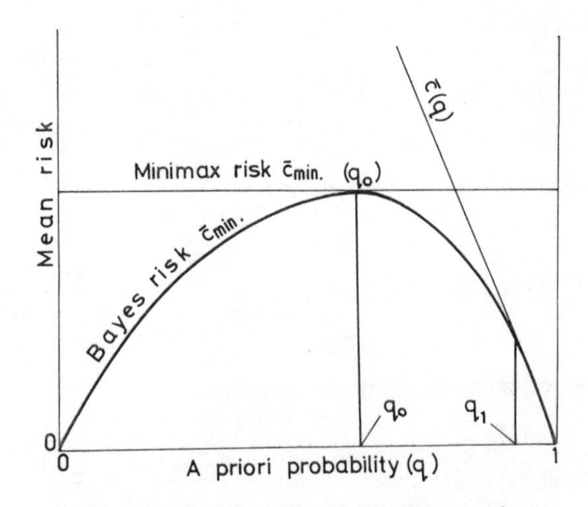

Fig. 6.11 The Bayes risk and the minimax risk.

Obviously, the risk $\bar{c}(q)$ can be very high when the *a priori* probabilities for the two hypotheses are unknown and the Bayes criterion is applied at an arbitrary value for q. If the Bayes criterion is applied for the value $q = q_0$ corresponding to the maximal Bayes risk, the tangent $\bar{c}_{min}(q_0)$ will be horizontal, so that certainly the risk will not be higher than this value.

Consequently, in order to use this criterion, the unknown probability q is taken as a variable; the Bayes risk is calculated for all values of q between 0 and 1, and finally that value of q is chosen which corresponds to the maximal risk, i.e. to the most unfavourable situation.

6.4.3 The criterion of the ideal observer

This criterion is derived from the Bayes criterion in the case when the cost of correct decisions is zero ($C_{00} = C_{11} = 0$), and the cost of erroneous decisions is unity ($C_{10} = C_{01} = 1$), so that the costs related to each type of error are equal:

$$C_{10} - C_{00} = C_{01} - C_{11}$$

Hence, by (6.19), the threshold for this criterion is:

$$k_0' = \frac{q}{1-q} \tag{6.33}$$

If the likelihood ratio is $k(y)$, the decision rule for this criterion will be:

$$k(y) \underset{H_1}{\overset{H_0}{\lesseqgtr}} k_0' \tag{6.34}$$

6.4.4 The criterion of maximal likelihood

This criterion is used in the absence of any information about both the *a priori* probabilities and the decision costs, and we set $q = 0.5$, $C_{00} = C_{11} = 0$, and $C_{01} = C_{10} = 1$. With these values in (6.19), the threshold value for this test is found to be:

$$k_0'' = 1 \tag{6.35}$$

so the decision rule is:

$$k(y) \underset{H_1}{\overset{H_0}{\lesseqgtr}} 1 \tag{6.36}$$

For the two-step model of Fig. 6.9, the decision level corresponding to this criterion will be derived from the condition:

$$p_0(y) = p_i(y) \tag{6.37}$$

From (6.21) and (6.37), the decision threshold is found to be:

$$y_{\mathrm{K}}'' = \frac{\hat{y}_i + \hat{y}_0}{2} \tag{6.38}$$

so the decision rule for a single measurement is:

$$y \underset{H_1}{\overset{H_0}{\lesseqgtr}} \frac{\hat{y}_i + \hat{y}_0}{2} \tag{6.39}$$

and, for N independent measurements:

$$(\bar{y})_N \underset{H_1}{\overset{H_0}{\lesseqgtr}} \frac{\hat{y}_i + \hat{y}_0}{2} \tag{6.39$'$}$$

Generally, in the case of several measurements $y_1, y_2, \ldots, y_N$, the decision rule is:

$$k(y_1, y_2, \ldots, y_N) \underset{H_1}{\overset{H_0}{\lesseqgtr}} 1 \tag{6.40}$$

The procedure for formulating detection decisions by using this criterion in the case of several observations is illustrated in Fig. 6.12.

The decision surface $S_{\mathrm{K}''}$ is the hyperplane perpendicular to the vector $\vec{Y}_i - \vec{Y}_0$ and intersecting it at its mid-point. This plane divides the domain of observations into two regions R_0 and R_1. If the observation vector $\vec{Y}_i$ with co-ordinates $(y_1, y_2, \ldots, y_N)$ is contained in the region R_0, we will choose hypothesis H_0, and if it is contained in R_1, we will choose hypothesis H_1. The vector $\vec{Y}_i$ will always include a certain noise, specified as $\Delta\vec{Y}$.

6.4.5 The Neyman-Pearson criterion

In most cases, it is difficult to know or estimate the *a priori* probabilities q and costs C_{ij}, so the Bayes test cannot be applied. In such cases the Neyman-Pearson test is used instead, and the critical value k_0 for the likelihood ratio

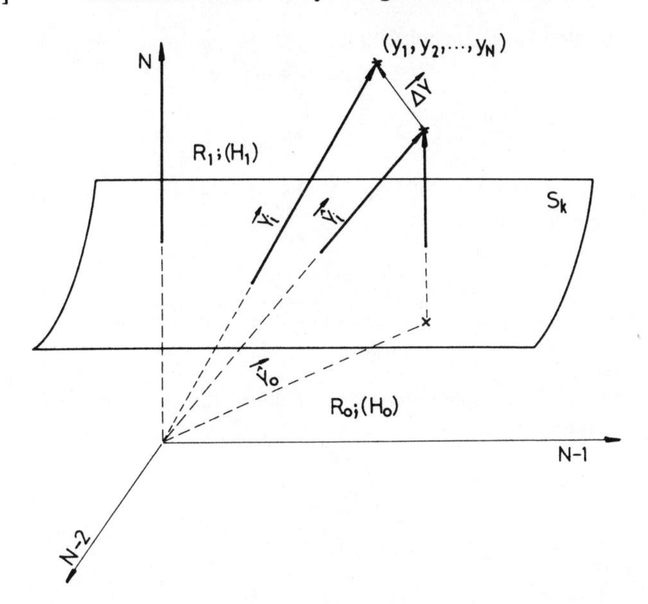

Fig. 6.12 Formulation of decision detections on the maximal likelihood criterion.

$k(y)$, the decision rule and the decision threshold are established on the basis of the probability of false detection $[(P_{10})_0]_{y_k}$.

In order to describe this criterion we shall again refer to Fig. 6.9, and take into account the probability density for background signals:

$$p_0(y) = \frac{1}{\sigma_y\sqrt{2\pi}} \exp\left[-\frac{(y-\hat{y}_0)^2}{2\sigma_y^2}\right] \quad (6.41)$$

According to the Neyman-Pearson criterion, the value of the decision threshold y_k corresponding to a probability of false detection preset at $[(P_{10})_0]_{y_k}$ is obtained from the condition:

$$[(P_{10})_0]_{y_k} = \frac{1}{\sigma_y\sqrt{2\pi}} \int_{y_k}^{+\infty} \exp\left[-\frac{(y-\hat{y}_0)^2}{2\sigma_y^2}\right] dy \quad (6.42)$$

With the normal deviate value $z = (y - \hat{y}_0)/\sigma_y$, the decision threshold will be given by:

$$y_k = \hat{y}_0 + z_k\sigma_y \quad (6.43)$$

where z_k is obtained from the condition:

$$[(P_{10})_0]_{y_k} = \frac{1}{\sqrt{2\pi}} \int_{z_k}^{+\infty} \exp\left[-\frac{z^2}{2}\right] dz \tag{6.44}$$

Hence, for the two-step model of Fig. 6.9, the Neyman-Pearson criterion leads to the following decision rule:

$$y \underset{H_1}{\overset{H_0}{\lesseqgtr}} \hat{y}_0 + z_k \sigma_y \tag{6.45}$$

Obviously, for N independent measurements the decision surface and the decision rule will be:

$$(y_k)_N = \hat{y}_0 + z_k \frac{\sigma_y}{\sqrt{N}} \tag{6.46}$$

and

$$(\bar{y})_N \underset{H_1}{\overset{H_0}{\lesseqgtr}} \hat{y}_0 + z_k \frac{\sigma_y}{\sqrt{N}} \tag{6.47}$$

where $(\bar{y})_N = \sum_{j=1}^{N} y_j/N$ is the mean of the N measurements.

Example 6.1. By using the Neyman-Pearson criterion for normally distributed background signals, evaluate the equation of the decision threshold and the decision rule corresponding to the probability of false detection $[(P_{10})_0]_{y_k} = 0.01$. In this case the value of the Laplace function will be: $\Phi(z_k) = 0.5 - 0.01 = 0.49$.

From the table of the normal distribution function, we find that the probability $P_{10} = 0.01$, i.e. $\Phi(z_k) = 0.49$, is obtained for $z_k = 2.32$. Hence, in the case of a single measurement, the decision threshold and the decision rule are:

$$y_k = \hat{y}_0 + 2.32\,\sigma_y \text{ and } y \underset{H_1}{\overset{H_0}{\lesseqgtr}} \hat{y}_0 + 2.32\,\sigma_y$$

and, in the case of N measurements:

$$(y_k)_N = \hat{y}_0 + 2.32\frac{\sigma_y}{\sqrt{N}} \text{ and } (\bar{y})_N \underset{H_1}{\overset{H_0}{\lesseqgtr}} \hat{y}_0 + 2.32\frac{\sigma_y}{\sqrt{N}}$$

In conclusion, the minimum risk is ensured by using the Bayes criterion. However, because establishing the values of the cost matrix elements C_{ij} and *a priori* probability q is usually tedious, there are many cases in which the Bayes criterion will not be applied. Accepting that the relative costs for each kind of error are equal, we obtain the criterion of the ideal observer. In the absence of any information about C_{ij} and q, the critical value of the likelihood ratio is taken equal to unity, i.e. we use the criterion of maximal likelihood.

The application of the Bayes test to the problem of detection in analytical systems was first discussed by Svoboda and Gerbatsch [5]. These authors used the criterion of maximal likelihood, because they accepted equal costs for the two kinds of errors and an *a priori* probability $q = 0.50$.

The problems of analytical detection are more suitably treated by using the Neyman-Pearson criterion. Liteanu and Rîcă [6–8] have applied this criterion in the formulation of a general strategy for analytical detection. This criterion has allowed quantitative formulation of several concepts for analytical detection, among which we may mention the decision threshold, decision rule, characteristic of detection, detection limit of a method of analysis, etc.

6.5 THE INFORMATIONAL DEFINITION OF THE DECISION THRESHOLD

The definition of the decision threshold obtained from the Neyman–Pearson criterion can be interpreted according to information theory, by taking into account the entropy of two events associated to the magnitude of the background signal with respect to the decision threshold y_k.

Thus, again considering Fig. 6.9, the field of background signals will be resolved into two sets, $y \leqslant y_k$ and $y > y_k$. The probabilities for these sets are easily obtained from the probability density of the background signals $p_0(y)$:

$$\begin{aligned} P(y \leqslant y_k) &= [(P_{00})_0]_{y_k} \\ P(y > y_k) &= [(P_{10})_0]_{y_k} \end{aligned} \tag{6.48}$$

According to (3.6), the entropy of this system is:

$$\begin{aligned} H[(P_{00})_0, (P_{10})_0]_{y_k} = &- [(P_{00})_0]_{y_k} \log [(P_{00})_0]_{y_k} \\ &- [(P_{10})_0]_{y_k} \log [(P_{10})_0]_{y_k} \end{aligned} \tag{6.49}$$

Since $[(P_{00})_0]_{y_k} = 1 - [(P_{10})_0]_{y_k}$, the entropy of the two events can be expressed as a function of the probability of false detection:

$$\begin{aligned} H[(P_{10})_0, (P_{00})_0]_{y_k} = &- \left(1 - [(P_{10})_0]_{y_k}\right) \log \left(1 - [(P_{10})_0]_{y_k}\right) \\ &- [(P_{10})_0]_{y_k} \log [(P_{10})_0]_{y_k} \end{aligned} \tag{6.50}$$

For instance, a false detection probability $[(P_{10})_0]_{y_k} = 0.001$ gives $\Phi(z_k) = 0.5 - 0.001 = 0.499$, for which $z_k = 3.1$. From this, the value of the decision threshold will be $y_k = \hat{y}_0 + 3.1\ \sigma_y$, and the entropy of the two sets of background signals, $y \leqslant \hat{y}_0 + 3.1\ \sigma_y$ and $y > \hat{y}_0 + 3.1\ \sigma_y$, will have the value

$$H[(P_{00})_0 = 0.999, (P_{10})_0 = 0.001]_{y_k} = 0.014 \text{ bit.}$$

Hence, the decision threshold defined by the Neyman-Pearson criterion is a well-specified value $y_k > y_0$ on the signal axis, with respect to which the probability of false detection $[(P_{10})_0]_{y_k}$ and the entropy $H[(P_{10})_0, (P_{00})_0]_{y_k}$ have well-established values.

6.6 SEQUENTIAL DETECTION

For all the decision criteria given so far, it has been assumed that the number of measurements used is constant. If this number is not preselected and as many measurements are made as are required in order to decide for one of the hypotheses, the procedure is called a sequential test.

The sequential method already has important technical applications in signal detection problems [9]. We shall now consider the likelihood ratio sequential test, as applied to analytical detection.

The theory of this test was formulated by Wald [10], and it has already been mentioned in Section 2.6.8, in relation to the general problem of hypothesis testing.

In this test, after each measurement a decision is taken to accept one of the hypotheses H_0 and H_1 or make further measurement(s).

To use the test, before starting the measurements we set the values acceptable for the probabilities of the first kind and second kind of error, $(P_{10})_0$ and $(P_{01})_i$. Next, using these probabilities, we establish two critical values (thresholds) for the likelihood ratio:

$$k_A = \frac{1 - (P_{01})_i}{(P_{10})_0} \tag{6.51}$$

and

$$k_B = \frac{(P_{01})_i}{1 - (P_{10})_0} \tag{6.52}$$

Thus, from the result y_1 obtained from the first measurement we calculate the likelihood ratio $k(y_1)$:

$$k(y_1) = \frac{p_i(y_1)}{p_0(y_1)} \tag{6.53}$$

The likelihood ratio $k(y_1)$ is compared with the two critical values k_A and k_B, and a decision is taken as follows:

(a) if $k(y_1) \leqslant k_B$, the test is stopped and the hypothesis H_0 is accepted;
(b) if $k(y_1) \geqslant k_A$, the test is stopped and the hypothesis H_1 is accepted;
(c) if $k_B < k(y_1) < k_A$, further measurement is necessary.

If a further measurement is made, the new likelihood ratio is calculated and compared with k_A and k_B, and so on. Thus, after N measurements, all the values obtained are used to calculate the likelihood ratio:

$$k(y_1, y_2, \ldots, y_N) = \frac{p_i(y_1, y_2, \ldots, y_N)}{p_0(y_1, y_2, \ldots, y_N)} \tag{6.54}$$

The decision is then taken on the same basis as for a single measurement.

For the two-step model of Fig. 6.9, the likelihood ratio after the Nth experiment is given by:

$$k(y_1, y_2, \ldots, y_N) = \frac{\left(\dfrac{1}{\sigma_y\sqrt{2\pi}}\right)^N \prod\limits_{i=1}^{N} \exp\left[-\dfrac{(y_j-\hat{y}_i)^2}{2\sigma_y^2}\right]}{\left(\dfrac{1}{\sigma_y\sqrt{2\pi}}\right)^N \prod\limits_{j=1}^{N} \exp\left[-\dfrac{(y_j-\hat{y}_0)^2}{2\sigma_y^2}\right]} \tag{6.55}$$

Substitution of (6.55) into the inequality

$$k_B \leqslant k(y_1, y_2, \ldots, y_N) \leqslant k_A \tag{6.56}$$

leads, after taking of logarithms, to the inequality:

$$\frac{\sigma_y^2 \ln k_B}{\hat{y}_i - \hat{y}_0} + N\frac{\hat{y}_i + \hat{y}_0}{2} \leqslant \sum_{j=1}^{N} y_j \leqslant \frac{\sigma_y^2 \ln k_A}{\hat{y}_i - \hat{y}_0} + N\frac{\hat{y}_i + \hat{y}_0}{2} \tag{6.57}$$

where $\sum\limits_{j=1}^{N} y_j$ is the sum of the N successive measurement results.

Hence, for this model, the decisions are taken as follows:

(a) if $\sum\limits_{j=1}^{N} y_j \leqslant \dfrac{\sigma_y^2 \ln k_B}{\hat{y}_i - \hat{y}_0} + N\dfrac{\hat{y}_i + \hat{y}_0}{2}$ (6.58)

the hypothesis H_0 is accepted;

(b) if $\sum\limits_{j=1}^{N} y_j \geqslant \dfrac{\sigma_y^2 \ln k_A}{\hat{y}_i - \hat{y}_0} + N\dfrac{\hat{y}_i + \hat{y}_0}{2}$ (6.59)

the hypothesis H_1 is accepted;

(c) if neither (6.58) nor (6.59) is satisfied, further measurements are made.

The two-step model is used again in Fig. 6.13, where we show the three decision domains $R_0^{(1)}$, $R_1^{(1)}$, and $R_2^{(1)}$ after the first measurement.

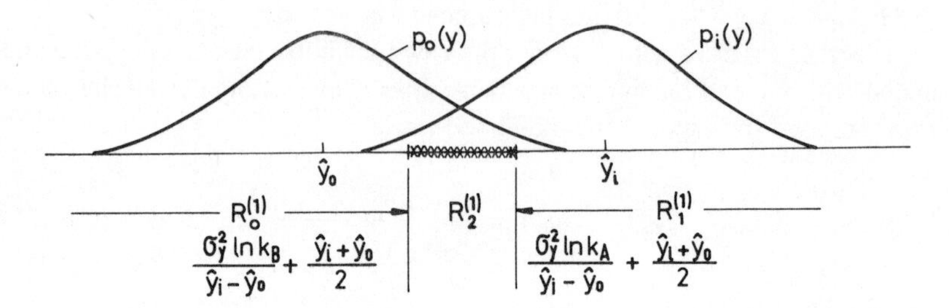

Fig. 6.13 The domains in measurement space for the sequential probability ratio test.

The procedure, applied to the two-step model of Fig. 6.9, is illustrated in Fig. 6.14. The decision functions (6.58) and (6.59), which give the dependence between the sequential sums of results and the number of measurements N, divide the co-ordinate plane into three regions. The sum of all the results is calculated after each measurement, and the corresponding position in the co-ordinate plane is established; the decision is formulated accordingly. The application of the test to analytical detection decisions is exemplified in Section 6.11.4.

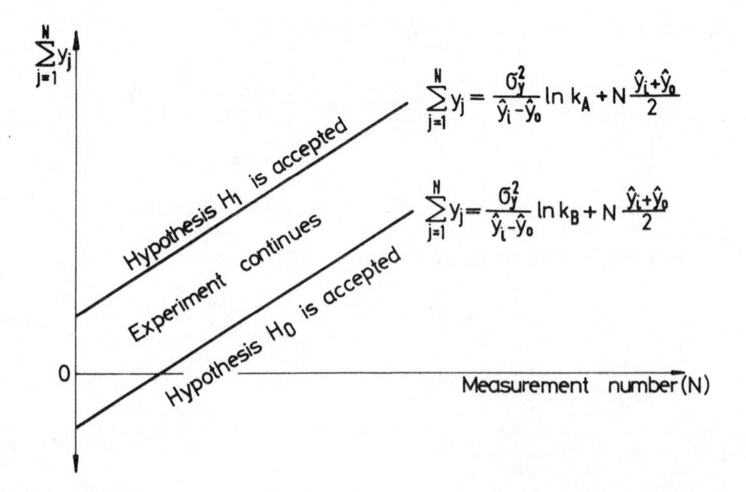

Fig. 6.14 Formulation of detection decisions on the sequential probability ratio test.

6.7 NON-PARAMETRIC DETECTION

In parallel with the development of the theory and technique of signal detection, the theory and practice of non-parametric detection has been developed in order to establish the optimal strategy for signal detection and aid in design of the instruments for detection [11].

Non-parametric detection, which is also known as free distribution detection, is based on the concepts of non-parametric statistics. Of the large literature available we shall mention here only three books [12–14] and that a chapter on non-parametric statistics is usually found in any book of statistics.

Several non-parametric tests have already been given in Section 2.6.7: the Smirnov test, the Wilcoxon test, and the X test. These are also known as order statistics tests, because the order relations satisfied by the results are taken into account. Applications of non-parametric tests to studies of stability of analytical systems were given in Section 4.3.2. Simple applications to problems of analytical detection will be considered in Section 6.11.3.

6.8 CRITERION OF THE LEAST-SQUARE MEAN ERROR

We shall once more use the model in Fig. 6.9 and denote by $\hat{y}_0$ and $\hat{y}_i$ the mean values of the two distributions (the true values of the signal) corresponding to the two hypotheses H_0 and H_1 respectively.

The criterion of the least-square mean error formulates the detection decision on the basis of the inequality:

$$\sum_{j=1}^{N} (y_j - \hat{y}_0)^2 \underset{H_1}{\overset{H_0}{\lessgtr}} \sum_{j=1}^{N} (y_j - \hat{y}_i)^2 \tag{6.60}$$

or, in another form,

$$\sum_{j=1}^{N} \hat{y}_0 y_j \gtrless \sum_{j=1}^{N} \hat{y}_i y_j + \frac{1}{2}\left(\sum_{j=1}^{N} \hat{y}_0^2 - \sum_{j=1}^{N} \hat{y}_i^2\right) \tag{6.61}$$

In the usual terminology employed in statistics, the operation of obtaining the sum of the products of the observations y_j with $\hat{y}_0$ or $\hat{y}_i$ is known as correlation of the observations with the possible signals. The technical instruments which perform such operations are called correlators.

The criterion of the least-square mean error can also be applied in the case of multiple detection, i.e. when we have to detect one signal amongst several possible signals.

Consider the case of $M > 2$ possible signals, $\hat{y}_0, \hat{y}_1, \hat{y}_2, \ldots, \hat{y}_{M-1}$, and let $H_0, H_1, \ldots, H_{M-1}$ be the hypotheses attached to them. Let $y_1, y_2, \ldots, y_N$ be a set of observations pertaining to one of the M possible hypotheses. In order to formulate the detection decision, the set of observations is correlated in turn with each possible signal, and the results are compared, two by two, as shown in (6.61); in this way, the detection decision is formulated. Graphically, the detection of one out of M possible components by using N observations is illustrated in Fig. 6.15. The chosen criterion divides the N-dimensional space of the observations into M regions: $R_0, R_1, R_2, \ldots, R_{M-1}$, one for each signal in turn. For the criterion of the least-square mean error, these regions are bounded by the hyperplanes bisecting the segments joining the true signals $(\hat{y}_0, \hat{y}_1, \ldots \hat{y}_{M-1})$ in pairs. The hypothesis corresponding to the region containing the vector of the observations $\vec{Y}$ will be the one selected.

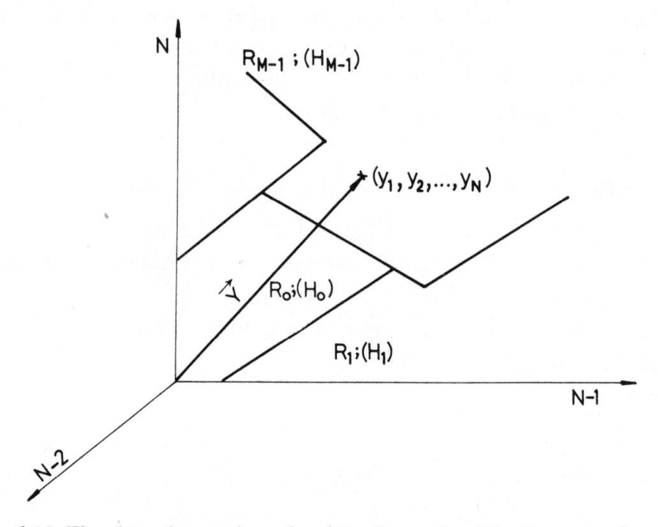

Fig. 6.15 The detection regions for detecting a signal in the case of multiple hypotheses.

Now, let us consider this criterion as applied to the detection of a continuous signal.

As shown in Section 6.1, the signals can be represented as a function $y(x)$, where the variable x may be either time or any other parameter which varies within an interval $(0, X)$.

We shall consider the case of two signals $\hat{y}_0(x)$ and $\hat{y}_i(x)$, and attach to them the hypotheses H_0 and H_1; we also consider the result $y(x)$ obtained from one experiment. By analogy with (6.60) and (6.61), the decision rules for this case will be written as:

$$\int_0^X [\hat{y}_0(x) - y(x)]^2 \mathrm{d}x \underset{H_1}{\overset{H_0}{\lessgtr}} \int_0^X [\hat{y}_i(x) - y(x)]^2 \, \mathrm{d}x \tag{6.62}$$

and

$$\int_0^X \hat{y}_0(x)y(x)\,dx \underset{H_1}{\overset{H_0}{\gtrless}} \int_0^X \hat{y}_1(x)y(x)\,dx + \frac{1}{2}\int_0^X [\hat{y}_0^2(x) - \hat{y}_1^2(x)]\,dx \tag{6.63}$$

The scheme of a detector based on the principle of signal correlation is given in Fig. 6.16.

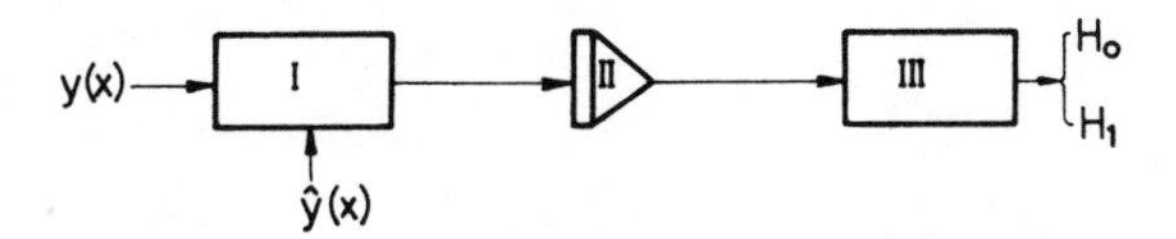

Fig. 6.16 A detector with a correlator.

The observed signal $y(x)$ is introduced into a multiplication circuit **I** together with the locally generated signal $\hat{y}(x)$, which must be identical with the signal which is going to be detected. Their product $y(x)\hat{y}(x)$ is integrated by an integration circuit **II**, where the correlation function is obtained. Next, the comparator **III** compares this with a decision threshold, and by this the decision is formulated.

The structure of a detector based on the correlation principle, for the case when one signal is to be detected out of M possible signals (multiple hypothesis case) is illustrated in Fig. 6.17.

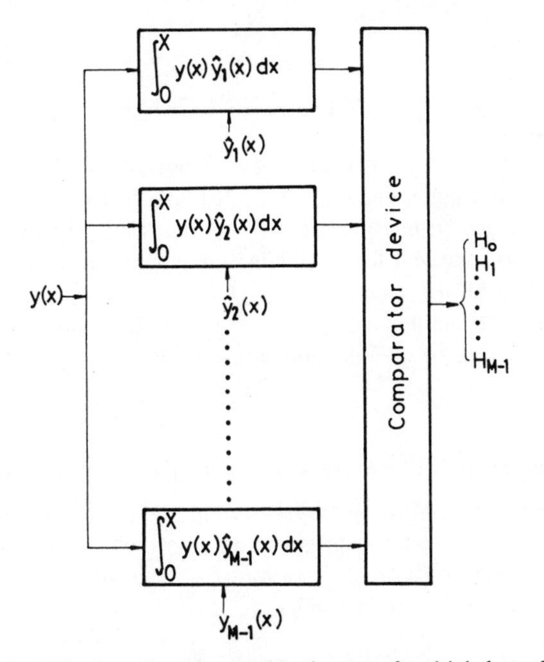

Fig. 6.17 The detection of a signal in the case of multiple hypotheses.

The correlation principle can be applied for each of the analytical detection criteria discussed above. We shall use the model in Fig. 6.9, with N independent observations, $y_1, y_2, \ldots, y_N$ belonging to one of the two possible hypotheses. The likelihood ratio is given by Eq. (6.55).

If k_0 is the critical value of the likelihood ratio for one of the decision criteria, the decision rule will be:

$$k(y_1, y_2, \ldots, y_N) \underset{H_1}{\overset{H_0}{\lesseqgtr}} k_0 \tag{6.64}$$

that is:

$$\sum_{j=1}^{N} \hat{y}_0 y_j \underset{H_1}{\overset{H_0}{\gtreqless}} \sum_{j=1}^{N} \hat{y}_i y_j + \frac{1}{2}\left(\sum_{j=1}^{N} \hat{y}_0^2 - \sum_{j=1}^{N} \hat{y}_i^2\right) - \sigma_y^2 \ln k_0 \tag{6.65}$$

For the case of a continuum, by analogy with (6.65), the decision rule will be:

$$\int_0^X \hat{y}_0(x)y(x)\mathrm{d}x \underset{H_1}{\overset{H_0}{\gtreqless}} \int_0^X \hat{y}_i(x)y(x)\mathrm{d}x + \frac{1}{2}\int_0^X [\hat{y}_0^2(x) - \hat{y}_i^2(x)]\,\mathrm{d}x - A \tag{6.66}$$

where A is a constant value determined from the critical value of the likelihood ratio and the noise.

The principle of signal correlation has important technical applications in communications and signal detection [15, 16]. It consists in correlating a signal either with itself (self-correlation) or with another signal. In analytical measurements it is used as a method for enhancing the signal-to-noise ratio [17–21], and also for extracting information from complex analytical signals, i.e. for deciding whether the signal to be detected is present or not. (Applications of the correlation principle to analytical detection and determination will be discussed in Chapter 9.)

6.9 THE DETECTION CHARACTERISTIC OF AN ANALYTICAL SYSTEM

Within the framework of the model given in Fig. 6.7 for the relation between signal and amount of component, there is a well-determined function which relates the probability of correct detection (P_{11}), the probability of false detection (P_{10}), and the amount of component ($\hat{c}$):

$$P_{11} = f(P_{10}, \hat{c}) \tag{6.67}$$

This functional relationship is called the **detection characteristic** of the analytical system [7].

This concept is introduced for analytical systems by analogy with the corresponding concept of the operating characteristic of a detection system defined in the statistical theory of signal detection.

Within the framework of the accepted model [Eq. (6.4)], for a certain analytical system of measurements there is a well-defined relationship between the amount of component to be determined ($\hat{c}$), and the signal-to-noise ratio, $r = (\hat{y}_i - \hat{y}_0)/\sigma_y$.

To generalize the concept of detection characteristic we shall first evaluate it as a relation between the probability of correct detection, the probability of false detection, and the signal-to-noise ratio, i.e.

$$P_{11} = f(P_{10}, r) \tag{6.68}$$

It can be evaluated deductively from the model given in Fig. 6.8. Thus, we can write:

$$r = \frac{\hat{y}_i - \hat{y}_0}{\sigma_y} = \frac{y_k - \hat{y}_0}{\sigma_y} - \frac{y_k - \hat{y}_i}{\sigma_y} \tag{6.69}$$

We shall write

$$\frac{y_k - \hat{y}_0}{\sigma_y} = z_k \text{ and } \frac{y_k - \hat{y}_i}{\sigma_y} = z_d$$

Hence

$$r = z_k - z_d \tag{6.70}$$

The z_k and z_d values are defined by the probabilities of false and correct detection respectively, from:

$$(P_{10})_0 = \frac{1}{\sqrt{2\pi}} \int_{z_k}^{+\infty} \exp\left[-\frac{z^2}{2}\right] dz \tag{6.71}$$

and

$$(P_{11})_i = \frac{1}{\sqrt{2\pi}} \int_{z_d}^{+\infty} \exp\left[-\frac{z^2}{2}\right] dz \tag{6.72}$$

Consequently, by using Eqs. (6.70) – (6.72), we can evaluate the detection characteristic (6.68).

Example 6.2. Consider the values $r = 1$ and $P_{10} = 0.01$. From (6.71), with $P_{10} = 0.01$ we obtain $\Phi\,(z_i) = 0.5 - 0.01 = 0.49$ and $z_k = 2.33$. By substituting $r = 1$ and $z_k = 2.33$ into (6.70) we obtain $z_d = 1.33$. From (6.74) we obtain $P_{11} = 0.092$ for $z_d = 1.33$. The results of several calculations of this kind are gathered into Table 6.1. These data have been used for drawing the detection characteristic $P_{11} = f(P_{10}, r)$ which is given in Fig. 6.18.

Table 6.1 Values of the detection characteristic P_{11} (the probability of correct detection) for various values of the false detection probability P_{10} and of the signal-to-noise ratio $r = z_k - z_d$.

$r = 1$		$r = 2$		$r = 3$	
P_{10}	P_{11}	P_{10}	P_{11}	P_{10}	P_{11}
0.01	0.092	0.01	0.37	0.01	0.75
0.05	0.26	0.05	0.64	0.05	0.94
0.10	0.39	0.10	0.76	0.10	0.96
0.20	0.56	0.20	0.88	0.20	0.98
0.30	0.68	0.30	0.93	0.30	0.99
0.40	0.77	0.40	0.96	0.40	0.997
0.50	0.84	0.50	0.98	—	—
0.60	0.89	0.60	0.99	—	—
0.70	0.94	0.70	0.994	—	—
0.80	0.97	—	—	—	—
0.90	0.99	—	—	—	—

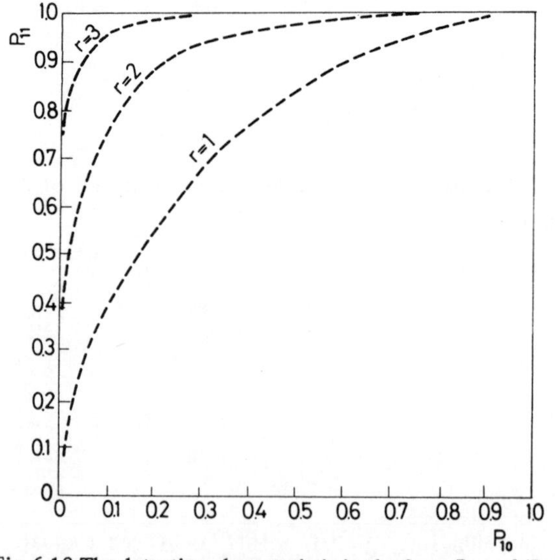

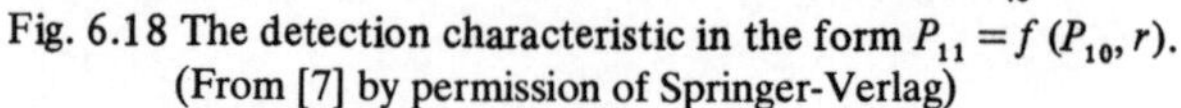
Fig. 6.18 The detection characteristic in the form $P_{11} = f\,(P_{10}, r)$. (From [7] by permission of Springer-Verlag)

By a similar treatment we obtain the detection characteristic expressed as a dependence between the probability of correct detection and the signal-to-noise ratio, $P_{11} = f(r)$. The corresponding curves, obtained at several discrete values of the probability of false detection P_{10}, are given in Figs. 6.19 and 6.20.

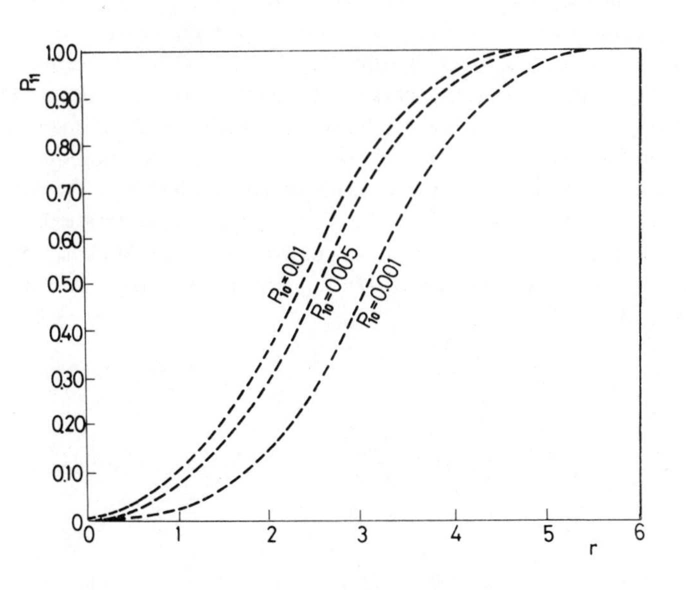

Fig. 6.19 The detection characteristic in the form $P_{11} = f(r)$. (From [7] by permission of Springer-Verlag)

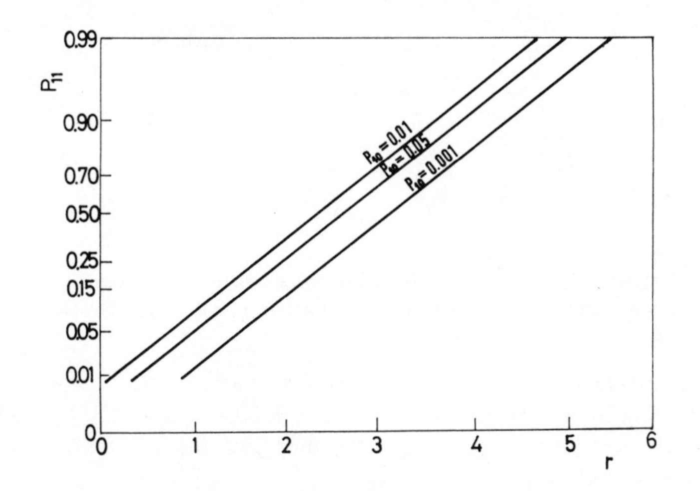

Fig. 6.20 The detection characteristic in the form $P_{11} = f(r)$ on normal probability paper. (From [7] by permission of Springer-Verlag)

If we use the mean of several individual measurements, the characteristic of detection will be modified because the model for the relation between amount and mean analytical signal is modified. If the model for the relation between $\hat{c}$ and y is $y \Rightarrow N(\hat{y} = \hat{y}_0 + b\hat{c};\ \sigma_y)$, then the model when $(\bar{y})_N$ is derived from N individual measurements will be $(\bar{y})_N \Rightarrow N(\hat{y} = \hat{y}_0 + b\hat{c};\ \sigma_y/\sqrt{N})$. This modification is shown graphically in Fig. 6.21, where we have represented the detection characteristic for $P_{10} = 0.01$ in the form $P_{11} = f(r)$, both for individual measurements and for mean measurements obtained from two and four individual measurements. Such curves are evaluated in exactly the same manner as the curves given in Figs. 6.18–6.20, with the only differences being the change in the standard deviation for the mean result $(\sigma_y)_N = \sigma_y/\sqrt{N}$ and the consequent change in the signal-to-noise ratio, $(r)_N = r/\sqrt{N}$, where σ_y and r are the standard deviation and signal-to-noise ratio for individual measurements, and $(\sigma_y)_N$ and $(r)_N$ the corresponding quantities for the mean results of N individual measurements.

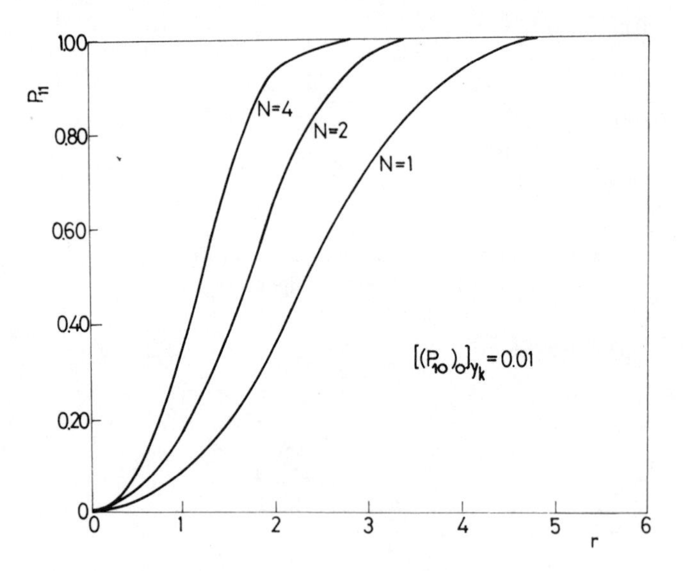

Fig. 6.21 Modifications of the detection characteristic in the form $P_{11} = f(r)$ when using results which are mean values of several measurements.

Example 6.3. Evaluate the detection characteristic in the form $P_{11} = f(c, P_{10})$ for the model $y \Rightarrow N(\hat{y} = 5.23 + 105.1\hat{c};\ \sigma_y = 2.35)$, established for the spectral emission signal (given in digits) and the amounts of manganese present in steel.

The calculations for a probability of false detection $[(P_{10})_0]_{y_k} = 0.001$ are given in Table 6.2 for the case of individual measurements.

Table 6.2 The evaluation of the detection characteristic $P_{11} = f(\hat{c})$ for the false detection probability $[(P_{10})_0]_{y_k} = 0.001$, by using the model $\hat{y} = f(\hat{c})$ characterized by $\sigma_y = 2.35$.

$[(P_{10})_0]_{y_k} = 0.001, z_k = 3.1$				
$\hat{c}$ (%)	$\hat{y} = 5.23 + 105.1\hat{c}$	$r = \dfrac{\hat{y} - 5.23}{2.35}$	$z_d = z_k - r$	P_{11}
0.01	6.28	0.45	+2.65	0.004
0.02	7.33	0.89	+2.21	0.014
0.03	8.38	1.34	+1.76	0.039
0.04	9.43	1.79	+1.31	0.095
0.05	10.48	2.24	+0.84	0.200
0.06	11.53	2.68	+0.42	0.377
0.07	12.82	3.13	−0.03	0.512
0.08	13.87	3.48	−0.38	0.648
0.09	14.92	3.92	−0.82	0.794
0.10	15.79	4.37	−1.27	0.898
0.11	17.02	4.82	−1.72	0.957
0.12	18.07	5.27	−2.17	0.985
0.13	19.12	5.71	−2.61	0.998
0.14	20.17	6.16	−3.06	0.998

The detection characteristics for individual measurements, and for mean values of two and four determinations are shown graphically in Fig. 6.22.

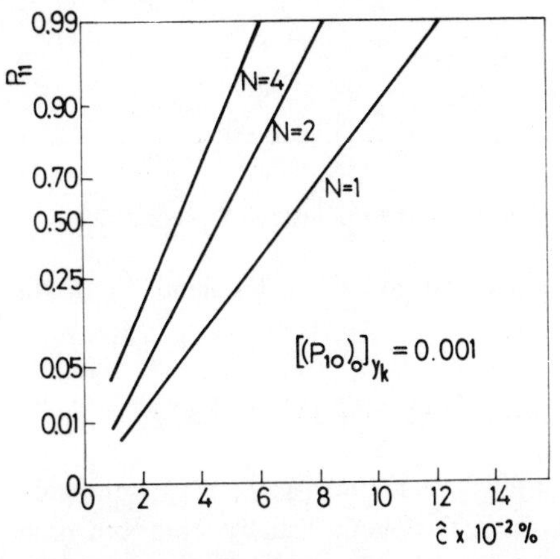

Fig. 6.22 The detection characteristic in the form $P_{11} = f(c)$ for $[(P_{10})_0]_{y_k} = 0.001$ when using mean results of several measurements.

The following conclusions can be derived.

(a) Because each analytical system has its own calibration model it will also have its own detection characteristic.

(b) The detection characteristic of an analytical system used for detection purposes is the fundamental parameter of the quality of that system.

(c) Generally, the detection characteristic can be improved by increasing the signal-to-noise ratio or by using the mean of several individual measurements.

6.10 RESOLUTION OF ANALYTICAL SIGNALS

We have considered so far only the case when the signal to be detected appears in the presence of Gaussian noise, and have not considered interference by other signals of the same type.

As shown in Section 5.2, interference may appear in any analytical process, and rigorous evaluation of an individual analytical signal is possible only when the interference is very low, i.e. when the overlap with other signals is negligible.

Consider the simple case of the two signals $\hat{Y}_1(x)$ and $\hat{Y}_2(x)$ in Fig. 6.23, which we will call signals **I** and **II**. Signal **II** may be a copy of signal **I**, the only difference (m_2-m_1) being a translation along the axis of the x parameter (time, frequency, wavelength, mass, volume, etc.). In a measurement process these signals may appear either individually or concomitantly. Let us assume that these signals have a Gaussian profile, which is the usual case in spectroscopy, chromatography, etc.

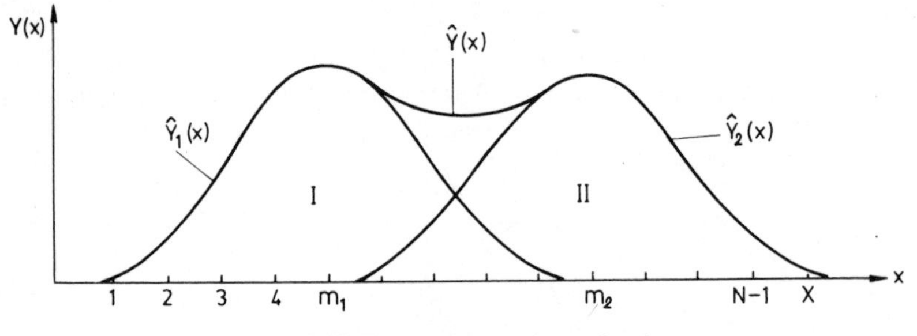

Fig. 6.23 The resolution of two signals.

The two signals add up to give the total signal $\hat{Y}(x)$, which, when there is no noise, is given by:

$$\hat{Y}(x) = \hat{Y}_1(x) + \hat{Y}_2(x) \tag{6.73}$$

whereas the experimentally evaluated signal $Y(x)$ obtained in real conditions will always include a noise ΔY, which is usually a random variable. Thus:

$$Y(x) = \hat{Y}_1(x) + \hat{Y}_2(x) + \Delta Y \tag{6.74}$$

A measure of the degree of overlap of the two signals is given by the quantity:

$$\Lambda = \int_0^X \hat{Y}_1(x)\hat{Y}_2(x)\mathrm{d}x \tag{6.75}$$

When the two signals are completely separated $|\Lambda| = 0$; the greater the overlap, the larger the value of $|\Lambda|$.

The resolution of two or several signals can be interpreted by using their representation in an N-dimensional space. Thus, we consider N points $(1, 2, 3, \ldots, N)$ on the abscissa of Fig. 6.23, belonging to the interval $(0, X)$ in which the signal is observed. As shown in Section 6.1, if an axis is associated with each one of these points, the two signals **I** and **II** can be individually represented as two vectors $\vec{\hat{Y}}_1$ and $\vec{\hat{Y}}_2$ with co-ordinates $(\hat{y}_{11}, \hat{y}_{12}, \ldots, \hat{y}_{1N})$ and $(\hat{y}_{21}, \hat{y}_{22}, \ldots, \hat{y}_{2N})$ (Fig. 6.24).

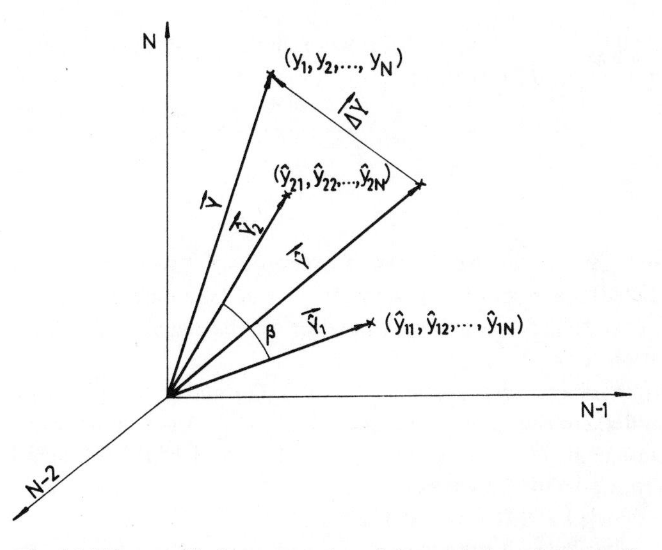

Fig. 6.24 The graphical interpretation of the resolution of two signals.

Because N can be made as large as we please, the signals can be described to any degree of rigour. The angle β between the two vectors is a direct measure of the degree of separation, and for two completely separated signals $\beta = 90°$. The total signal $\hat{Y}(x)$ is represented by the vector $\vec{\hat{Y}}$ resulting from vectorial addition of the two vectors, and is contained in the plane defined by them.

A measurement of $Y(x)$ under real conditions will include a noise $\vec{\Delta Y}$ and is represented in the figure as a vector $\vec{Y}$ with co-ordinates $(y_1, y_2, \ldots, y_N)$.

By analogy with the case of two signals, the resolution of several signals can also be interpreted graphically.

To avoid interference the parameter Λ must be minimized. In the limit, to avoid interference completely, Λ must be reduced to 0 for all the possible signals.

Some elements of resolution strategy will now be considered. To clarify the concept we shall consider the simple case of two signals **I** and **II**, given in Fig. 6.23. These signals can conveniently be written as:

$$\hat{Y}_1(x) = A\hat{y}_1(x) \tag{6.76}$$

$$\hat{Y}_2(x) = B\hat{y}_2(x) \tag{6.77}$$

and the functions $\hat{y}_1(x)$ and $\hat{y}_2(x)$ will be taken to fulfil the conditions:

$$\int_0^X [\hat{y}_1(x)]^2 \mathrm{d}x = \int_0^X [\hat{y}_2(x)]^2 \mathrm{d}x = 1 \tag{6.78}$$

Generally, $\hat{y}_1(x)$ and $\hat{y}_2(x)$ are known functions, and the unknown quantities are A and B. Then, the quantity:

$$\Lambda_0 = \int_0^X \hat{y}_1(x)\hat{y}_2(x)\mathrm{d}x \tag{6.79}$$

is a measure for the relative degree of overlap of the two signals. This quantity will be called the resolution parameter. It can take values between zero and unity, $0 \leqslant \Lambda_0 \leqslant 1$; for complete separation of the signals $\Lambda_0 = 0$, for complete superposition $\Lambda_0 = 1$.

When there are two possible signals $\hat{Y}_1(x)$ and $\hat{Y}_2(x)$ which may appear either simultaneously or only one at a time, the decision taken on the basis of one measurement $Y(x)$ within the observation domain $(0, X)$ must be chosen from the four possible hypotheses:

H_0 — signals **I** and **II** are both absent;
H_1 — only signal **I** is present;
H_2 — only signal **II** is present;
H_3 — signals **I** and **II** are both present.

The methodology for making the choice is the resolution strategy.

In the absence of noise, a correct decision (i.e. without error) can be made by evaluating the quantities:

$$A' = \frac{1}{(1-\Lambda_0^2)} \int_0^X [\hat{y}_1(x) - \Lambda_0\hat{y}_2(x)]\, Y(x)\mathrm{d}x \tag{6.80}$$

and

$$B' = \frac{1}{(1-\Lambda_0^2)} \int_0^X [\hat{y}_2(x) - \Lambda_0 \hat{y}_1(x)]\, Y(x) \mathrm{d}x \tag{6.81}$$

The decision rule will be:

H_0 is true if $A' = B' = 0$;
H_1 is true if $A' = A > 0$, and $B' = 0$;
H_2 is true if $A' = 0$, and $B' = B > 0$;
H_3 is true if $A' = A > 0$, and $B' = B > 0$.

Hence, if either A' or B' is zero, it is decided that the corresponding signal is absent.

Equations (6.80) and (6.81) are dealt with in practice by correlating the measured result $Y(x)$ with the functions $[\hat{y}_1(x) - \Lambda_0 \hat{y}_2(x)]/(1 - \Lambda_0^2)$ and $[\hat{y}_2(x) - \Lambda_0 \hat{y}_1(x)]/(1 - \Lambda_0^2)$, and this is done by using appropriate technical equipment (correlators).

Under real conditions, the observation $Y(x)$ made in the interval $(0, X)$ will include a noise ΔY which will give the quantities A' and B' by a certain degree of uncertainty. Hence, statistical criteria must be used in formulating detection decisions.

The statistical criteria already described for the detection of a signal in the presence of noise can also be used in the case of resolution of two or more signals. As explained before, for the application of the Bayes criterion there are difficulties even in the case of the detection of one signal in the presence of noise. However, the Neyman-Pearson criterion is easy to use.

Let us denote by $p_0(A')$ and $p_0(B')$ the probability density functions for A' and B' when the signals **I** and **II** are absent. The decision thresholds A_k and B_k of the two signals, corresponding to the probabilities of false detection $(P_{10})_\mathrm{I}$ and $(P_{10})_\mathrm{II}$, are obtained from the equations:

$$(P_{10})_\mathrm{I} = \int_{A_\mathrm{k}}^{+\infty} p_0(A') \mathrm{d}A' \tag{6.82}$$

and

$$(P_{10})_\mathrm{II} = \int_{B_\mathrm{k}}^{+\infty} p_0(B') \mathrm{d}B' \tag{6.83}$$

For a normal distribution of the quantities A' and B', without systematic deviation, in the absence of signals, we have:

$$A' \Rightarrow N(0; \sigma_{A'}) \tag{6.84}$$

$$B' \Rightarrow N(0; \sigma_{B'}) \tag{6.85}$$

so the decision thresholds for the two signals are:

$$\begin{aligned} A_k &= z_k \, \sigma_{A'} \\ B_k &= z_k \, \sigma_{B'} \end{aligned} \tag{6.86}$$

where z_k is obtained from the condition

$$P_{10} = \frac{1}{\sqrt{2\pi}} \int_{z_k}^{+\infty} \exp\left[-\frac{z^2}{2}\right] dz.$$

The decision rule is formulated as follows:

hypothesis H_0 is chosen if $A' \leqslant A_k$ and $B' \leqslant B_k$;
hypothesis H_1 if $A' > A_k$ and $B' \leqslant B_k$;
hypothesis H_2 if $A' \leqslant A_k$ and $B' > B_k$;
hypothesis H_3 if $A' > A_k$ and $B' > B_k$.

Consequently, the choice from the four possible hypotheses is reduced to two independent choice operations – one for each signal – and these choices are made by comparing the quantities A' and B' with the decision thresholds A'_k and B'_k.

The procedure is illustrated graphically in Fig. 6.25, where the probability densities $p_0(A')$ and $p_0(B')$ for the quantities A' and B' are shown for the case when the signals are absent. By using the decision thresholds A'_k and B'_k evaluated as shown above, the plane A', B' is divided into four regions, R_0, R_1, R_2 and R_3, in one-to-one correspondence with the hypotheses H_0, H_1, H_2 and H_3. One hypothesis is chosen according to the position of the experimental point (A', B') in the co-ordinate plane.

Obviously, by analogy with the case for two signals, a resolution strategy can also be developed for the case of a larger number of signals.

Example 6.4. Consider the general problem of the resolution of two Gaussian signals:

$$\hat{Y}_1(x) = A_1 \frac{1}{\sigma\sqrt{2\pi}} \exp\left[-\frac{(x-m_1)^2}{2\sigma^2}\right] \tag{6.87}$$

$$\hat{Y}_2(x) = B_1 \frac{1}{\sigma\sqrt{2\pi}} \exp\left[-\frac{(x-m_2)^2}{2\sigma^2}\right] \tag{6.88}$$

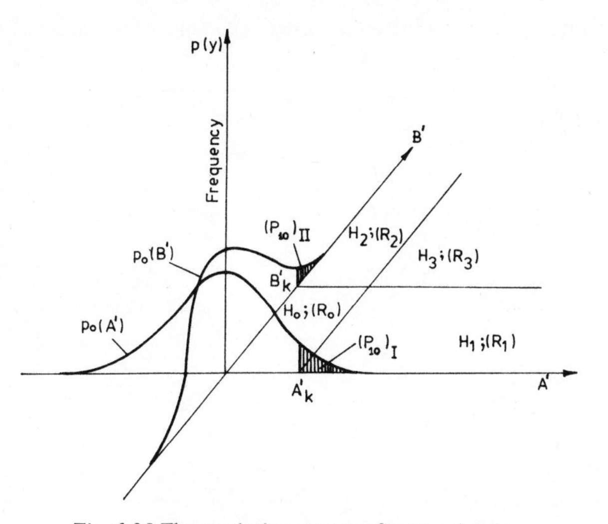

Fig. 6.25 The resolution strategy for two signals.

This problem may appear in various cases, for instance in chromatography, spectroscopy, etc. By normalization of the signals (6.87) and (6.88) in agreement with (6.76) the following results are obtained for the two signals:

$$A = \frac{A_1}{\sqrt{2\sigma\sqrt[4]{\pi}}} \text{ and } \hat{y}_1(x) = \frac{1}{\sqrt{\sigma\sqrt[4]{\pi}}} \exp\left[-\frac{(x-m_1)^2}{2\sigma^2}\right]$$

$$B = \frac{B_1}{\sqrt{2\sigma\sqrt[4]{\pi}}} \text{ and } \hat{y}_2(x) = \frac{1}{\sqrt{\sigma\sqrt[4]{\pi}}} \exp\left[-\frac{(x-m_2)^2}{2\sigma^2}\right]$$

According to (6.79), the resolution parameter Λ_0 is given by:

$$\Lambda_0 = \int_{-\infty}^{+\infty} \hat{y}_1(x)\hat{y}_2(x)\mathrm{d}x = \exp\left[-\left(\frac{m_2-m_1}{2\sigma}\right)^2\right] \qquad (6.89)$$

Several values of the resolution parameter Λ_0, corresponding to various values of the difference $(m_2 - m_1)$, expressed as a multiple of the standard deviation, are given in Table 6.3. The relation between Λ_0 and $(m_2 - m_1)$ is given graphically in Fig. 6.26.

For $m_2 = m_1$, $\Lambda_0 = 1$, i.e. the two signals are completely superimposed.

The classical condition $m_2 - m_1 = 4\sigma$, which ensures that the two Gaussian signals are practically resolved (see Section 5.2), leads to the value $\Lambda_0 = 0.02$ for the degree of overlap.

Table 6.3 Values of the resolution parameter Λ_0 for several values of the quantity $(m_2 - m_1)$.

$m_2 - m_1$	Λ_0	$m_2 - m_1$	Λ_0
0	1.00000	4σ	0.01832
1σ	0.77880	5σ	0.00193
2σ	0.36788	6σ	0.00012
3σ	0.10540	7σ	0.00000

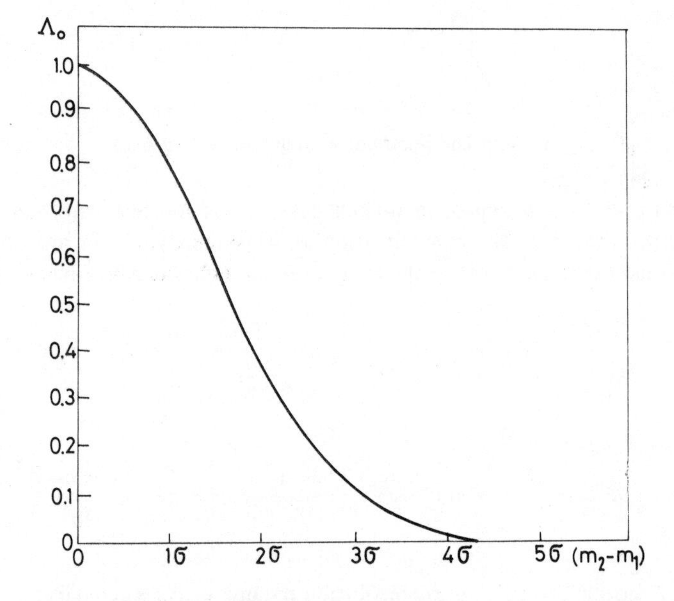

Fig. 6.26 Model of the dependence between the resolution parameter Λ_0 and the quantity $(m_2 - m_1)$.

6.11 CLASSICAL STATISTICAL TESTS APPLIED TO DECISION FORMULATION IN ANALYTICAL DETECTION

This section will be devoted to some simple and practical questions in connection with the analytical detection of one component. We shall begin with the practical method for evaluating the decision threshold and formulating analytical detection decisions. The application of the classical statistical tests to analytical detection will then be exemplified.

6.11.1 The decision threshold

For a normal distribution of background signals, in agreement with the Neyman-Pearson criterion for a detection decision taken on a single measurement, the decision threshold is (see 6.4.5):

$$\hat{y}_k = \hat{y}_0 + z_k \sigma_y \tag{6.90}$$

The decision will be for the hypothesis H_1 that the component is present, if the analytical measurement is $y > \hat{y}_k$. However, because H_1 is a complex hypothesis, a measurement $y \leqslant \hat{y}_k$ will not necessarily mean that the component is absent. The measurement $y \leqslant \hat{y}_k$ may be due either to the absence of the component (hence, the absence of the signal), or to the presence of the component in such a small amount that the probability of obtaining a signal $y > \hat{y}_k$ is also very small. Consequently, if $y \leqslant \hat{y}_k$ it would be more correct to say that the hypothesis H_1 is not significant (i.e. it cannot be verified).

From (6.90) the practical evaluation of the decision threshold requires the *a priori* evaluation of the mean background signal $\hat{y}_0$ and the standard deviation σ_y for signal fluctuations. The simplest method is to estimate $\hat{y}_0$ and σ_y on a set of results $y_1, y_2, \ldots, y_N$ obtained by repeated measurements of the background signal, for a material not containing the component to be detected. The estimated values of these quantities are:

$$\bar{y}_0 = \sum_{j=1}^{N} y_j/N \quad \text{and} \quad s_y = \sqrt{\frac{\sum_{j=1}^{N} (y_j - \bar{y}_0)^2}{N-1}}$$

and are used for obtaining an estimation for the decision threshold:

$$y'_k = \bar{y}_0 + z_k s_k \tag{6.91}$$

Obviously, the greater N, the closer the estimated value y'_k to the true value $\hat{y}_k$, approaching it asymptotically. Because in practice we can use only estimated values y'_k for the decision threshold, a statistical confidence interval must be evaluated for the true value $\hat{y}_k$. This can be done by means of the statistics of the variable y'_k, which in turn will be determined by the statistics of the variables $\bar{y}_0$ and s_y.

As is well known, the sample mean $\bar{y}_0$ estimated for the set of N data $y \Rightarrow N(\hat{y}_0; \sigma_y)$ will itself belong to a set $\bar{y}_0 \Rightarrow N(\hat{y}_0; \sigma_y/\sqrt{N})$. Similarly, the sample standard deviation s_y will belong to a set $s_y \Rightarrow N(\sigma_y; \sigma_y/\sqrt{2(N-1)})$. Consequently, for large N and the variables $\bar{y}_0$ and s_y independent, we can accept that the variable y'_k belongs to a set $y'_k \Rightarrow N(\hat{y}_k; \sigma_{y_k})$, where

$$\sigma_{y_k} = \sqrt{\frac{\sigma_y^2}{N} + z_k^2 \frac{\sigma_y^2}{2(N-1)}} \tag{6.92}$$

For sample values, the variable $(y'_k - \hat{y}_k)/s_k$ has a t-distribution with $\nu = N-1$ degrees of freedom. Hence, for a confidence probability P, the confidence interval for the true value of the decision threshold is:

$$y'_k - |t_{\alpha/2;N-1}|s_{y_k} < \hat{y}_k < y'_k + |t_{\alpha/2;N-1}s_{y_k}| \tag{6.93}$$

where $\alpha = 1 - P$.

The sample standard deviation for the decision threshold, s_{y_k}, will be determined from (6.92), by using the sample value s_y instead of σ_y in this equation.

Hence, the quantity on the right-hand side of inequality (6.93) has the significance of the decision threshold value, evaluated from a finite number of data, for a false detection probability $[(P_{10})_0]\,_{y_k}$ and a confidence probability $(1 + P)/2$. We can guarantee [with a confidence probability equal to $(1 + P)/2$] that the event of observing a background signal $y_0 > y'_k + |t_{\alpha/2;N-1}|s_{y_k}$ occurs with a probability which is either equal to or smaller than $[(P_{10})_0]\,_{y_k}$. Consequently, with respect to the decision threshold defined in this way, the hypothesis H_1 is accepted if the measured result is $y > y'_k + |t_{\alpha/2;N-1}|s_{y_k}$.

If the number of data used for the estimation of $\bar{y}_0$ and s_y is increased, the evaluated decision threshold will tend towards the true value (6.90), defined on the basis of probability of false detection.

With a finite number of measurements of the background signal, for a certain confidence probability we can evaluate the decision threshold by using the t-distribution.

The variable $t = (y_0 - \bar{y}_0)/s_y$ has a t-distribution with $N - 1$ degrees of freedom, so that the statistical confidence interval for the background signal y_0, corresponding to a confidence probability P, will be:

$$\bar{y}_0 - |t_{\alpha/2;N-1}|s_y < y_0 < \bar{y}_0 + |t_{\alpha/2;N-1}|s_y \tag{6.94}$$

The right-hand side of this inequality, i.e. the upper confidence limit for the background fluctuations, can be taken as the decision threshold, in the sense that it guarantees, with a confidence probability equal to $(1 + P)/2$, that the background signals will never exceed this limit. Hence, from a single result we will decide for the hypothesis H_1 if the analytical signal is $y > \bar{y}_0 + |t_{\alpha/2;\,N-1}|s_y$.

As N is increased, the decision threshold given by the right-hand side of (6.94) tends towards the true decision threshold corresponding to a probability of false detection $[(P_{10})_0]\,_{y_k} = (1 - P)/2$.

Example 6.5. To establish the decision threshold for the detection of an element with an emission spectrometer, the background signals (in digits) have been measured. The values are: 160, 165, 162, 151, 157, 156, 163, 167, 158, 156, 160, 161, 158, 166, 165. Calculate the decision threshold and the corresponding confidence interval, for a probability of false detection $[(P_{10})_0]\,_{y_k} = 0.05$ and a confidence probability $P = 0.90$.

From the series of results we obtain the estimated values $\bar{y}_0 = 160.3$ digits and $s_y = 4.4$ digits.

For a false detection probability $[(P_{10})_0]_{y_k} = 0.95$ we find $z_k = 1.65$. From this and (6.90), the estimated value of the decision threshold is $y'_k = 168$ digits.

From (6.92), the sample standard deviation for the estimated value of the decision threshold is $s_{y_k} = 1.77$ digits. The tables of the t-distribution function give $t_{0.05;14} = 1.76$. Hence, the confidence interval for the decision threshold rounded off to an integral number of digits will be:

$$164 \text{ digits} < y_k < 171 \text{ digits}$$

Hence, the hypothesis H_1 will be accepted for $y > 171$ digits.

According to (6.94), at a confidence probability $P = 0.90$, the upper limit for background fluctuations is 168 digits. With this as the detection threshold, the hypothesis H_1 will be accepted if $y > 168$ digits.

6.11.2 The t-test in detection

We shall now consider two applications of the t-test to decision formulation for analytical detection.

(a) *Difference between two sample means* (see Section 2.6.2.2)

Consider the following two series of analytical measurements made for detecting a component: $y_{01}, y_{02}, \ldots, y_{0N_0}$ and $y_{11}, y_{12}, \ldots, y_{1N_1}$. The first series represents the measurements for a material not containing the component, and the second series represents those for the test material. Let $\bar{y}_0$, s_0^2, $\bar{y}_1$, and s_1^2 be the means and variances of the two sets.

Assuming that the two series of measurements belong to normal distributions which have a common variance, and using a significance level α, we will decide for the hypothesis H_1 if $t_{\text{exp}} > t_{(\alpha;N_0+N_1-2)}$, where

$$t_{\text{exp}} = \frac{\bar{y}_1 - \bar{y}_0}{\sqrt{\dfrac{N_0+N_1}{N_0N_1}\left[\dfrac{(N_0-1)s_0^2+(N_1-1)s_1^2}{N_0+N_1-2}\right]}} \tag{6.95}$$

If the two series of measurements belong to sets which have different variances, the approximate t-test (see Section 2.6.2.2) is used.

Example 6.6. The results of emission spectrometry measurements on three samples of steel, for detection of titanium, are given in Table 6.4. It is known that titanium is not present in the first sample (0). The results in the table are the density differences (ΔS) between the sites on the photographic plate that correspond to the titanium line $\lambda_{\text{Ti}} = 3103.8$ Å. and those that correspond to the internal standard line $\lambda_{Fe} = 3102.87$ Å. The hypothesis of the presence of titanium in samples 1 and 2 will be verified for a significance level $\alpha = 0.05$.

According to (6.95), for samples (1) and (0) $t_{1exp} = 0.37$. Similarly, for samples (2) and (0) $t_{2exp} = 2.85$.

Because $t_{1exp} < t_{(0.05;20)}$ and $t_{2exp} > t_{(0.05;20)}$, at the significance level $\alpha = 0.05$, the hypothesis that titanium is present is confirmed only for sample (2).

Table 6.4 The results of analytical measurements for titanium detection on three samples of steel: 0 – without titanium; 1 and 2 – with titanium.

Measurement no.	Steel samples 0	1	2
	$\Delta S \times 10^3$		
1	−265	−290	−240
2	−262	−227	−236
3	−254	−252	−232
4	−266	−252	−253
5	−225	−272	−222
6	−246	−251	−243
7	−298	−227	−210
8	−251	−239	−241
9	−248	−250	−243
10	−287	−280	−271
11	−242	−268	−231

(b) *Difference between two frequencies of occurrence* (see Section 2.6.4)

We assume that for detection purposes N_0 analytical measurements have been made on a material without the component to be detected, and N_1 on the material to be investigated.

We choose a level y_r on the signal axis y and divide each set of data into two sets with respect to it. Let n_0 and n_1 be the number of measurements for which $y > y_r$ for the two sets. Hence, the probability of occurrence of a signal $y > y_r$ for the reference material is estimated by $p_0 = n_0/N_0$, while for the test material it is estimated by $p_1 = n_1/N_1$.

By using the t-test for N_0 and N_1 measurements, we will decide for hypothesis H_1 if p_1 is significantly larger than p_0.

Thus, if $t_{exp} > t_{(\alpha;N_0 + N_1 - 2)}$, it will be decided that the component to be detected is present. The experimental value t_{exp} is obtained from:

$$t_{exp} = \frac{(p_1 - \frac{1}{2n_1}) - (p_0 + \frac{1}{2n_0})}{\sqrt{\bar{p}(1-\bar{p})(\frac{1}{n_0} + \frac{1}{n_1})}} \tag{6.96}$$

and

$$\bar{p} = \frac{n_0 + n_1}{N_0 + N_1}$$

Example 6.7. We give in Table 6.5 three series of spectral measurements made on three samples of steel (0, 1 and 2) for the detection of molybdenum. Molybdenum is known to be absent from sample (0). Verify the hypothesis that molybdenum is present in samples (1) and (2), at a significance level $\alpha = 0.05$.

Table 6.5 Values of the analytical signal (digits) for molybdenum as obtained by emission spectrometry in 33 runs of the experiment for three samples of steel.

Digits y	Signal frequency		
	0	1	2
27	–	–	–
28	–	–	–
29	1	–	–
30	2	–	–
31	5	1	1
32	9	5	2
33	9	7	5
34	6	11	10
35	1	6	8
36	–	2	5
37	–	1	2

With respect to a level $y_r = 32.5$ digits, the occurrence probabilities for $y > y_r$ are estimated by the values:

$$p_0 = 16/33, p_1 = 27/33 \text{ and } p_2 = 30/33.$$

According to (6.96), we find $t_{1\text{exp}} = 2.64$. Because $t_{1\text{exp}} > t_{0.05;64} = 2.01$, the hypothesis that molybdenum is present in sample (1) must be accepted.

As $p_2 > p_1$, the same hypothesis can be accepted for sample (2) without repeating the calculations.

6.11.3 Non-parametric tests in detection

We shall consider here the practical procedure for the application of the Wilcoxon test, and also van der Waerden's X-test, to detection. These tests have already been presented in Section 2.6.7, so we shall directly consider some practical examples.

Example 6.8. We shall consider again the data of example 6.6. For a significance level $\alpha = 0.05$, we shall use the two tests mentioned above in order to establish the presence of titanium in samples (1) and (2).

Let us write the measurement results for samples (0), (1), and (2) as y_0, y_1, and y_2. The results arranged in increasing order, give the two series:

$$\text{I} - y_0 y_1 y_0 y_1 y_1 y_1 y_0 y_0 y_0 y_0 y_1 y_1 y_0 y_1 y_1 y_0 y_0 y_0 y_1 y_1 y_1 y_0$$

$$\text{II} - y_0 y_0 y_2 y_0 y_0 y_0 y_0 y_2 y_0 y_0 y_0 y_2 y_2 y_0 y_2 y_2 y_2 y_2 y_2 y_0 y_2 y_2$$

According to the Wilcoxen test (Section 2.6.7.2), if the results y_0 and y_1, (or y_0 and y_2) belong to the same set, for $N_0 = N_1 = N_2 = 11$ measurements the mean and variance of the number of inversions would be, for each series: $\bar{u} = 60$ and $\sigma_u^2 = 232$.

In the experimental series I, the number of inversions is $u_1 = 63$. Hence, $z_{1\,\text{exp}} = (u_1 - \bar{u})/\sigma_u = 0.18$; the experimental series II contains $u_2 = 98$ inversions, so $z_{2\,\text{exp}} = (u_2 - \bar{u})/\sigma_u = 2.50$.

Because $z_{1\,\text{exp}} < z_{0.05} = 1.96$, and $z_{2\,\text{exp}} > z_{0.05} = 1.96$, the hypothesis that titanium is present is confirmed only for sample (2), i.e. the same result as obtained with the t-test.

According to the van der Waerden X-test (Section 2.6.7.3), from the experimental series I and II the values for the variable X are: $X_{1\,\text{exp}} = 0.38$ and $X_{2\,\text{exp}} = 5.04$. From the table of values of the variable X, we find $X_{(0.05;N_0=N_1=N_2=11)} = 4.08$.

Because $X_{1\,\text{exp}} < X_{(0.05;11)}$, and $X_{2\,\text{exp}} > X_{(0.05;11)}$, again the hypothesis that titanium is present is confirmed only for sample (2).

6.11.4 Sequential probability ratio test in detection

The theory of this test was given in Sections 2.6.8 and 6.6. Here we shall exemplify its application to the formulation of analytical detection decisions for both the normal and the binomial distribution.

Example 6.9. The following model is found for the relation between the spectral emission signal y (digits) and the molybdenum concentration c_{Mo} (%), for low molybdenum contents:

$$y \Rightarrow N(32.36 + 848\, c_{Mo}; \sigma_y^2 = 1.85)$$

Accepting for the first and second kind of error the values $P_{10} = P_{01} = 0.05$, establish, by using the likelihood ratio sequential test, the detection rule for testing the hypotheses H_0 that molybdenum is absent ($c_{Mo} = 0.000\%$) and H_1 that molybdenum is present ($c_{Mo} \geqslant 0.001\%$).

(a) *Decision formulation by using the normal distribution.*

From the experimentally established relation the following models are derived for the distribution of the signals given by samples with molybdenum concentrations of $c_{Mo} = 0.000\%$, and $c_{Mo} = 0.001\%$, respectively:

$$y_0 \Rightarrow N(\hat{y}_0 = 32.36; \sigma_y^2 = 1.85)$$

and

$$y_1 \Rightarrow N(\hat{y}_1 = 33.21; \sigma_y^2 = 1.85)$$

In order to simplify the calculations, we shall use the variables $(y_0 - \hat{y}_0)$, and $(y_1 - \hat{y}_0)$. For these variables, the following models are obtained:

$$(y_0 - \hat{y}_0) \Rightarrow N(0, \sigma_y^2 = 1.85),$$

$$(y_1 - \hat{y}_0) \Rightarrow N(0.85; \sigma_y^2 = 1.85).$$

From (6.58) and (6.59), these two models lead to the following decision rule.

The hypothesis H_0 is accepted if:

$$\sum_{j=1}^{N} (y_j - \hat{y}_0) \leqslant -6.42 + 0.42N;$$

the hypothesis H_1 is accepted if:

$$\sum_{j=1}^{N} (y_j - \hat{y}_0) \geqslant 6.42 + 0.42N;$$

if neither inequality is satisfied, the experiment is continued.

Here y_j is the result of the jth analytical measurement made, and N is the order number of the last measurement. The results obtained with this test on two samples of steel are given in Table 6.6, and displayed graphically in Fig. 6.27, from which it is seen that the hypothesis H_1 ($c_{Mo} \geqslant 0.001\%$) is accepted after the 11th measurement for sample 1, and the 7th measurement for sample 2.

Table 6.6 The sequential detection (on a normal distribution) of molybdenum by using the sequential probability ratio test with respect to the hypothesis H_0: $c_{Mo} = 0.000\%$, and the hypothesis H_1: $c_{Mo} = 0.001\%$.

N	Sample 1		Sample 2		Decision level for hypothesis H_1
	y	$\sum_{j=1}^{N} (y_j - 32.36)$	y	$\sum_{j=1}^{N} (y_j - 32.36)$	
1	35	2.64	33	0.64	6.84
2	31	1.28	34	2.28	7.26
3	33	1.92	34	3.92	7.68
4	32	1.56	32	3.56	8.10
5	34	3.20	34	5.20	8.52
6	32	2.84	34	6.84	8.94
7	35	5.48	37	11.48	9.36
8	34	7.12	35	14.12	9.78
9	33	7.76	34	15.76	10.20
10	35	10.40	36	19.40	10.62
11	34	12.04	34	21.04	11.04
12	32	11.68	34	22.68	11.46
13	33	12.35	35	25.32	11.88
14	34	13.96	35	27.96	12.30
15	32	13.60	36	31.60	12.72
16	35	16.24	35	34.24	13.14
17	35	18.88	35	36.88	13.56
18	32	18.52	33	37.52	13.98

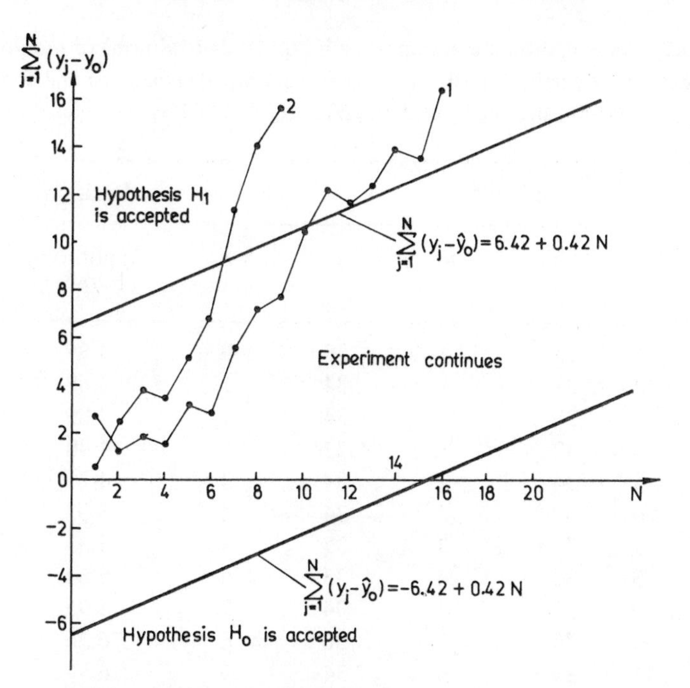

Fig. 6.27 The formulation of detection decisions for molybdenum in two samples of steel, by using the sequential probabilistic ratio test for the case of the normal distribution.

(b) *Decision formulation by using the binomial distribution.*

Here we set a level y_r = 33.5 digits and calculate the probability of occurrence of a background signal $y_0 > y_r$ (at c_{Mo} = 0.000%), and also for a signal $y_1 > y_r$ when c_{Mo} = 0.001%.

From the distribution of background signals, $y_0 \Rightarrow N(\hat{y}_0 = 32.36; \sigma_y^2 = 1.85)$, the occurrence probability for a signal $y_0 > y_r = 33.5$ digits is found to be $p_0 = 0.21$, and the distribution of the signals corresponding to the concentration c_{Mo} = 0.001%, $y \Rightarrow N(\hat{y}_1 = 33.21; \sigma_y^2 = 1.85)$ gives the probability $p_1 = 0.41$ for signals $y_1 > y_r = 33.5$ digits.

Let m be the number of occurrences of an analytical signal $y > y_r = 33.5$ digits in a sequence of N experiments. Then, as shown in Section 2.6.8.2, for $p_0 = 0.21, p_1 = 0.41, P_{10} = P_{01} = 0.05$, the decision rule will be:

hypothesis H_0 is accepted if $m \leqslant -3.06 + 0.305N$;
hypothesis H_1 is accepted if $m \geqslant 3.06 + 0.305N$;
if neither inequality is satisfied, the experiment will be continued.

The results, for the data of example 6.9, are given in Table 6.7 and Fig. 6.28. The hypothesis that molybdenum is present is verified after the 17th experiment for sample 1, and the 8th experiment for sample 2.

Table 6.7 The sequential detection (on a binomial distribution) of molybdenum, by using the sequential probability ratio test with respect to the hypothesis H_0: $c_{Mo} = 0.000\%$, and the hypothesis H_1: $c_{Mo} = 0.001\%$.

N	Sample 1		Sample 2		Decision level for hypothesis H_1
	y	m	y	m	
1	35	1	33	0	3.37
2	31	1	34	1	3.68
3	33	1	34	2	3.99
4	32	1	32	2	4.40
5	34	2	34	3	4.71
6	32	2	34	4	5.02
7	35	3	37	5	5.33
8	34	4	35	6	5.64
9	33	4	34	7	5.95
10	35	5	36	8	6.26
11	34	6	34	9	6.57
12	32	6	34	10	6.88
13	33	6	35	11	7.19
14	34	7	35	12	7.50
15	32	7	36	13	7.81
16	35	8	35	14	8.12
17	35	9	35	15	8.43
18	32	9	33	16	8.74
19	35	10	34	17	8.86
20	36	11	35	18	9.16

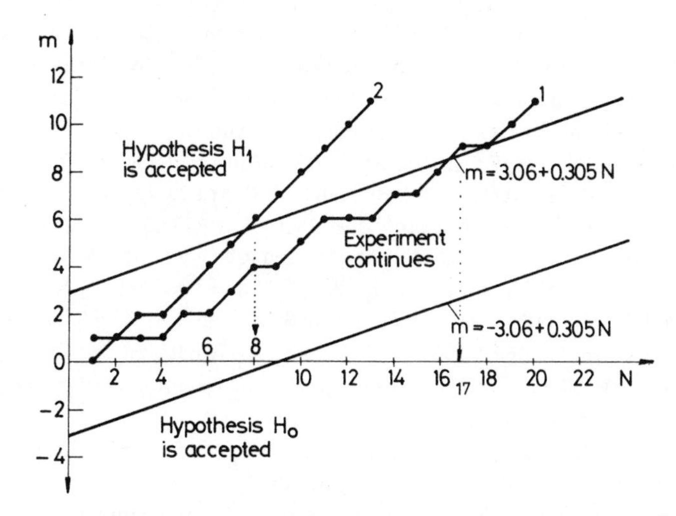

Fig. 6.28 The formulation of detection decisions for molybdenum in two samples of steel, by using the sequential probabilistic ratio test for the case of the binomial distribution.

6.12 ANALYTICAL APPLICATIONS OF PATTERN RECOGNITION METHODS

It was stated in Section 6.1.3 that analytical identification, which needs further development in analytical chemistry, has the purpose of establishing the physical and chemical nature of a material (in whole or in part). In contrast to analytical detection, where we are looking for answers of the form 'the component A is present', or 'the components B and C are absent', analytical identification requires a complete as possible specification of the nature of the investigated material. Consequently, analytical identification can be regarded, in a general way, as detection of a number of unknown signals. Usually, analytical identification based on **pattern recognition** implies both detection and determination.

In its chemical application, the concept of pattern recognition means the classification of materials into distinct classes, on the basis of multidimensional experiments.

Thus, by using diffraction patterns, an unknown mineral can be assigned to a certain mineral class, with a certain structure and composition. From the various spectra (infrared, mass, NMR, emission, etc.) we can place an unknown substance in a certain chemical class.

Generally, chemical classification is possible owing to the existence of some relation between the class of the material and the result of the analytical experiment. Almost always this relation is so complex that it cannot be evaluated theoretically, for example, the relation between the chemical structure of a substance and its spectrum. Thus an empirical approach is invariably required.

Electronic computers have made it possible to design automatic systems for pattern recognition. In this way, data obtained from one or several instruments can be processed, and very intricate problems approached and solved.

The tremendous interest and importance of this procedure of data-processing in obtaining chemical information is shown by the large number of recent works devoted to it. Although the application of these methods in analytical chemistry started only in about 1964, Shoenfeld and DeVoe [22] have reviewed approximately a hundred works which appeared on this subject in only five journals between October 1971 and January 1976. It can be seen from the published work that the widest application is in structure determination and identification of compounds from spectral data.

Generally, there are three distinct steps in pattern recognition: measurement, data processing, and decision formulation.

The measurements (the data) are represented as a multidimensional vector:

$$Y_j = (y_{j1}, y_{j2}, \ldots, y_{ji}, \ldots, y_{jN}) \tag{6.97}$$

Thus, no matter how complex, the spectrum of a substance can be represented to any degree of rigour. In (6.97) the symbols $y_{j1}, y_{j2}, \ldots, y_{jN}$ stand for the amplitudes of the jth spectrum at N points. Combination of several such spectra can be represented by a matrix as follows:

$$Y = (Y_1, Y_2, \ldots, Y_j, \ldots, Y_M) = \begin{bmatrix} y_{11}, & y_{12}, \ldots, & y_{1i}, \ldots, & y_{1N} \\ y_{21}, & y_{22}, \ldots, & y_{2i}, \ldots, & y_{2N} \\ \vdots & \vdots & \vdots & \vdots \\ y_{j1}, & y_{j2}, \ldots, & y_{ji}, \ldots, & y_{jN} \\ \vdots & \vdots & \vdots & \vdots \\ y_{M1}, & y_{M2}, \ldots, & y_{Mi}, \ldots, & y_{MN} \end{bmatrix} \tag{6.98}$$

where y_{ji} is the ith variable of the jth model ($i = 1, 2, \ldots, N; j = 1, 2, \ldots, M$). In the case of spectroscopic data, the matrix (6.98) will represent the spectrum of M components, each spectrum represented by its signal amplitudes at N points.

The aim of the data processing is to diminish the number of measurements required and to increase their informational significance. Adequate preprocessing generally results in increased correctness of pattern recognition and in important reductions of the cost of identification.

The techniques of pattern recognition are usually based on empirical principles of decision formulation. The measurements are made under real conditions, so they are affected by noise, and an important method for increasing their efficiency is to formulate the decision (or classification) criteria on statistical principles. The decision criteria given above for analytical detection can also find application in identification.

We shall consider two methods, the **file search method**, and the **learning machine method**, which form the basis of most of the works published on this subject.

6.12.1 The file search method

This method is used in many automatic systems for structure determination and identification of chemical compounds by means of mass, infrared or emission spectroscopy, etc. [23-42]. The success of automatic systems of this kind, which use spectrochemical data, is due to the simplicity and generality of the principle of the method, the existence of libraries rich in known spectra, and the extended use of computers for processing chemical measurements.

The principle is comparison of the unknown spectrum with the known spectra in a library. The library spectra are encoded into a form suitable for the computer available and stored. For identification, the test spectrum is encoded in the same way and compared either with each spectrum encoded and stored in the library, or with a set of spectra which satisfy some preselected conditions [33]. By use of an adequate criterion of agreement, the spectrum or spectral class which most closely resembles the test spectrum is finally established.

The encoding must be done so as to satisfy several requirements. Thus, each encoded spectrum must contain the largest possible amount of information, occupy the smallest possible volume of store, be retrievable with maximum speed, and give the maximum degree of correctness of identification. A binary encoding system is almost always used in conjunction with searching through the files of spectra [23-40]. This simple encoding system ensures that the data are compressed to the utmost, and allows a high searching speed. The encoded spectrum is represented as a sequence of digits 0 or 1. The simplest variant is the 1-bit encoding of the amplitude of a maximum, in which 1 bit is used for one dimension of the spectrum. In this method, all the maxima in a spectrum are specified by 1 if the amplitudes are above a certain threshold, and by 0 if they are below it.

Consider a hypothetical spectrum covering an interval $(a-b)$ μm, and split it into intervals of length Δx μm; at each subinterval, encode a maximum with 1 bit (Fig. 6.29). The spectrum vector is $Y = (y_1, y_2, \ldots, y_N)$, where $y_1, y_2, \ldots, y_N$ stand for the maximum amplitudes in the $N = (b-a)/\Delta x$ successive intervals. Figure 6.29 shows the encoding with respect to an amplitude threshold y_k.

A spectrum encoded in this way is represented by a series of N digits 0 or 1. Obviously, the maximum number of different spectra which can be encoded in

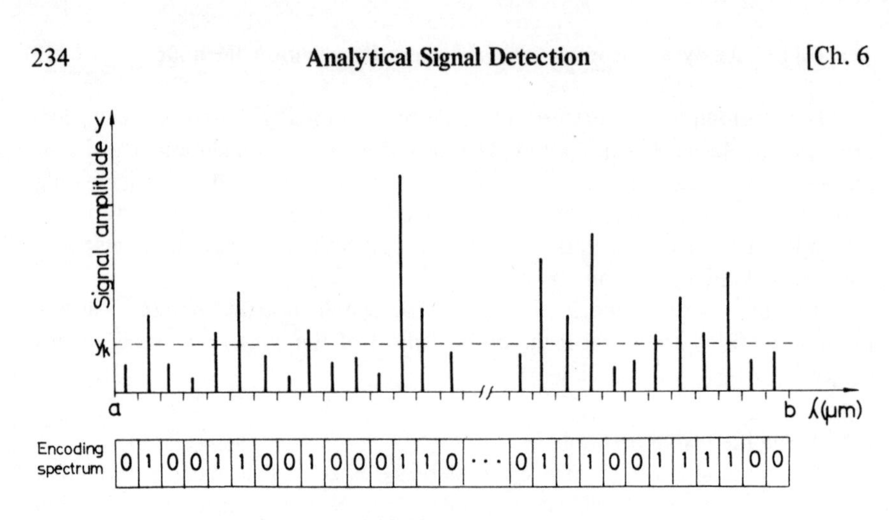

Fig. 6.29 The 1-bit encoding.

this way is 2^N. In N-dimensional space, these spectra will be represented by the vertices of a hypercube with edges equal to unity and one vertex at the origin. For example, for $N = 3$ (three-dimensional space) the 2^3 encoded spectra are represented as the vertices of an ordinary cube, as shown in Fig. 6.30.

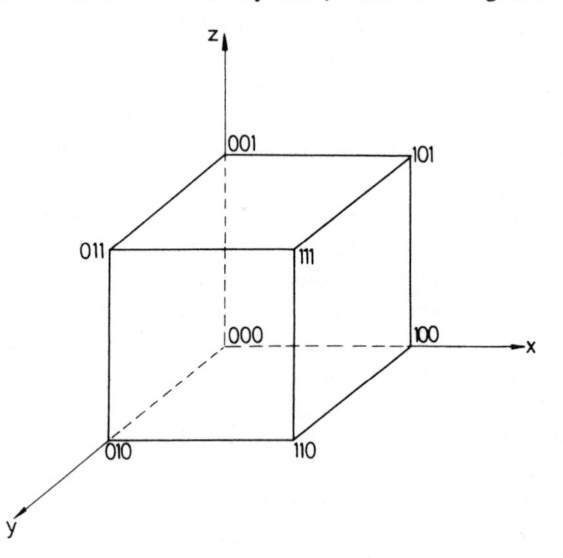

Fig. 6.30 A three-dimensional representation for spectra encoded with 1 bit per dimension.

In order to increase the informational content of the encoded spectrum, 2 or more bits can be used per dimension. If the amplitude is encoded by using 2 bits per dimension, then we assume that it is known at four levels, symbolically written as 0, 1, 2, 3. When 3 bits per dimension are used, the amplitude is assumed to be known at 8 levels, symbolically written as 0, 1, 2, 3, 4, 5, 6, 7. An example where a spectrum is encoded with 3 bits per dimension is given in Fig. 6.31, where the 8 levels are taken as equidistant. Before each group of 3 bits there is a free cell in which 1 bit is inscribed in order to distinguish between the library (known) spectra and the unknown spectrum.

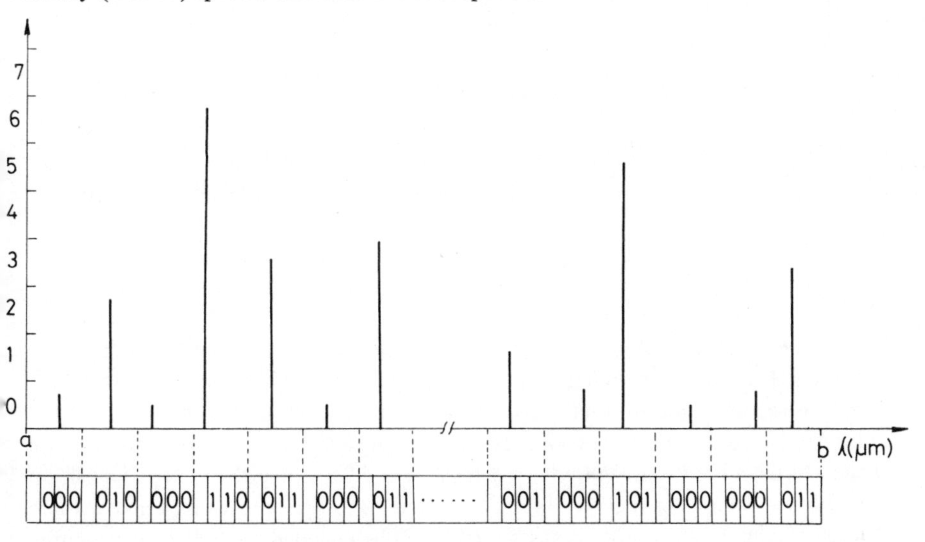

Fig. 6.31 Encoding with 3 bits per dimension.

Some libraries of spectra have the amplitudes of the maxima registered starting from a minimum of 0.01% and going up to 100% at 0.01% intervals. Hence, in this case the amplitude is considered to be known at 10^4 levels so approximately 13 bits are needed for encoding one dimension of the spectrum. Obviously, a spectrum coded in this way will contain a very large amount of information.

In several works [29, 30, 34], the spectra are encoded by the procedure of 'spectral abbreviation'. In this case, the spectrum is divided into N successive regions (windows), each containing the same number of mass or frequency units. For each window the first one, two or three of the most intense maxima will be encoded. Generally, by the procedure of spectral abbreviation, the store required is diminished, the library search speed is increased, and the decrease in correctness of identification is negligible. The abbreviation procedure has been further developed by Grotch [34], by using the so-called 'compressed code' for encoding the position of the maxima. We shall consider a hypothetical mass spectrum (Fig. 6.32), divided into successive windows of 14 mass units each. As $14 < 2^4$, 4 bits per window are enough for encoding the position of the highest maximum in the window. Again, for the spectrum of Fig. 6.32 we assume that the amplitude is known at 4 levels (0, 1, 2, 3), so 2 supplementary bits are required for encoding the amplitude of a maximum.

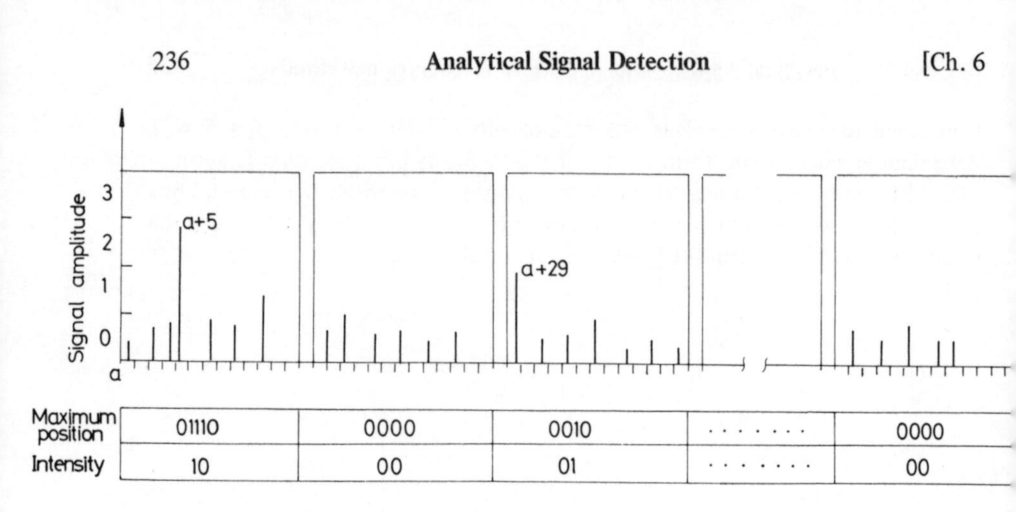

Fig. 6.32 Encoding the position of a maximum by using 4 bits.

The most prominent maximum from the first window corresponds to the mass $(a + 5)$, so the encoded position will be written as $(a+5)-a+1=6=0110$. In the second window there is no maximum with amplitude greater than the threshold, so it will be encoded as 0000. The most prominent maximum from the third window corresponds to a mass $(a + 29)$, hence the encoded position will be written as $(a + 29) - (a + 28) + 1 = 2 = 0010$. In the last window there is no maximum above the threshold, so it will be encoded as 0000. Consequently, the encoded spectrum will be written as 0110 0000 0010 . . . 0000.

Several criteria can be used in order to compare the unknown spectrum with the known spectra from the library. Thus, the **least-squares criterion** uses the relation:

$$D_j = \sum_{i=1}^{N} (y_i - y_{ji})^2 \tag{6.99}$$

where y_i are the successive amplitudes corresponding to the ith dimension, $i = (1, 2, \ldots, N)$, of the unknown spectrum $Y = (y_1, y_2, \ldots, y_i, \ldots, y_N)$, and y_{ji} are the amplitudes of the known spectrum $Y_j = (y_{j1}, y_{j2}, \ldots, y_{ji}, \ldots, y_{jN})$ corresponding to the same dimensions. Obviously, by the least-square criterion we will choose (identify) from the library that spectrum (or those spectra) which correspond to the smallest value of the quantity D_j.

Another criterion used for formulating identification decisions uses the quantity:

$$D_j = \sum_{i=1}^{N} y_i \, y_{ji} \tag{6.100}$$

In this case the identification search will pick out the spectrum or spectra for which the quantity D_j from (6.100) has the largest value.

In formulating identification decisions we also use the criterion based on the absolute deviations between the unknown spectrum and the library spectra. In this case we evaluate the quantity:

$$D_j = \sum_{i=1}^{N} |y_i - y_{ji}| \tag{6.101}$$

The lower this sum, the better the agreement between the unknown spectrum Y and the known spectrum Y_j. Consequently, by this criterion we choose the spectrum or spectra for which the quantity (6.101) is smallest.

The automatic systems of identification operate by using (6.99), (6.100) and (6.101) with encoded spectra.

For encoded spectra, relation (6.99) will correspond to the sum of the logical operation 'exclusive OR' (XOR), which is written as:

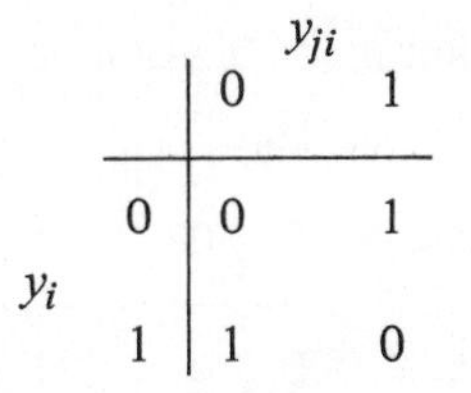

		y_{ji} 0	1
y_i	0	0	1
	1	1	0

Similarly, relation (6.100) will correspond to the sum of the logical operation 'AND', which is written as:

		y_{ji} 0	1
y_i	0	0	0
	1	0	1

Let us consider two hypothetical spectra: the unknown spectrum $Y = (y_1, y_2, \ldots, y_i, \ldots, y_{20})$, and the known spectrum $Y_j = (y_{j1}, y_{j2}, \ldots, y_{ji}, \ldots, y_{j20})$. By encoding them with 1 bit per dimension we shall obtain two sequences of 20 figures which are 0 or 1, one sequence for each spectrum. The significance of the logical operations AND and XOR is:

spectrum Y: 01011101010000100110

spectrum Y_j: 01011001010000100010

XOR $(y_i; y_{ji})$: 00000100000000000100

AND $(y_i; y_{ji})$: 01011001010000100010

By using the criterion (6.99) for the encoded spectra, we have $D_j = \sum_{i=1}^{20} (\text{XOR})_i = 2$, which represents the number of channels (dimensions) in disagreement between the two spectra. From the same encoded data, the criterion (6.100) gives $D_j = \sum_{i=1}^{20} (\text{AND})_i = 7$, i.e. the number of channels (dimensions) in which both spectra are encoded as 1.

In order to increase the correctness of the identification, Grotch [33, 36] has introduced a new criterion resulting from a linear combination of the logical operations XOR and AND, that is:

$$D_j = \sum_{i=1}^{N} [(\text{XOR})_i - \mu(\text{AND})_i] \tag{6.102}$$

The significance of the linear combination of the two logical operations is:

		y_{ji} = 0	y_{ji} = 1
y_i	0	0	1
	1	1	$-\mu$

By using the criterion (6.102) for the identification of several mass spectra, Grotch [33] has concluded that the best results are obtained with $\mu = 2$.

By using (6.99) for the case of two spectra encoded with 1 bit per dimension, we find:

$$D_j = b + b_j - 2a \tag{6.103}$$

where b and b_j are the number of channels (dimensions) from the unknown spectrum and from the known spectrum j, where the amplitude is above the threshold, i.e. which are encoded as 1, and a represents the number of channels where the amplitude is above the threshold in both spectra.

By use of (6.103) Woodruff *et al.* [37] have formulated a new criterion for measuring the agreement between two spectra. By normalizing the quantity D_j from (6.103) with respect to the number of channels where amplitudes of the two spectra are above the threshold, $n = b + b_j - a$, they obtain

$$\frac{D_j}{n} = \frac{b + b_j - 2a}{b + b_j - a} = 1 - \frac{a}{b + b_j - a} \tag{6.104}$$

The ratio D_j/n gives a measure for the degree by which the two spectra differ. The complementary quantity defined by:

$$S_j = 1 - \frac{D_j}{n} = \frac{a}{b + b_j - a} \tag{6.105}$$

obviously gives the measure of the resemblance of the two spectra.

By use of logical quantities, the measure of resemblance S_j is given by the ratio of the sum of the logical operation AND to the sum of the logical operation 'inclusive OR' (ORI), that is:

$$S_j = \sum_{i=1}^{N} (\mathrm{AND})_i / \sum_{i=1}^{N} (\mathrm{ORI})_i \tag{6.106}$$

where the logical operation 'inclusive OR' is defined by:

		y_{ji} 0	1
y_i	0	0	1
	1	1	1

The analytical measurements are made under real conditions, i.e. in the presence of noise, so both the library spectra (which have also been obtained experimentally) and the unknown spectra will have an inherent uncertainty. Consequently, the decisions of identifications and for estimating system performance must all be based on statistical principles. The noise may lead to the following situations in the case of the file search method.

1. The two encoded spectra which are compared belong to the same chemical type but appear to be different because of measurement errors.

2. The two encoded spectra which are compared differ because they pertain to different chemical categories and also because of measurement errors.

Obviously, the decision between these two situations must be based on statistical criteria. The theory of statistical decision presented in connection with the analytical detection can be applied to these problems of identification, and for this reason we shall not repeat it. Statistical consideration concerning hypothesis testing for the identification of spectra encoded in the binary system can be found in references [35] and [36].

6.12.2 The learning machine method

The learning machine method is an empirical method for processing experimental data. The term 'learning' is used here to indicate that the process of decision gradually improves performance in solving problems as it accumulates experience in solving them. With negative feedback the decision process may be modified in order to diminish the percentage of erroneous answers and to increase the system performance of pattern recognition.

The application of this method for extracting chemical information (for identification) from experimental data is based on the premise that the experimental data are related to the chemical category of the substance investigated. The decision function is established empirically, on the basis of a training set of data (data which correspond to known chemical categories) by taking into account the relation between the input (the data) and the output (the categories). The data to be processed by this method are given in vectorial form, $Y = (y_1, y_2, \ldots, y_N)$, as in the case of the previous method. The data are processed before the decision process with a view to reduction of the dimensions of the data, simplification of the decision process, and increase in correctness of the decision. In this respect adequate processing of the original spectra allows extraction from them of a higher amount of chemical information, i.e. improvement in the results of the chemical classification.

To increase the amount of chemical information which can be extracted and to improve correspondingly the process of pattern recognition, Jurs [43] and Wangen *et al.* [44] have used a preliminary Fourier transform of mass spectra. For the same purpose, Kowalski and Reilly [45] have subjected NMR spectra to a preliminary self-correlation transform. Further, Kowalski and Bender [46] have processed mass spectra by using the Hadamar transformation, which has resulted in an improved process of pattern recognition. The Hadamar transformation has also been used by Brunner and co-workers [47] for NMR spectra.

In the case of the learning machine method, the decision function is obtained empirically. This method is usually based on the procedure of the 'threshold logical unity' (TLU) described by Nillson [48]. The TLU principle is one of binary classification, which can classify a model by choosing the corresponding category out of two possible categories. Many problems of structure and chemical composition can be solved by using this method of binary classification [49-70]. In particular, there is one variant of the TLU method which is largely used because it is very convenient because of its simplicity.

Several chemical problems, for instance, of establishing whether an atomic species or a functional group is present or absent in a compound, are solved by a simple binary decision which must answer either 'yes', or 'no' to a question such as 'does the compound contain oxygen?'.

There is also another class of problems, for example establishing the number of atoms of a certain element or the number of functional groups of a certain kind in a given molecule. In such cases we have a problem of multiple classification (there are several categories), and in order to classify the investigated material we use the 'array of binary classifiers'.

We shall assume that the original data are given in the vectorial form $Y = (y_1, y_2, \ldots, y_N)$. The TLU procedure consists in evaluating a function of decision (discrimination), geometrically represented as a decision hyperplane (decision surface) having the same dimensions, N, as the measurement vector. The measurement space is divided by this surface into two separate regions, each corresponding to one of the two categories into which the data can be classified. For the two-dimensional case, this is shown in the left side of Fig. 6.33, where the points corresponding to the two categories (1) and (2) can be separated by a decision straight line (or plane), which belongs to the set of possible lines (or planes). In this case we say that the two categories are linearly separable, because their points are separated by a straight line, in such a way that the points on one side belong to one category, and the points on the other side belong to the other category.

Now, generally, for a linearly separable N-dimensional case, the equation of the decision plane, or in other words, the linear decision function, is:

$$w_1y_1 + w_2y_2 + \ldots + w_iy_i + \ldots + w_Ny_N = a \qquad (6.107)$$

This linear function can be changed to the form:

$$w_1y_1 + w_2y_2 + \ldots + w_iy_i + \ldots + w_Ny_N - a = 0 \qquad (6.108)$$

By use of the notations $-a = w_{N+1}$ and $1 = y_{N+1}$, (6.108) can be written:

$$\sum_{i=1}^{N+1} w_iy_i = 0 \qquad (6.109)$$

Equation (6.109) is satisfied by the points in the decision plane. The points situated on opposite sides of the decision plane satisfy the following inequalities:

$$\sum_{i=1}^{N+1} w_iy_i > 0, \text{ for category (1)} \qquad (6.110)$$

$$\sum_{i=1}^{N+1} w_iy_i < 0, \text{ for category (2)} \qquad (6.111)$$

These inequalities give the decision rule by which the data, and correspondingly, the test object are classified into one of the two possible categories.

The coefficients $w_1, w_2, \ldots, w_i, \ldots, w_N, w_{N+1}$ from (6.109) define the so-called weight vector:

$$W = (w_1, w_2, \ldots, w_i, \ldots, w_N, w_{N+1}) \tag{6.112}$$

In another form, the equation of the decision plane is written as a scalar product:

$$W \cdot Y = 0 \tag{6.113}$$

By multiplying each component of the weight vector W by a positive constant, a new weight vector will be obtained, but it will lead to the same decision plane and decision rule. In order to remove this degeneracy of the weight vector, it is convenient to divide all its components by the absolute value of w_{N+1}, and to use the unique vector of components $w_i/|w_{N+1}|$. Keeping the same notations for the first N components, the unique weight vector will be written as:

$$W = (w_1, w_2, \ldots, w_i, \ldots, w_N, -1) \tag{6.114}$$

The significance of the supplementary dimension for the data and weight vectors, the $(N+1)$th dimension, is shown in Fig. 6.33, where the change from two to three dimensions is considered. Owing to this change, the decision plane will contain the origin of the co-ordinate axes and the weight vector will be perpendicular to the decision plane, at the origin of the co-ordinate axes. The points belonging to categories (1) and (2) will be found on opposite sides of the decision plane. In Fig. 6.33, the points which belong to category (1) are found

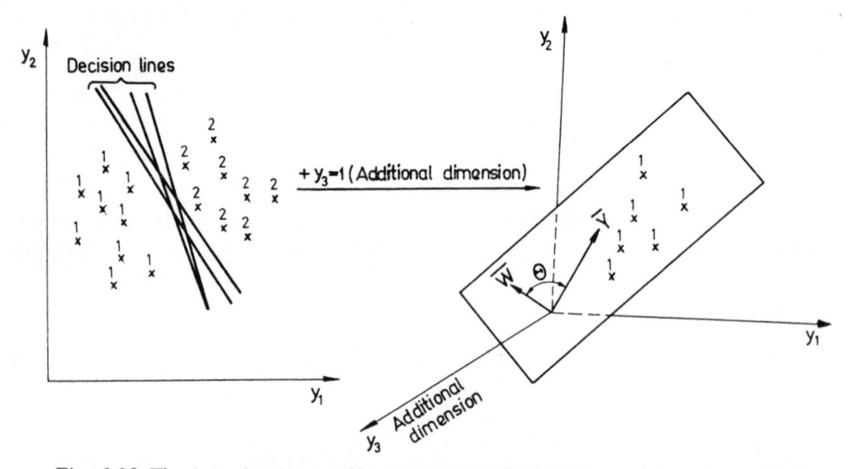

Fig. 6.33 The introduction of $N + 1$ dimensions in order to move the decision surface to the origin of the coordinate axis.

in front of the decision plane and are marked, while the points of category (2) are behind the decision plane and are unmarked.

Generally, if Y_i is an observation vector, and W the weight vector, the decision, that is, the classification of observations, is made according to the rule:

$$\text{if } s = W \cdot Y_i = |W| \cdot |Y_i| \cos\theta > 0, \tag{6.115}$$

the data belong to category (1);

$$\text{if } s = W \cdot Y_i = |W| \cdot |Y_i| \cos\theta < 0, \tag{6.116}$$

the data belong to category (2).

Here θ is the angle between the weight vector W and the data vector Y_i.

The scheme of the principle of binary linear classification is given in Fig. 6.34 [49].

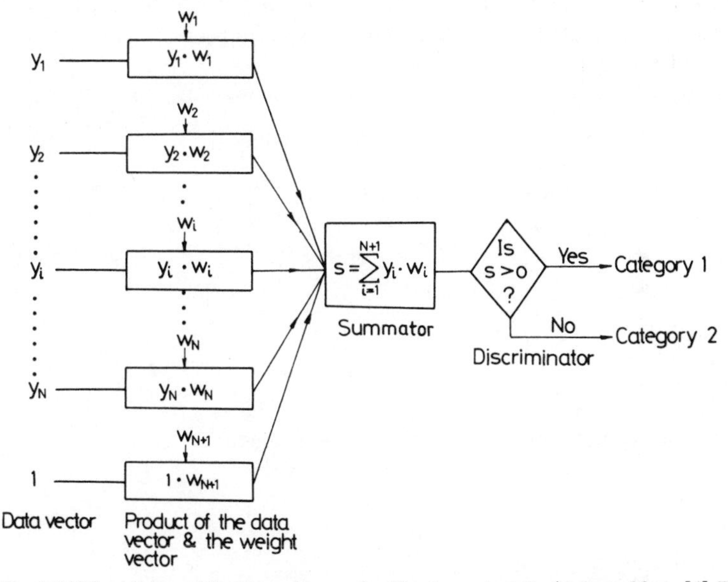

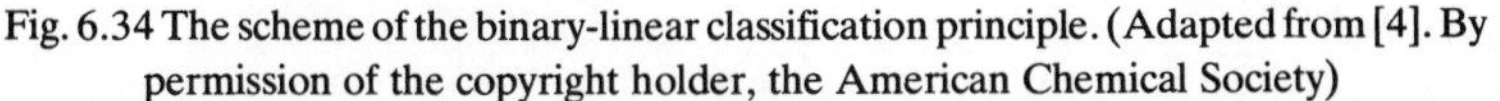
Fig. 6.34 The scheme of the binary-linear classification principle. (Adapted from [4]. By permission of the copyright holder, the American Chemical Society)

To evaluate the weight vector, we use an empirical procedure with error correction, i.e. with negative feedback for an erroneous answer [60]. For this we use a training set, containing data for which the corresponding categories or classes are known. A suitable starting value for the weight vector is first established. Then the classification procedure is tried on the data of the training set, as if the corresponding categories were unknown. If it is found that there are errors of classification, the weight vector is corrected correspondingly through negative feedback. Obviously, if the data of the two categories are linearly separable, the algorithm must converge towards the value of the weight vector for which classification errors are eliminated. There are several feedback methods for evaluating the weight vector. A simple and efficient correction method consists in moving the decision plane along the perpendicular to the plane, taken from the erroneously classified point, to a new position on the other side of the erroneously classified point (doing this ensures the correct classification of the point). The new position is chosen in such a way that the distance between the plane and the point, before and after the correction, is the same.

Let us write the scalar product of the weight vector W with the erroneously classified data vector Y_i as:

$$s = W \cdot Y_i \tag{6.117}$$

Hence, we assume that s has the wrong sign, i.e. opposite to the sign that should have been obtained for the class containing the vector Y_i. Let W' be the weight vector corrected as described above: its scalar product with Y_i must differ from (6.117) only in sign, so:

$$W' \cdot Y_i = -s \tag{6.118}$$

If we write the corrected weight vector as

$$W' = W + cY_i \tag{6.119}$$

then

$$-s = Y_i(W + cY_i) \tag{6.120}$$

and the value of the correction coefficient is obtained:

$$c = \frac{-2s}{Y_i \cdot Y_i} \tag{6.121}$$

Thus the corrected weight vector will be given by:

$$W' = W - \left(\frac{2s}{Y_i \cdot Y_i}\right) Y_i \tag{6.122}$$

This procedure is continued until all the training set is correctly classified.

In order to improve the linear binary pattern classifiers, Wangen *et al.* [59] have used a 'width' parameter and defined a 'null region' in the pattern space. For the case of two classes which are not linearly separable, i.e. classes with overlapping distributions, the null region contains the subsets of data which prevent the linear separation (these subsets will contain data from the two classes). Obviously, all the data which are not included in the null region will be linearly separable.

The null region for the two classes of data (1) and (2), which are not linearly separable, is illustrated in Fig. 6.35.

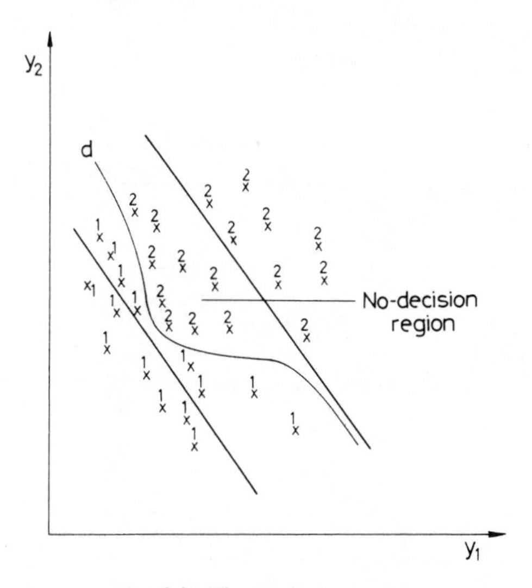

Fig. 6.35 The no-decision region.

Such a problem would be optimally solved by using a non-linear decision surface d, which would separate the two classes completely. Generally speaking, solving such a problem is a very difficult task.

In the case of overlap of the ensembles of data belonging to the two classes, there is also the problem of finding the optimal value of the weight vector. That is, we have to find that value of the weight vector for which the pattern recognition, expressed as the percentage of correct decisions, is maximal. Ritter *et al.* [67] have used the simplex optimization method for this problem. The investigated variables are the weight vector components: $w_1, w_2, \ldots, w_N, w_{N+1}$, and the registered response is the number of correctly classified data from the training set. The simplex is displaced along the response surface until the optimal value of the weight vector is obtained: this value will correspond to the maximal value of pattern recognition for the training set.

Many chemical problems require a classification into more than two classes, i.e. multiple classification. Such problems can be solved by an array of several binary classifiers. A widely used method of this kind is the 'branching tree method' [49]. We give in Fig. 6.36 the structure of an array of binary classifiers [49] used for establishing the molecular formula, by mass spectrometry, for a compound belonging to the categories $C_{1-7}H_{1-16}O_{0-3}N_{0-2}$. The weight vectors for each of the binary classifications 1, 2, . . . , 26 entering into the array structure are calculated from training sets.

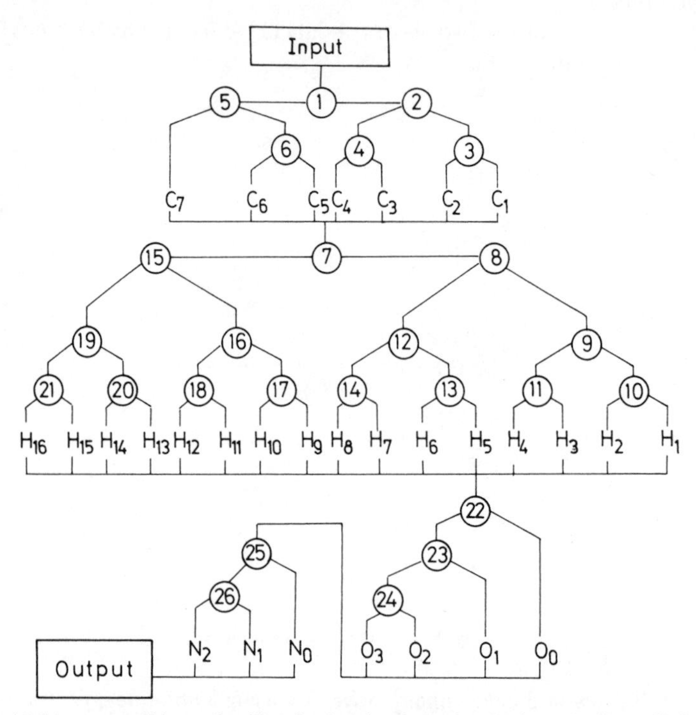

Fig. 6.36 An array of binary classifiers for deriving the molecular formula. (From [49] by permission of the copyright holder, the American Chemical Society)

The problems of multiple classification can also be solved by using a parallel array of binary classifications [55, 59], or by using the Haming type of binary codes [58, 59].

The quality of performance of the binary classification systems based on the learning machine principle is characterized by the parameters of recognition, reliability, convergence rate, and prediction.

The recognition refers to the capacity of the system to recognize the members of the training set used in the learning process. Obviously, when the training set contains two linearly separable classes of data and the weight vector (the decision plane) is well chosen, the recognition must be equal to 100%, which means that all the training set data are correctly classified.

The reliability refers to the capacity of the system to classify correctly the training set data when these data are distorted by errors due to noise in the measurement process.

The convergence rate refers to the rate at which the decision function (i.e. the weight vector) is obtained through the applied feedback. When the training set can be divided into two linearly separable classes, the set will be correctly classified after a finite number of iterations; hence, a correct value will be obtained for the weight vector, and the learning process can then be stopped.

The predictive ability refers to the ability of the system to correctly classify patterns which are not members of the training set.

Data on the convergence rate and predictive ability for adaptive systems of binary classification of the learning machine type are given by Jurs *et al.* [50].

With respect to the learning machine method presented above, Kowalski and Bender [70] have shown that there are three drawbacks: (a) the method is linear, so it is difficult to use it in non-linear cases; (b) the method does not lead to a unique solution; (c) the risk of an erroneous classification remains unknown.

Hence, the next problem is to find the possibilities for minimizing or removing these drawbacks. A solution to the problem of the risk of classification might be obtained by applying statistical decision strategy to the problem of classification.

6.12.3 Other methods of pattern recognition

Besides file search and learning machine methods, other methods for extracting chemical information from experimental data have been developed.

(a) *The K nearest neighbour method (KNN)* [71–76]

This method is conceptually different from the pattern recognition methods given so far, but is simple to apply. For the classification of an unknown compound, it measures the Euclidean distances from the data for the unknown to the known data for a training set belonging to several known classes. The Euclidean distance D_j from the 'unknown' data $Y = (y_1, y_2, \ldots, y_N)$, to the data for the jth member of the training set, $Y_{ji} = (y_{j1}, y_{j2}, \ldots, y_{jN})$, is given by:

$$D_j = \left[\sum_{i=1}^{N} (y_i - y_{ji})^2 \right]^{1/2} \qquad (6.123)$$

The simplest variant is the method of the nearest neighbour (NN), where the unknown compound is assigned to the class of its nearest neighbour in the set of known data.

In the KNN variant, one vote is given for each of the K closest neighbours, and the unknown is assigned to the class which obtains the largest number of votes.

To improve the significance of this method, the votes are suitably weighted [76]. For example, each of the K votes is weighted with respect to the distance

D_j from the unknown to the corresponding known member, or with respect to the square of this distance, D_j^2.

For a binary classification with the classes written as (1) and (2), by counting the votes v_j as +1 for the members of class (1), and as −1 for the members of class (2), and by weighting these votes by D_j, or D_j^2, the sign of the sum $\sum_{j=1}^{K} v_j/D_j$, or $\sum_{j=1}^{K} v_j/D_j^2$, respectively, will indicate the class to which we must assign the unknown: when the sums are positive the unknown will be classified in class (1), and when they are negative it will be classified in class (2).

The KNN method of classification does not generally have the disadvantages mentioned for the learning machine method. It can also be used for non-linear classifications, i.e. linear separability of the classes is no longer a restriction. In contradistinction to the file search method and the learning machine method the KNN method of pattern recognition is a statistically based method.

(b) *The least-squares method* [51]

This method applies the statistical method of least squares to the evaluation of a weight vector which will allow an unknown to be assigned to one of several possible classes, and this in a single step of the calculation. This classification method is applicable to linearly separable classes.

We shall consider the evaluation of a weight vector which would allow us to differentiate chemical compounds with 0, 1, 2 or 3 oxygen atoms per molecule; thus, we have four classes. For the product of the weight vector (W) and the observation vector (Y), we shall prescribe the following theoretical values corresponding to the four classes: $s_j^* = 0, 1, 2, 3$, i.e. its value is equal to the number of oxygen atoms per molecule. The coefficients of the weight vector, which will finally be used to effect the classifications, are obtained by requiring, in accordance with the least-squares method, that the sum of $(s_j - s_j^*)^2$ should be a minimum. To evaluate such a weight vector we shall consider the sum of the squares of the deviations for a set of data from a known classification:

$$S = \sum_{j=1}^{M} (s_j - s_j^*)^2 = \sum_{j=1}^{M} \left(\sum_{i=1}^{N} w_i y_{ji} - s_j^* \right)^2 \tag{6.124}$$

where M is the number of classes, N the number of dimensions, and S the sum of the squares.

The w_i coefficients of the weight vector are obtained from the system of equations:

$$\frac{\partial S}{\partial w_k} = 2 \sum_{j=1}^{M} \sum_{i=1}^{N} (w_i y_{ji} - s_j^*)(y_{jk}) = 0 \tag{6.125}$$

where $k = 1, 2, \ldots, N$.

The classification is then made according to the value of the product of the vector of the unknown and the weight vector so derived.

For the example considered above, if we obtain $s = 1$, the unknown will be assigned to the class of compounds with one oxygen atom per molecule. If the experimentally obtained value is, say, $s = 1.71$, by the least deviation principle the unknown will be included in the class of compounds with two oxygen atoms per molecule.

(c) *The method of the class with the largest number of votes* [65]

We shall consider the case of M classes. To separate each class from all the others we need $M(M-1)/2$ weight vectors. There will be $(M-1)$ positive votes for the classification of the unknown in its own class, and each of the other classes will have a smaller number of positive votes. The general procedure will be easily clarified by the following example. Consider the three classes, (1), (2) and (3), given in Fig. 6.37 [65]. The decision surface I separates class (1) from class (2), decision surface II separate class (1) from class (3), and surface III separates class (2) from class (3). Hence, the total number of weight vectors is $3(3-1)/2 = 3$. Let us take as the unknown a component from class (1): on the one hand it will be classified in class (1) on the basis of decision surface I, and also on the basis of decision surface II; on the other hand, it will be classified in class (2) on the basis of decision surface III. Hence, by a majority of votes the unknown is classified in its own class, class (1).

(d) *The spectrum sum method* [76, 77]

This method for extracting chemical information from spectra consists in comparing the spectrum of the unknown chemical compound with the mean spectra of certain classes of known chemical compounds.

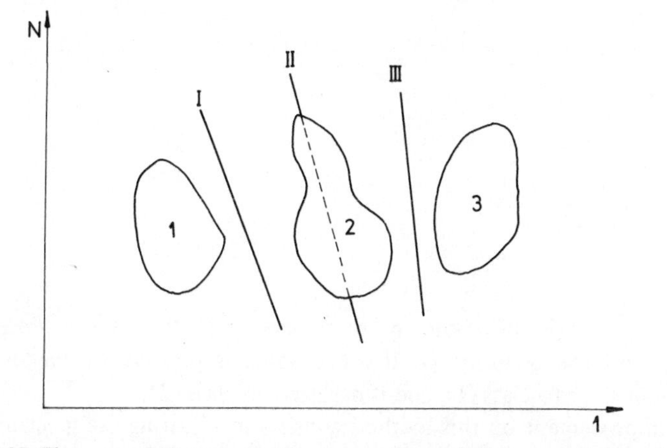

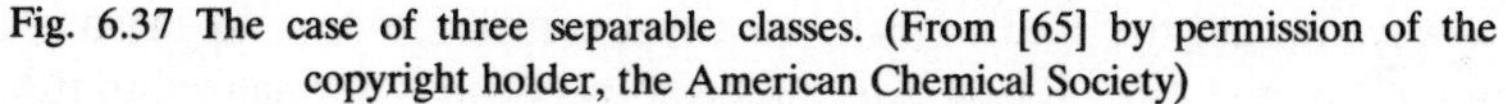
Fig. 6.37 The case of three separable classes. (From [65] by permission of the copyright holder, the American Chemical Society)

We shall consider the simple case of two classes of substances: for example, class (1) contains substances with hydroxyl groups, and class (2) contains substances without hydroxyl groups. The mean spectra of the two classes of substances will be written as:

$$\bar{Y}_1 = \frac{\sum_{j=1}^{m_1} y_{ji}^{(1)}}{m_1} \tag{6.126}$$

and

$$\bar{Y}_2 = \frac{\sum_{j=1}^{m_2} y_{ji}^{(2)}}{m_2} \tag{6.127}$$

where $y_{ji}^{(1)}$ and $\bar{Y}_1$ are, respectively, an individual spectrum and the mean of all the spectra in class (1); $y_{ji}^{(2)}$ and $\bar{Y}_2$ the corresponding entities for class (2); m_1 and m_2 are the numbers of spectra in classes (1) and (2), respectively.

To establish the *a priori* probabilities of the two classes, the following quantities will be evaluated instead of the data given by (6.126) and (6.127):

$$\bar{Y}_1' = \frac{m_1 \bar{Y}_1}{m_1 + m_2} \tag{6.128}$$

$$\bar{Y}_2' = \frac{m_2 \bar{Y}_2}{m_1 + m_2} \tag{6.129}$$

We shall write:

$$D = \bar{Y}_1' - \bar{Y}_2' \tag{6.130}$$

In order to classify an unknown compound we shall take the product of its spectrum and the quantity D. If the product is positive we reckon that the unknown belongs to class (1), and if negative, to class (2).

An improvement on this method consists in adjusting the spectrum [76] in such a way that all the spectra make an equal contribution to the mean vectors ($\bar{Y}_1'$ and $\bar{Y}_2'$) of the two classes. For this, each spectrum is normalized to the sum of the total intensity of the spectrum. Thus, instead of the spectrum Y_{ji}, we use the spectrum:

$$Y'_{ji} = \frac{Y_{ji}}{\sum_{i=1}^{N} y_{ji}} \tag{6.131}$$

where N is the total number of points in the spectrum, and y_{ji} is the intensity of point i in the spectrum j.

In addition to the methods discussed above we may also mention the *discriminant function method* [78], *the correlation set method* [79], and the *factor analysis method* [80, 81]. These methods will certainly have great importance in the future. Their development on statistical lines should increase their efficiency in extracting chemical information from experimental data.

REFERENCES

[1] Davenport, W. B. and Root, W. L., *An Introduction to the Theory of Random Signals and Noise,* McGraw-Hill, New York (1958).
[2] Helstrom, C. W., *Statistical Theory of Signal Detection,* Pergamon, London (1960).
[3] Hancock, J. C, and Wintz, P. A., *Signal Detection Theory,* McGraw-Hill, New York (1966).
[4] Root, W. L., *Proc. IEEE,* **58**, 611 (1970).
[5] Svoboda, V. and Gerbatsch, R., *Z. Anal. Chem.,* **242**, 1 (1968).
[6] Liteanu, C. and Rîcă, I., *Mikrochim. Acta,* **1973**, 745.
[7] Liteanu, C. and Rîcă, I., *Mikrochim. Acta,* **1975**, 311.
[8] Liteanu, C. and Rîcă, I., *Pure Appl. Chem.,* **44**, 535 (1975).
[9] Bussgang, J. J., *Proc. IEEE,* **58**, 731 (1970).
[10] Wald, A., *Sequential Analysis*, Wiley, New York (1947).
[11] Thomas, J. B., *Proc. IEEE,* **58**, 623 (1970).
[12] Walsh, J. E., *Handbook of Nonparametric Statistics,* Van Nostrand, Princeton, Vol. 1 (1962), Vol. 2 (1965), Vol. 3 (1968).
[13] Fraser, D. A. S., *Nonparametric Methods in Statistics,* Wiley, New York (1957).
[14] Noether, G. E., *Elements of Nonparametric Statistics,* Wiley, New York (1957).
[15] Lee, Y. W., *Statistical Theory of Communication,* Wiley, New York (1960).
[16] Lange, F. H., *Correlation Technique,* Van Nostrand, Princeton (1967).
[17] Hieftje, G. M., *Anal. Chem.,* **44**, (7), 69A (1972).
[18] Hieftje, G.M., Bystroff, R. I. and Lim, R., *Anal. Chem.,* **45**, 253 (1973).
[19] Horlick, G., *Anal. Chem.,* **44**, 943 (1972).

[20] Horlick, G. and Codding, E. G., *Anal. Chem.*, **45**, 1749 (1973).
[21] Horlick, G., *Anal. Chem.*, **45**, 319 (1973).
[22] Shoenfeld, P. S. and DeVoe, J. R., *Anal. Chem.*, **48**, 403R (1976).
[23] Anderson, D. H. and Covert, G. L., *Anal. Chem.*, **39**, 1288 (1967).
[24] Erley, D. S., *Anal. Chem.*, **40**, 894 (1968).
[25] Erley, D. S., *Appl. Spectry.*, **25**, 200 (1971).
[26] Lytle, F. E., *Anal. Chem.*, **42**, 355 (1970).
[27] Lytle, F. E. and Brazie, T. L., *Anal. Chem.*, **42**, 1532 (1970).
[28] Jurs, P. C., *Anal. Chem.*, **43**, 364 (1971).
[29] Knock, B. A., Smith, I. C., Wright, D. E., Ridley, R. G. and Kelly, W., *Anal. Chem.*, **42**, 1516 (1970).
[30] Hertz, H. S., Hites, R. A. and Biemann, K., *Anal. Chem.*, **43**, 681 (1971).
[31] Grotch, S. L., *Anal. Chem.*, **42**, 1214 (1970).
[32] Wangen, L. E., Woodward, W. S. and Isenhour, T. L., *Anal. Chem.*, **43**, 1605 (1971).
[33] Grotch, S. L., *Anal. Chem.*, **43**, 1362 (1971).
[34] Grotch, S. L., *Anal. Chem.*, **45**, 2 (1973).
[35] Grotch, S. L., *Anal. Chem.*, **46**, 526 (1974).
[36] Grotch, S. L., *Anal. Chem.*, **47**, 1285 (1975).
[37] Woodruff, H. B., Lowry, S. R. and Isenhour, T. L., *Anal. Chem.*, **46**, 2150 (1974).
[38] Woodruff, H. B., Lowry, S. R., Ritter, G. L. and Isenhour, T. L., *Anal. Chem.*, **47**, 2027 (1975).
[39] Naegeli, P. R. and Clerc, J. T., *Anal. Chem.*, **46**, 739A (1974).
[40] Varmuza, K. and Krenmayr, P., *Z. Anal. Chem.*, **266**, 274 (1973).
[41] Crawford, L. R. and Morrison, J. D., *Anal. Chem.*, **40**, 1464 (1968).
[42] Abramson, F. P., *Anal. Chem.*, **47**, 45 (1975).
[43] Jurs, P. C., *Anal. Chem.*, **43**, 1812 (1971).
[44] Wangen, L. E., Frew, N. M., Isenhour, T. L. and Jurs, P. C., *Appl. Spectry.*, **25**, 203 (1971).
[45] Kowalski, B. R. and Reilley, C. N., *J. Phys. Chem.*, **75**, 1402 (1971).
[46] Kowalski, B. R. and Bender, C. F., *Anal. Chem.*, **45**, 2234 (1973).
[47] Brunner, T. R., Williams, R. C., Wilkins, C. L. and McCombie, P. J., *Anal. Chem.*, **46**, 1798 (1974).
[48] Nilsson, N. J., *Learning Machines*, McGraw-Hill, New York (1965).
[49] Jurs, P. C., Kowalski, B. R. and Isenhour, T. L., *Anal. Chem.*, **41**, 21 (1969).
[50] Jurs, P. C., Kowalski, B. R., Isenhour, T. L. and Reilley, C. N., *Anal. Chem.*, **41**, 690 (1969).
[51] Kowalski, B. R., Jurs, P. C., Isenhour, T. L. and Reilley, C. N., *Anal. Chem.*, **41**, 695 (1969).
[52] Kowalski, B. R., Jurs, P. C., Isenhour, T. L. and Reilley, C. N., *Anal. Chem.*, **41**, 1945 (1969).

[53] Jurs, P. C., Kowalski, B. R., Isenhour, T. L. and Reilley, C. N., *Anal. Chem.*, **41**, 1949 (1969).
[54] Wangen, L. E. and Isenhour, T. L., *Anal. Chem.*, **42**, 737 (1970).
[55] Jurs, P. C., Kowalski, B. R., Isenhour, T. L. and Reilley, C. N., *Anal. Chem.*, **42**, 1387 (1970).
[56] Jurs, P. C., *Anal. Chem.*, **42**, 1633 (1970).
[57] Jurs, P. C., *Anal. Chem.*, **43**, 22 (1971).
[58] Sybrandt, L. B. and Perone, S. P., *Anal. Chem.*, **43**, 382 (1971).
[59] Wangen, L. E., Frew, N. M. and Isenhour, T. L., *Anal. Chem.*, **43**, 845 (1971).
[60] Isenhour, T. L. and Jurs, P. C., *Anal. Chem.*, **43** (10), 20A (1971).
[61] Kowalski, B. R., *Anal. Chem.*, **47**, 1152A (1975).
[62] Abe, H. and Jurs, P. C., *Anal. Chem.*, **47**, 1829 (1975).
[63] Zander, C. S., Stuper, A. J. and Jurs, P. C., *Anal. Chem.*, **47**, 1085 (1975).
[64] Zander, C. S. and Jurs, P. C., *Anal. Chem.*, **47**, 1562 (1975).
[65] Bender, C. F. and Kowalski, B. R., *Anal. Chem.*, **46**, 294 (1974).
[66] Bender, C. F. and Kowalski, B. R., *Anal. Chem.*, **45**, 590 (1973).
[67] Ritter, G. L., Lowry, S. R., Wilkins, C. L. and Isenhour, T. L., *Anal. Chem.*, **47**, 1951 (1975).
[68] Lytle, F. E., *Anal. Chem.*, **44**, 1867 (1972).
[69] Felty, W. L. and Jurs, P. C., *Anal. Chem.*, **45**, 885 (1973).
[70] Kowalski, B. R. and Bender, C. F., *Anal. Chem.*, **44**, 1405 (1972).
[71] Cover, T. M. and Hart, P. E., *IEEE Trans. Inf. Theory,* IT-13, 21, (1967).
[72] Cover, T. M. *IEEE Trans. Inf. Theory,* IT-14, 50 (1968).
[73] Peterson, D. W., *IEEE Trans. Inf. Theory,* IT-16, 26 (1970).
[74] Pichler, M. A. and Perone, S. P., *Anal. Chem.*, **46**, 1790 (1974).
[75] Clark, H. A. and Jurs, P. C., *Anal. Chem.*, **47**, 374 (1975).
[76] Justice, J. B. and Isenhour, T. L., *Anal. Chem.*, **46**, 223 (1974).
[77] Justice, J. B., Anderson, D. N., Isenhour, T. L. and Marshall, J. C., *Anal. Chem.*, **44**, 2087 (1972).
[78] Kawahara, F. K., Santner, J. F. and Julian, E. C., *Anal. Chem.*, **46**, 266 (1974).
[79] Smith, D. H., *Anal. Chem.*, **44**, 536 (1972).
[80] Justice, J. B., Jr. and Isenhour, T. L., *Anal. Chem.*, **47**, 2286 (1975).
[81] Rozett, W. and Petersen, E. McL., *Anal. Chem.*, **47**, 2377 (1975).

Chapter 7

The Detection Limit in Chemical Analysis

7.1 INTRODUCTION

With the growing importance of trace analysis, it became necessary to characterize rigorously the detection capacity of a method of analysis. This was necessary in order to evaluate and compare the detection capacity of particular methods and to formulate the problem of improving the detection capacity or performance of a method. It has become one of the most debated problems in analytical chemistry. There is a very large number of works devoted to this problem, among which we mention [1-52]; these works have shown up various misunderstandings, confusions, and different viewpoints and terminologies.

The problem first arose in qualitative analysis, in connection with the use of colour reagents for detection. It was only later, however, with the extensive use of physico-chemical methods and instrumentation, that the problem became really urgent.

The basis for the rigorous solution of the problem was given by Kaiser [1-8] and Wilson [9] from the concepts of mathematical statistics. The statistically based concept of 'detection limit' (Nachweisgrenze), as introduced by Kaiser, is defined and evaluated as shown in Fig. 7.1. [4].

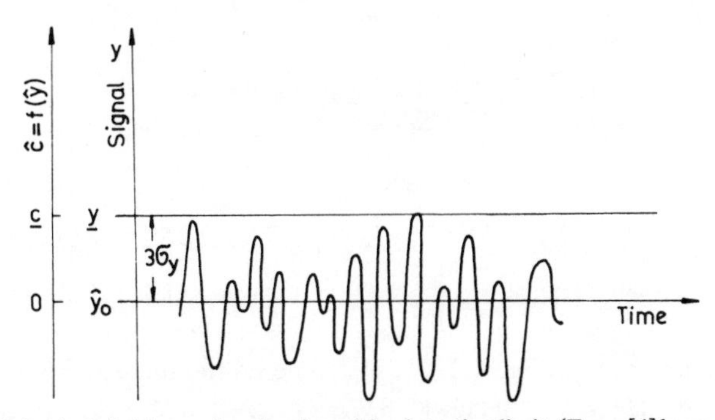

Fig. 7.1 Kaiser's definition and evaluation of the detection limit. (From [4] by permission of Springer-Verlag)

Here $\hat{y}_0$ and σ_y are the mean and standard deviation for normally distributed background signals. The mean background signal is evaluated with a blank sample, i.e. a sample of test material which does not contain the component to be detected. In Kaiser's definition the symbol $\underline{y}$ stands for the upper level of background signal, and the proposed value for this threshold is

$$\underline{y} = \hat{y}_0 + 3\sigma_y \tag{7.1}$$

The probability for obtaining a background signal smaller than this threshold, $y < \underline{y}$, is $P_{00} = 0.9985$.

Again according to Kaiser, the detection limit $\underline{c}$ is obtained from the analytical calculation function written as:

$$\hat{c} = f(\hat{y}), \tag{7.2}$$

by interpolating the threshold value $\underline{y}$, so that:

$$\underline{c} = f(\underline{y}) \tag{7.3}$$

The quantity $\underline{c}$ defined in this way has been accepted by many authors as a measure of the detection capacity in chemical analysis [10–34].

The interpretation of $\underline{c}$, defined as above, has given rise to various discussions and contradictions. Thus, it has been shown by Ehrlich and Gerbatsch [35] that $\underline{c}$ must not be regarded as the smallest amount which can be detected with certainty. This results clearly from Fig. 7.2, which shows the probability density $p_0(y)$ for background signals and the probability density $p_{\underline{c}}(y)$ for signals from samples containing the amount $\underline{c}$ of the component to be detected.

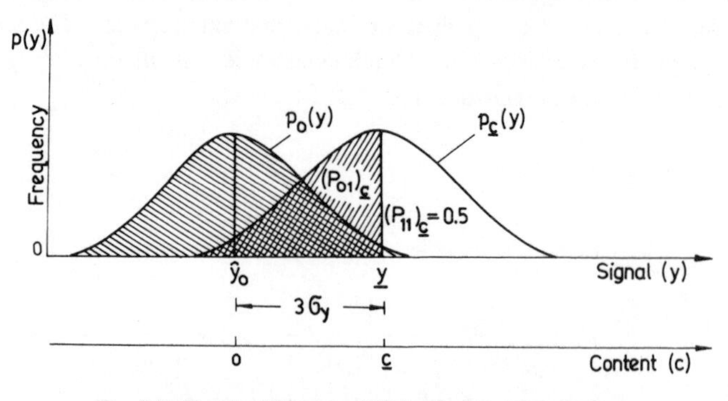

Fig. 7.2 The significance of Kaiser's detection limit $\underline{c}$.

As can be seen in Fig. 7.2, for a sample containing the amount $\underline{c}$ of component to be detected, a signal $y > \underline{y}$ is obtained with probability $P_{11} = 0.50$. Hence under these conditions it is only in 50% of cases that we will be able to distinguish the signal from the background noise and detect the component.

As explained by Nalimov *et al.* [36], this is due to the fact that errors of the second kind (i.e. the probability P_{01} for misinterpreting a signal as a background signal when the component is actually present) are not taken into account in Kaiser's definition.

Other authors [37-40, 51] have defined the detection limit as the minimum amount of component which can be detected with certainty by a single analytical measurement. When this definition is adopted we must take into account the errors of the first and second kinds which may appear in the detection process. When the detection limit is defined in this way, the analytical signal $\hat{y}_d$ corresponding to it is usually obtained from the equation:

$$\hat{y}_d = \hat{y}_0 + 6\sigma_y \tag{7.4}$$

It seems clear that the detection limit should be defined as the least amount which is detectable with a given degree of certainty. Liteanu and Florea [46] have developed a method based on frequency measurements for its rigorous evaluation.

As shown before [48-50], this general problem of defining and evaluating detection capacity can be completely solved only by means of the statistical theory of signal detection.

7.2 DEFINING THE DETECTION LIMIT

We shall consider the model given in Fig. 2.9, of a normally distributed analytical signal y, for which the mean (i.e. the true value) $\hat{y} = \hat{y}_0 + b\hat{c}$ is linearly dependent on the amount of component to be detected, while the standard deviation σ_y is independent of it, that is:

$$y \Rightarrow N(\hat{y} = \hat{y}_0 + b\hat{c};\, \sigma_y) \tag{7.5}$$

We shall further assume that the analytical system is in a strictly stationary stable state.

We now refer to Fig. 7.3, where we have represented by $p_0(y)$ the probability density for background signals (that is, for blank sample signals), and by $p_d(y)$ the probability density for signals from samples containing the component to be detected in an amount exactly equal to the detection limit $\hat{c}_d$, and shall proceed to the definition of $\hat{c}_d$.

There are four kind of signals with respect to the decision level y_k, and the corresponding probabilities satisfy the following equations:

$$\begin{aligned} [(P_{10})_0]_{y_k} + [(P_{00})_0]_{y_k} &= 1 \\ [(P_{01})_d]_{y_k} + [(P_{11})_d]_{y_k} &= 1 \end{aligned} \tag{7.6}$$

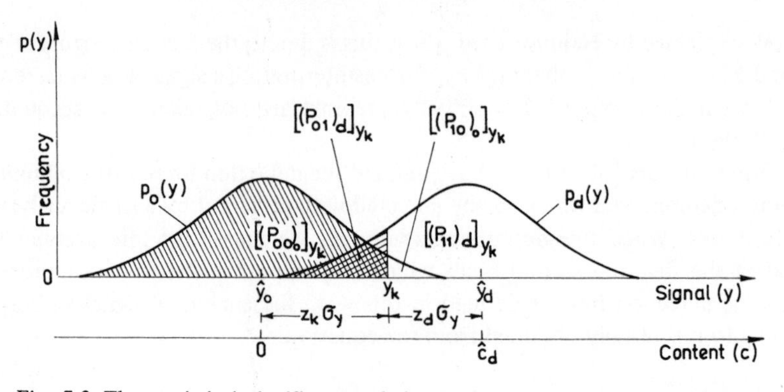

Fig. 7.3 The statistical significance of the detection limit and of the defining quantities.

The consequence of this system of equations is that for the two-step model, the values of only two probabilities, one from each of the equations, are required to completely characterize a detection process for one component, and consequently the detection capacity of a given method of analysis.

From Fig. 7.3 the signal-to-noise ratio corresponding to the detection limit $\hat{c}_d$ is:

$$r_d = \frac{\hat{y}_d - \hat{y}_o}{\sigma_y} = \frac{\hat{y}_k - \hat{y}_o}{\sigma_y} - \frac{\hat{y}_k - \hat{y}_d}{\sigma_y} \tag{7.7}$$

Let us write:

$$z_k = \frac{\hat{y}_k - \hat{y}_o}{\sigma_y} \tag{7.8}$$

and

$$z_d = \frac{\hat{y}_k - \hat{y}_d}{\sigma_y} \tag{7.9}$$

The values of z_k and z_d are obtained as the values of the argument of the normal distribution function corresponding to the given values $[(P_{10})_o]_{y_k}$ and $[(P_{11})_d]_{y_k}$; that is:

$$[(P_{10})_o]_{y_k} = \frac{1}{\sqrt{2\pi}} \int_{y_k}^{\infty} \exp\left[-\frac{z_k^2}{2}\right] dz \tag{7.10}$$

$$[(P_{11})_d]_{y_k} = \frac{1}{\sqrt{2\pi}} \int_{y_k}^{\infty} \exp\left[-\frac{z_d^2}{2}\right] dz \tag{7.11}$$

Hence, the signal-to-noise ratio corresponding to the detection limit

$$r_{\mathrm{d}} = z_{\mathrm{k}} - z_{\mathrm{d}} \tag{7.12}$$

is a function of the probability of false detection $[(P_{10})_{\mathrm{o}}]_{y_{\mathrm{k}}}$ and the probability of true detection $[(P_{11})_{\mathrm{d}}]_{y_{\mathrm{k}}}$.

Hence, within the limits of this model, when only a single measurement is used the analytical signal corresponding to the detection limit is

$$\hat{y}_{\mathrm{d}} = \hat{y}_{\mathrm{o}} + r_{\mathrm{d}}\,\sigma_y \tag{7.13}$$

Now, by using the analytical calculation function:

$$\hat{c} = f(\hat{y}) \tag{7.14}$$

the detection limit $\hat{c}_{\mathrm{d}}$ is obtained from $\hat{y}_{\mathrm{d}}$:

$$\hat{c}_{\mathrm{d}} = f(\hat{y}_{\mathrm{d}}) \tag{7.15}$$

When the signal is a linear function of amount,

$$\hat{y} = \hat{y}_{\mathrm{o}} + b\hat{c} \tag{7.16}$$

we have

$$\hat{c}_{\mathrm{d}} = r_{\mathrm{d}}\frac{\sigma_y}{b} \tag{7.17}$$

When the detection is based not on a single measurement but on the mean of several the standard deviation of the mean signal will obviously be smaller, and this will entail a corresponding change in the value of the detection limit. If the individual measurements correspond to the model (7.5), with standard deviation σ_y, the mean $(\bar{y})_n$ of n independent measurements will have a standard deviation

$$\sigma_{(\bar{y})_n} = \frac{\sigma_y}{\sqrt{n}} \tag{7.18}$$

so the corresponding model is:

$$(\bar{y})_n \Rightarrow N(\hat{y} = \hat{y}_{\mathrm{o}} + b\hat{c};\, \sigma_{(\bar{y})_n}) \tag{7.19}$$

Hence, the analytical signal corresponding to the detection limit becomes

$$(\bar{y}_{\mathrm{d}})_n = \hat{y}_{\mathrm{o}} + r_{\mathrm{d}}\frac{\sigma_y}{\sqrt{n}} \tag{7.20}$$

and the detection limit is

$$(\hat{c}_\mathrm{d})_n = r_\mathrm{d}\frac{\sigma_y}{b\sqrt{n}} \tag{7.21}$$

or

$$(\hat{c}_\mathrm{d})_n = \frac{\hat{c}_\mathrm{d}}{\sqrt{n}} \tag{7.22}$$

This equation shows that the detection limit can be lowered by increasing the number of measurements (see Chapter 9).

From this discussion it follows that, by choosing the values for the probability of false and correct detection and the number of analytical measurements, we can vary the detection limit of a given method within a certain range.

If the detection is based on a single measurement, the detection limit is the amount for which the probability of correct detection (that is, the probability that the measurement result y is greater than the decision threshold y_k, where y_k is defined by the probability of false detection $[(P_{10})_\mathrm{o}]_{y_\mathrm{k}}$) has a preselected value $[(P_{11})_\mathrm{d}]_{y_\mathrm{k}}$ of the detection characteristic. This procedure for defining the detection limit is represented in Fig. 7.4. The detection characteristic for single results is shown in this figure in the form $P_{11} = f(r)$, that is, $P_{11} = f(\hat{c})$, and the chosen decision level corresponds to a probability of false detection $[(P_{10})_\mathrm{o}]_{y_\mathrm{k}}$.

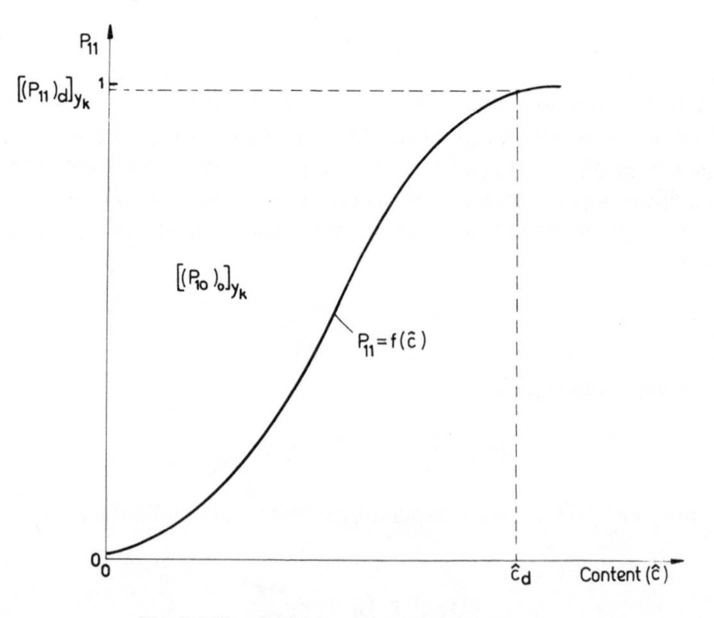

Fig. 7.4 The definition of the detection limit.

The detection limit is evaluated as the amount $\hat{c}_d$ corresponding to the probability the correct detection $[(P_{11})_d]_{y_k}$, on the curve for the detection characteristic.

7.3 THE INFORMATIONAL DEFINITION OF DETECTION LIMIT

With respect to the decision level y_k, defined from the probability of false detection $[(P_{10})_0]_{y_k}$ as shown in Section 6.4.5, we shall consider the two events $y \leqslant y_k$ and $y > y_k$ for the case when the probability density $p_d(y)$ is the particular density corresponding to signals for an amount equal to the detection limit $\hat{c}_d$. In this case the respective probabilities of these events are:

$$P(y \leqslant y_k) = [(P_{01})_d]_{y_k} \tag{7.23}$$

and

$$P(y > y_k) = [(P_{11})_d]_{y_k} \tag{7.24}$$

For this system the entropy is given by:

$$H[(P_{01})_d, (P_{11})_d]_{y_k} = -(P_{01})_d \log (P_{01})_d - (P_{11})_d \log (P_{11})_d \tag{7.25}$$

Using the identity:

$$(P_{01})_d = 1 - (P_{11})_d \tag{7.26}$$

we can write (7.25) as

$$H[(P_{01})_d, (P_{11})_d]_{y_k} = -[1-(P_{11})_d] \log [1-(P_{11})_d] - (P_{11})_d \log (P_{11})_d \tag{7.27}$$

Coming back to the definition of the detection limit for a single analytical measurement, and taking into account (7.27), we can make the following statement: the detection limit is the particular value $\hat{c}_d$ of the amount of component at which the probability of correct detection $[(P_{11})_d]_{y_k}$ or the entropy $H\,[(P_{01})_d, (P_{11})_d]_{y_k}$ evaluated with respect to the decision level y_k (defined on the basis of a probability of false detection $[(P_{10})_d]_{y_k}$ or an entropy of the background signals $H\,[(P_{00})_0, (P_{10})_0\,{}_{y_k}]$) is equal to a preselected value.

Hence, to define the value of the detection limit, we must select suitable values of $[(P_{10})_0]_{y_k}$, $[(P_{11})_d]_{y_k}$, and hence of the entropies $H\,[(P_{00})_0, (P_{10})_0]_{y_k}$ and $H\,[(P_{01})_d, (P_{11})_d]_{y_k}$.

The maximum entropy (the maximum uncertainty), before performing an analytical experiment of detection for one component, is equal to 1 bit. The experiment removes this prior entropy (uncertainty) to an extent depending on the amount of component to be detected. The higher this amount, the larger the uncertainty which is removed by performing the experiment.

If the amount of component to be detected is equal to the detection limit, the uncertainty which is removed by performing the experiment, or in other words the amount of information derived, will be given by:

$$I_d = (1 - H\,[(P_{01})_d, (P_{11})_d]_{y_k})\ \text{bit} \tag{7.28}$$

Example 7.1. Let us choose the values $[(P_{10})_o]_{y_k} = 0.001$ and $[(P_{11})_d]_{y_k} = 0.998$, respectively. $H\,[(P_{10})_o = 0.001, (P_{00})_o = 0.999]_{y_k} = 0.014$ bit and $H\,[P_{01})_d = 0.002, (P_{11})_d = 0.998]_{y_k} = 0.0208$ bit. For these values $z_k = 3.1$ and $z_d = -2.9$, so the signal-to-noise ratio is $r = 6$.

For these conditions, the analytical signal corresponding to the detection limit is calculated either from (7.13) if the detection is made on a single experiment, or from (7.20) if we use the mean of n results. Hence

$$\hat{y}_d = \hat{y}_o + 6\,\sigma_y \tag{7.29}$$

for detection made on one experiment, and

$$(\hat{y}_d)_n = \hat{y}_o + 6\frac{\sigma_y}{\sqrt{n}} \tag{7.30}$$

for detection by use of the mean of n experiments.

For these conditions, if the amount of component to be detected is equal to the detection limit, the uncertainty removed by the experiment will be:

$$I_d = (1 - 0.0208) = 0.9792\ \text{bit} \approx 1\ \text{bit}$$

Hence, the detection limit in this case means the minimum amount for which a single detection experiment can remove practically all the uncertainty existing prior to the experiment.

7.4 CONCLUSION ABOUT DEFINITION OF THE DETECTION LIMIT

Because of the importance of proper definition and clear understanding of the detection limit, it might be useful to stress some special points from the previous discussion. The conclusions given below refer mostly to a situation described by model (7.5).

(1) Kaiser's value $\underline{c}$ for the detection limit does not correspond to the minimum amount detectable with certainty, so in this sense it cannot be accepted as a measure for the detection capacity of an analytical method.

(2) The definition of a quantity characterizing the detection capacity must be based on the statistical theory of signal detection, taking into account the real model for the dependence between the analytical signal and the amount of component to be detected.

(3) The detection capacity can be completely evaluated by the **detection characteristic** which gives the dependence between the probability of correct detection P_{11}, the probability of false detection P_{10}, the number of measurements n on which the detection is based, and the amount $\hat{c}$ of the component to be detected:

$$P_{11} = f(P_{10}, n, \hat{c}) \tag{7.31}$$

(4) The **detection limit** is the amount of component to be detected which corresponds to a probability of false detection $[(P_{10})_o]_{y_k}$, a probability of correct detection $[(P_{11})_d]_{y_k}$, and the number of measurements n on which the detection decision is to be taken:

$$\hat{c}_d = f\left([(P_{10})_o]_{y_k}, [(P_{11})_d]_{y_k}, n\right) \tag{7.32}$$

Hence, to avoid misunderstanding and to make comparisons possible the parameters must be specified along with a complete set of analytical instructions.

(5) The following quantities have been introduced and defined in connection with the problem of defining the detection limit.

The decision threshold y_k, which is defined either for a single measurement, or for the average of n measurements, by the equations

$$y_k = \hat{y}_o + z_k \sigma_y \tag{7.33}$$

and

$$(y_k)_n = \hat{y}_o + z_k \frac{\sigma_y}{\sqrt{n}} \tag{7.34}$$

The value of z_k is established on the basis of a probability of false detection $[(P_{10})_o]_{y_k}$, or an entropy $H\,[(P_{10})_o, (P_{00})_o]_{y_k}$ for background signals, with respect to the decision threshold.

The hypothesis that the component to be detected is present is accepted if the single measurement result satisfies the condition:

$$y > y_k\ , \tag{7.35}$$

or the mean result satisfies the condition:

$$(\bar{y})_n > (y_k)_n \tag{7.36}$$

The analytical signal corresponding to the detection limit, which is found to be

$$\hat{y}_d = \hat{y}_o + r_d\,\sigma_y \qquad (7.37)$$

in the case of a single measurement, or

$$(\hat{y}_d)_n = \hat{y}_o + r_d \frac{\sigma_y}{\sqrt{n}} \qquad (7.38)$$

if the signal is the mean of n measurements.

The signal-to-noise ratio corresponding to the detection limit, r_d, which is established from the values selected for the two probabilities (false detection $[(P_{10})_o]_{y_k}$ and correct detection $[(P_{11})_d]_{y_k}$), or the background signal entropy $H\,[(P_{00})_o, (P_{10})_o]_{y_k}$ and the entropy for signals corresponding to the detection limit $H\,[(P_{11})_d, (P_{01})_d]_{y_k}$, where both are evaluated with respect to the decision threshold.

(6) For the analytical calculation function

$$\hat{c} = f(\hat{y}) \qquad (7.39)$$

the detection limit is obtained directly from the corresponding signal:

$$\hat{c}_d = f(\hat{y}_d) \qquad (7.40)$$

If the function is linear, the detection limit can be written as:

$$\hat{c}_d = r_d \frac{\sigma_y}{b} \qquad (7.41)$$

or for the mean of n experiments,

$$(\hat{c}_d)_n = r_d \frac{\sigma_y}{b\sqrt{n}} \qquad (7.42)$$

Hence the detection limit is proportional to the standard deviation (i.e. to the noise) and to the reciprocal of the sensitivity of the method (the greater the sensitivity, the lower the detection limit).

The detection capacity is increased by lowering the detection limit, by (a) decreasing the standard deviation, (b) increasing the sensitivity, or (c) increasing the number of experiments. (See also Chapter 9.)

(7) The value $\hat{c}_d$ of the detection limit, where this is understood to be the smallest amount which is detectable with practically complete certainty by one

analytical experiment, will be obtained with the following values for the defining parameters:

$$[(P_{10})_0]_{y_k} = 0.001 \text{ and } [(P_{11})_d]_{y_k} = 0.998$$

$$H\,[(P_{10})_0, (P_{00})_0]_{y_k} = 0.014 \text{ bit and } H\,[(P_{01})_d, (P_{11})_d]_{y_k} = 0.0208 \text{ bit}$$

The corresponding signal-to-noise ratio is $r_d = 6$. (This is the basis of the '6-sigma' rule.)

When the amount of component to be detected is equal to the detection limit defined in this way, the uncertainty existing prior to the experiment is practically completely removed by performing a single experiment. The meaning of the detection limit defined in this way is clarified by comparing it in Fig. 7.5 with the detection limit as defined by Kaiser.

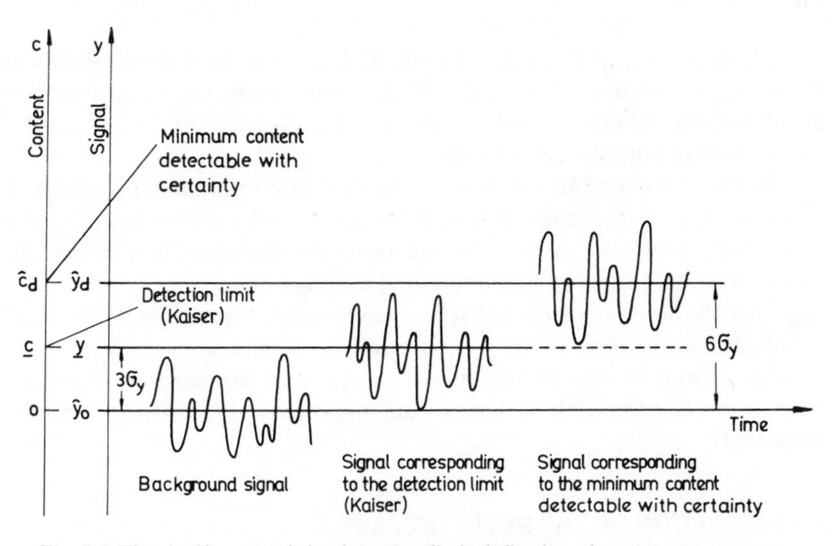

Fig. 7.5 The significance of the detection limit defined as the minimum amount detectable with certainty.

(8) If the standard deviation is not a constant, but varies with the amount of component, the standard deviation $(\sigma_y)_d$ corresponding to the detection limit $\hat{c}_d$ will be different from the standard deviation $(\sigma_y)_0$ for background fluctuations. In this case, the analytical signal corresponding to the detection limit is given by:

$$\hat{y}_d = \hat{y}_0 + z_k(\sigma_y)_0 - z_d(\sigma_y)_d \qquad (7.43)$$

For this case, the significance of the analytical signal corresponding to the detection limit can be seen in Fig. 7.6.

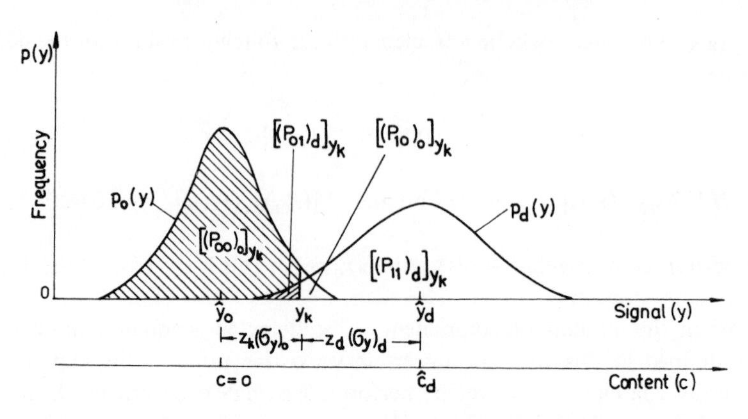

Fig. 7.6 The analytical signal corresponding to the detection limit, when the standard deviation for signal fluctuations varies with the amount.

(9) As already said, the detection limit depends on both the sensitivity and the standard deviation. Hence, the detection limit is associated with a particular set of working instructions, and provides, in fact, a criterion of the quality of the instructions and of their performance.

Hence, the detection characteristic or detection limit cannot be assigned a single value for a particular analytical technique, and the same set of working instructions may lead to different values of the detection limit or detection characteristic in different laboratories or at different times.

(10) The noise is generally of very complex origin that cannot be completely specified theoretically. Further, the sensitivity is also a function of several variables. Owing to these circumstances, the detection characteristic or detection limit cannot be completely evaluated from theory alone, but has to be determined empirically.

7.5 ESTIMATION OF THE DETECTION LIMIT

7.5.1 Introduction

The detection limits reported in the literature are usually evaluated according to Kaiser's definition. The calibration function and the analytical signal corresponding to the detection limit are evaluated, and from the estimation value of this signal

$$\bar{y}_d = \bar{y}_o + r_d \, \sigma_y \tag{7.44}$$

and the analytical calculation function:

$$\bar{c} = f(\bar{y}) \tag{7.45}$$

the estimation value of the detection limit is obtained:

$$\bar{c}_d = f(\bar{y}_d) \tag{7.46}$$

The principle of this method is shown in Fig. 7.7.

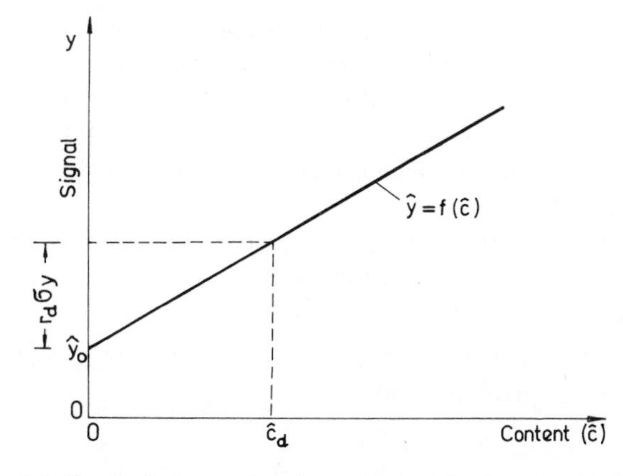

Fig. 7.7 The classical procedure for estimating the detection limit.

Examination of the method leads to the following points, which must be taken into account.

(1) A wide variety of expressions are used for evaluating the analytical signal corresponding to the detection limit.

A large number of works are based on Kaiser's theory of the detection limit, and use the value $r_d = 3$ (or close to it) for the signal-to-noise ratio corresponding to the detection limit. There is another series of works where the detection limit is taken as the smallest amount detectable with certainty; these take into account the errors of the first and second kind and use a signal-to-noise ratio of $r_d = 6$ (or close to 6).

(2) Any value for the detection limit is only an estimation, as it must be obtained by using estimation values for the corresponding analytical signal and also for the analytical calculation function. Hence, for rigour, we must give not only an estimation value for the detection limit but also the corresponding statistical confidence interval for the true value of the detection limit. To do so it is necessary to know the statistics of the estimation values of the analytical signal corresponding to the detection limit and of the calibration function.

(3) Generally speaking, there may be various departures from the model [Eq. (7.5)] which is most usually assumed for estimating the detection limit. For example, the calibration function may be non-linear, the distribution may not be normal or the standard deviation may depend on the amount of component. In order to obtain a rigorous evaluation of the detection limit, all these factors must be carefully investigated and the real model taken into account.

(4) The detection capacity is affected by the stability of the analytical system (Chapter 4), so its evaluation must involve a study of the stability.

(5) Because the value of the detection limit is specific to a certain set of conditions, these must be specified in any report of the result.

7.5.2 Estimation of the detection limit from a set of calibration data

As shown before, a set of calibration data is a set of pairs of values, of amount and signal, that is $(\hat{c}_i, y_i)$. We assume that the $\hat{c}_i$ values are rigorously known, while the analytical signals y_i are linearly dependent on $\hat{c}_i$ and are subject to a gaussian noise which is independent of $\hat{c}_i$ [in other words the calibration data belong to model (7.5)].

We shall describe two methods for estimating the detection limit and the statistical confidence interval for its true value.

Method I. The calibration function derived from the set of calibration data $(\hat{c}_i, y_i)$ by the least-squares method is

$$\bar{y} = \bar{y}_0 + b\hat{c} \tag{7.47}$$

The statistical confidence interval for the true value of the calibration function corresponding to a confidence probability P is confined by the two hyperbolic arcs given by:

$$y = \bar{y} \pm \frac{t_{(P;n-2)} \sqrt{\text{SSD}} \sqrt{S_{cc} + n(c - \bar{c})^2}}{\sqrt{n(n-2)\, S_{cc}}} \tag{7.48}$$

(see Section 2.7.1.1 for the meaning of the symbols, noting that c replaces x; $\text{SSD} = S_{yy} - b'S_{cy}$).

According to (7.13), the analytical signal corresponding to the detection limit is estimated by:

$$\bar{y}_{\text{d}} = \bar{y}_0 + r_{\text{d}} s_y \tag{7.49}$$

so to estimate it we need the mean background signal and the standard deviation for fluctuations of the signal.

The estimation mean background signal $\bar{y}_0$ is obtained from the intercept of the calibration function (7.47). The standard deviation s_y is estimated from the calibration data as follows:

$$s_{\text{y}} = \sqrt{\frac{1}{n-2} (S_{\text{yy}} - \frac{(S_{\text{cy}})^2}{S_{\text{cc}}})} \tag{7.50}$$

A statistical confidence interval for $\bar{y}_d$ can be evaluated on the following theoretical basis. From (7.49), the statistics of $\bar{y}_d$ will be determined by the statistics of the variables $\bar{y}_0$ and s_y.

According to Section 2.7.1.1, the variable $\bar{y}_0$, obtained as above, is normally distributed with mean $\hat{y}_0$ (the true, but unknown value) and a standard deviation $\sigma_{\bar{y}_0}$, that is:

$$\bar{y}_0 \Rightarrow N(\hat{y}_0; \sigma_{\bar{y}_0}) \tag{7.51}$$

where

$$\sigma_{\bar{y}_0} = \sqrt{\frac{\sigma_y^2 \sum_{i=1}^{n} c_i}{n \sum_{i=1}^{n} (c_i - \bar{c})^2}} \tag{7.52}$$

The variable s_y (the sample standard deviation for background signals) is assumed to be evaluated from a large number of data, and normally distributed with mean σ_y (the true value of the standard deviation) and standard deviation σ_{s_y} (see 2.5.2), so

$$s_y \Rightarrow N(\sigma_y; \sigma_{s_y}) \tag{7.53}$$

where σ_{s_y} is

$$\sigma_{s_y} = \sqrt{\frac{\sigma_y^2}{2(n-2)}} \tag{7.54}$$

If the variables $\bar{y}_0$ and s_y are independent, it follows from the law of addition of variances that $\bar{y}_d$ is normally distributed with mean $\hat{y}_d$ (the true, but unknown value) and standard deviation $\sigma_{\bar{y}_d}$:

$$\bar{y}_d \Rightarrow N(\hat{y}_d; \sigma_{\bar{y}_d}) \tag{7.55}$$

where

$$\sigma_{\bar{y}_d}^2 = \sigma_{\bar{y}_0}^2 + r_d^2\, \sigma_{s_y}^2 \tag{7.56}$$

and

$$\sigma_{\bar{y}_d} = \sigma_y \sqrt{\frac{\sum_{i=1}^{n} c_i^2}{n \sum_{i=1}^{n} (c_i - \bar{c})^2} + r_d^2 \frac{1}{2(n-2)}} \tag{7.57}$$

Hence, we can assume that the random variable

$$\frac{(\hat{y}_d - \bar{y}_d)}{s_{\bar{y}_d}} = t \qquad (7.58)$$

follows a Student distribution with $n-2$ degrees of freedom. Here $s_{\bar{y}_d}$ is the estimated standard deviation of the variable $\bar{y}_d$, which is the estimated value of the analytical signal corresponding to the detection limit, obtained from (7.57) by replacing σ_y with the estimation value s_y.

In agreement with the Student distribution, the confidence interval for the true value of the analytical signal corresponding to the detection limit is

$$\bar{y}_d - t_{(P;n-2)}s_{\bar{y}_d} < \hat{y}_d < \bar{y}_d + t_{(P;n-2)}s_{\bar{y}_d} \qquad (7.59)$$

where $t_{(P;n-2)}$ is the Student parameter corresponding to the confidence probability P and $n-2$ degrees of freedom.

Hence the detection limit is estimated by interpolating the signal $\bar{y}_d$, given by (7.49), in the calibration function (7.47), and the confidence interval for the detection limit is evaluated from the intersection between the confidence band (7.48) and the confidence interval (7.59). This is shown graphically in Fig. 7.8 (the symbols used in this section are the same as in Section 2.7.1.1, with c replacing x).

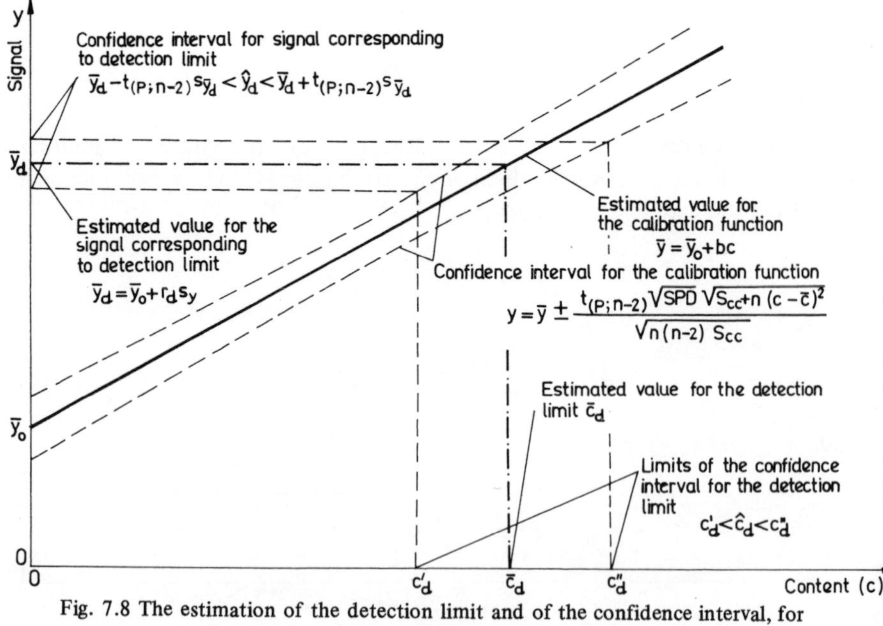

Fig. 7.8 The estimation of the detection limit and of the confidence interval, for a set of calibration data. (From [49] by permission of Pergamon Press)

Method II. We use the equation:

$$\hat{c}_d = r_d \frac{\sigma_y}{b} \tag{7.60}$$

A pair of experimental values s_y and b (the estimated standard deviation for signal fluctuations, and slope of the calibration function evaluated from calibration data) will lead to an estimation value for the detection limit, as follows:

$$\bar{c}_d = r_d \frac{s_y}{b} \tag{7.61}$$

Hence, the estimation value of the detection limit is a non-linear function of two random variables, s_y and b. The partial derivatives of (7.61) with respect to the variables s_y and b are, respectively:

$$\frac{\partial \bar{c}_d}{\partial s_y} = \frac{r_d}{b} \tag{7.62}$$

and

$$\frac{\partial \bar{c}_d}{\partial b} = r_d \frac{s_y}{b^2} \tag{7.63}$$

From addition of the variances, if s_y and b are independent, it follows from (7.62) and (7.63) that the standard deviation of $\bar{c}_d$ is

$$\sigma_{\bar{c}_d} = \frac{r_d\,\sigma_y}{b} \sqrt{\frac{1}{2(n-2)} + \frac{\sigma_y^2}{b^2 \sum_{i=1}^{n} (c_i - \bar{c})^2}} \tag{7.64}$$

Hence, we can assume that the variable

$$\frac{\hat{c}_d - \bar{c}_d}{s_{\bar{c}_d}} = t \tag{7.65}$$

has a Student distribution with $n-2$ degrees of freedom.

The Student distribution leads to the following statistical confidence interval for the true value of the detection limit:

$$\bar{c}_d - t_{(P;n-2)} s_{\bar{c}_d} < \hat{c}_d < \bar{c}_d + t_{(P;n-2)} s_{\bar{c}_d} \tag{7.66}$$

The estimated standard deviation $s_{\bar{c}_d}$ for the variable $\bar{c}_d$ is obtained from (7.64), by substituting the estimation value s_y for the standard deviation σ_y.

Example 7.2. The methods given above for estimating the detection limit and the statistical confidence interval will now be exemplified with the calibration data of Table 7.1. These are emission spectrometry calibration data for tungsten, obtained by repeated recording for seven standard samples of steel (indicated as 0, 1, 2, 3, 4, 5, 6).

The two methods given are rigorously valid for a calibration function described by the model (7.5). Hence, we shall first verify that this model holds for the calibration data.

Table 7.1 Calibration data

Standard sample	0	1	2	3	4	5	6
W%	0.000	0.012	0.105	0.001	0.006	0.077	0.048
	111	138	276	117	119	231	183
	113	133	276	112	124	235	186
	112	133	274	115	118	231	188
	113	133	275	113	120	236	186
Signal	112	135	278	116	120	232	189
digits	115	135	273	118	123	233	186
	116	138	276	116	120	231	187
	115	134	273	117	120	233	187
	115	133	273	116	120	233	185
	111	130	270	114	118	230	184
	111	133	272	115	117	231	186
	110	130	275	115	118	232	186
Sample variance	$s_0^2 =$ 3.97	$s_1^2 =$ 6.39	$s_2^2 =$ 4.75	$s_3^2 =$ 2.97	$s_4^2 =$ 4.20	$s_5^2 =$ 3.15	$s_6^2 =$ 2.72

The hypothesis that the standard deviation of the analytical signal is independent of the amount of tungsten can be verified with either the F test, or Bartlett's test (see Sections 2.6.3.2 and 2.6.3.3), by comparing the sample variances s_0^2, s_1^2, s_2^2, s_3^2, s_4^2, s_5^2, s_6^2, where each pertains to a particular level of tungsten. The ratio between any two of them is smaller than $F_{(0.05;11,\ 11)} = 2.82$. Hence, for the chosen significance level, we cannot reject the hypothesis that these sample variances are homogeneous, i.e. that all of them estimate the same quantity. Consequently, we cannot reject the hypothesis that the standard deviation is independent of the amount of tungsten.

We have still to verify the hypothesis that the dependence of signal on amount is linear, and also the hypothesis that there is no systematic error (a random interference term in the signal). These two hypotheses are simultaneously verified by comparing, by the F test, the estimation variance $s_r^2 = 4.02$ (i.e. the overall variance, resulting from all the estimated variances s_0^2, s_1^2, s_2^2, s_3^2, s_4^2, s_5^2, and s_6^2) with the estimated variance $s_y^2 = 5.73$ for fluctuations of the analytical signal around the straight line of calibration. We find that their ratio satisfies the inequality:

$$\frac{s_y^2}{s_r^2} = \frac{5.73}{4.02} = 1.43 < F_{(0.05;82,77)} = 1.45$$

so the hypothesis concerning the linearity and the absence of systematic errors cannot be rejected for the chosen significance level.

The hypothesis that the signal fluctuations are normally distributed is verified with the χ^2 test. The fluctuation of the analytical signal with respect to the estimation value corresponding to the calibration function gives the histogram given in Fig. 7.9. The results obtained by the χ^2 test are given in Table 7.2.

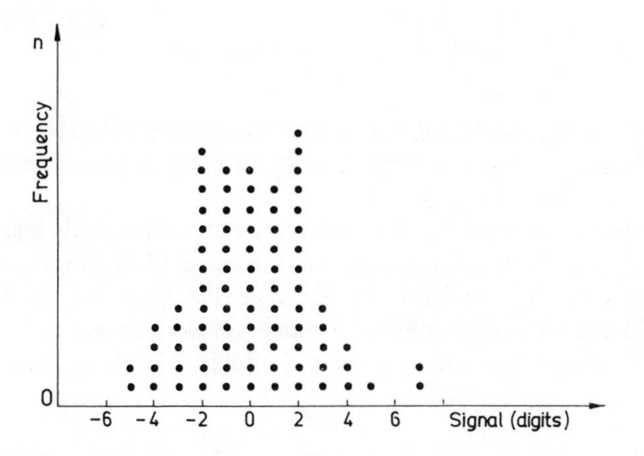

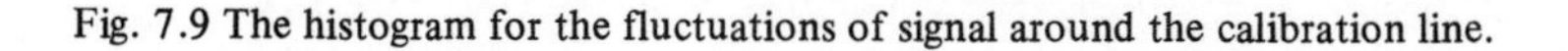
Fig. 7.9 The histogram for the fluctuations of signal around the calibration line.

Table 7.2 Testing of the hypothesis of normal distribution of analytical signal fluctuations by using the χ^2 test (Section 2.6.5)

The interval for signal fluctuations (digits) around calibration line		Absolute frequency: Experimentally obtained n		Absolute frequency: Estimated on the hypothesis of normal distribution n'		$\frac{(n-n')^2}{n'}$
−6.5	−5.5	0	11	0.68	12.41	0.167
−5.5	−4.5	2		1.71		
−4.5	−3.5	4		3.58		
−3.5	−2.5	5		6.44		
−2.5	−1.5	13		9.75		1.083
−1.5	−0.5	12		10.90		0.111
−0.5	+0.5	12		17.24		1.590
+0.5	+1.5	11		10.90		0.001
+1.5	+2.5	14		9.75		1.872
+2.5	+3.5	5	11	6.44	12.61	0.206
+3.5	+4.5	3		3.58		
+4.5	+5.5	1		1.71		
+5.5	+6.5	0		0.68		
+6.5	+7.5	2		0.20		
	Sum	84		83.59		
		$\nu = 6$			$\chi^2 = 5.030$	

The experimental value of χ^2 satisfies the inequality $\chi^2 = 5.030 < 12.59 = \chi^2_{(0.95;16)}$, so the hypothesis that the signal fluctuations are normally distributed cannot be rejected.

Hence, summing up, we cannot reject the hypothesis that the calibration data of Table 7.1 correspond to the model (7.5). Consequently, we can proceed to the estimation of the detection limit and of the statistical confidence interval by either of the methods given above.

Processing the calibration data in Table 7.1 (see Sections 2.7.1.1 and 5.6) gives

$$S_{cc} = 0.1267;\ S_{cy} = 195.4;\ S_{yy} = 301181;\ SSD = 470.2;\ \bar{y}_0 = 113;$$
$$b = 1539;\ s_y^2 = 5.734;\ s_{\bar{y}}^2 = 0.1290;\ s_{s_y}^2 = 0.0349;\ s_{\bar{y}_d}^2 = 1.421$$

The estimated calibration function is:

$$\bar{y} = 113 + 1539\,c$$

With the defining values $[(P_{10})_o]_{y_k} = 0.001$ and $[(P_{11})_d]_{y_k} = 0.998$, and a signal-to-noise ratio $r_d = 6$, we find from (7.49) the estimated value of the analytical signal corresponding to the detection limit:

$$\bar{y}_d = 127.4 \text{ digits}$$

By substituting this value in the calibration function, the detection limit is estimated to be:

$$\bar{c}_d = 0.009\% \text{ W}$$

The confidence intervals for the calibration function and for the detection limit, obtained from (7.48) and (7.59) with the same data as above, are

$$\bar{y} - t_{(P;82)}0.741\sqrt{0.1267 + 84(c-\bar{c})^2} < \hat{y} <$$

$$\bar{y} + t_{(P;82)}0.741\sqrt{0.1267 + 84(c-\bar{c})^2}$$

and

$$127.4 - t_{(P;82)}1.19 < \hat{y}_d < 127.4 + t_{(P;82)}1.19$$

For a confidence probability $P = 0.98$, we obtain $t_{(0.98;82)} = 2.40$, so that these two confidence intervals become respectively:

$$113 + 1539\,c - 1.778\sqrt{0.1267 + 84(c-0.0356)^2} < \hat{y} <$$

$$113 + 1539\,c + 1.778\sqrt{0.1267 + 84(c-0.0356)^2}$$

and

$$124.5 < \hat{y}_d < 130.2$$

Their intersection gives the confidence interval for the true value of the detection limit:

$$0.007\% \text{ W} < \hat{c}_d < 0.011\% \text{ W}$$

Let us now consider the second method. In this case, from (7.60), the detection limit will be:

$$\bar{c}_d = 0.009\%\ \mathrm{W}$$

From (7.64), the estimated standard deviation associated with the variable $\bar{c}_d$ will be:

$$s_{\bar{c}_d} = 0.00074,$$

and from (7.66), the confidence interval for the true value of the detection limit will be:

$$0.009 - t_{(P;82)}\, 0.00074 < \hat{c}_d < 0.009 + t_{(P;82)}\, 0.00074$$

With the confidence probability $P = 0.98$, i.e. $t = 2.40$, we obtain

$$0.0072\%\ \mathrm{W} < \hat{c}_d < 0.0108\%\ \mathrm{W}$$

7.5.3 Evaluation of a confidence value for the decision threshold and detection limit for a set of calibration data

A set of calibration data conforming to model (7.5) can be used to obtain a confidence value for the detection limit of a single measurement.

As shown before (Section 5.6), we can evaluate, by using the Student distribution, a confidence band (or, synonymously, a band variance) for the signal fluctuations. This band is bounded by two hyperbolic arcs:

$$\bar{y}_0 + bc - t_{(P;n-2)}\, s_y \sqrt{1 + \frac{1}{n} + \frac{(c-\bar{c})^2}{S_{cc}}} < y < \bar{y}_0 + bc + t_{(P;n-2)} s_y \sqrt{1 + \frac{1}{n} + \frac{(c-\bar{c})^2}{S_{cc}}} \qquad (7.67)$$

(the meaning of the symbols is given in Section 5.6).

The meaning of the confidence band and the method for evaluating the confidence value for the decision threshold and detection limit are illustrated in Fig. 7.10.

The intercept of the two hyperbolic arcs on the signal axis is the statistical confidence interval for the background signals, evaluated from the calibration data. Hence, the value y_k of the upper limit of this confidence interval has the significance of a confidence value for the detection threshold corresponding to a confidence probability of $(1 + P)/2$, or a significance level $(1 - P)/2$.

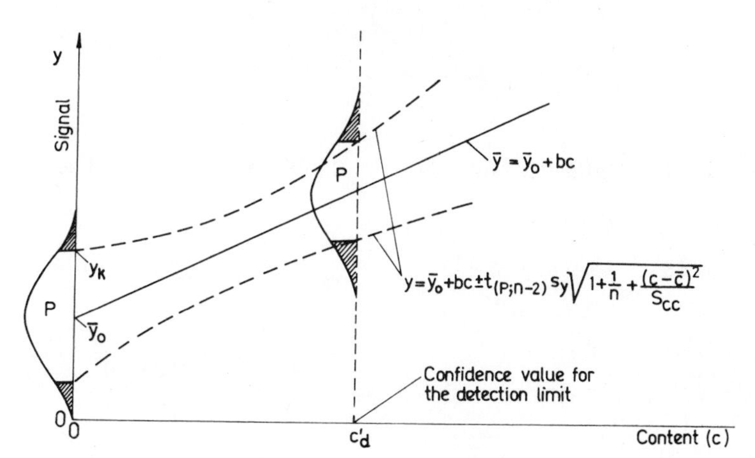

Fig. 7.10 The significance of various confidence values for the detection level, and for the detection limit.

The same value y_k on the lower hyperbola bounding the variance band corresponds to c'_d. This can be taken as a confidence value for the detection limit, such that this is the lowest amount that can be detected by one measurement with a confidence probability of $(1 + P)/2$. That is, a single measurement will result in a signal $y > y_k$.

Obviously, if the number of calibration data is increased, the estimated values $\bar{y}_0$, b, and s_y will tend, in the limit, towards the true values. In turn, the confidence value for the detection limit will tend towards the true value corresponding to the false detection probability $[(P_{10})_0]_{y_k} = (1 - P)/2$, and the confidence value for the detection limit will tend towards the true value corresponding to the correct detection probability $[(P_{11})_d]_{y_k} = (1 + P)/2$.

The problem of evaluating a confidence value for the decision threshold and for the detection limit from a set of calibration data is treated in detail by Hubaux and Vos [43]. These authors have investigated the effect of the following factors: the precision of the method, the number of standard samples, the range covered by the amount of component in the standard samples, the distribution of the amount of component in the standard samples, and the number of measurements.

Example 7.3. The method of evaluation of a confidence value for the decision threshold and detection limit will be exemplified with the calibration data from Table 7.1, at a confidence probability of $P = 0.998$.

From (7.67), the confidence interval for signal fluctuations is

$$113 + 1539\,c - t_{(P;82)}2.472 \sqrt{1 + \frac{1}{84} + \frac{(c - 0.0356)^2}{0.1269}} < y <$$

$$113 + 1539\,c + t_{(P;82)}2.472\sqrt{1 + \frac{1}{84} + \frac{(c - 0.0356)^2}{0.1269}}$$

For the confidence probability $P = 0.998$, i.e. $t_{(0.998;82)} = 3.40$,

$$113 + 1539\,c - 8.405\sqrt{1 + \frac{1}{84} + \frac{(c - 0.0356)^2}{0.1269}} < y <$$

$$113 + 1539\,c + 8.405\sqrt{1 + \frac{1}{84} + \frac{(c - 0.0356)^2}{0.1269}}$$

Taking $c = 0$ in the equation for the upper hyperbola, the confidence value for the decision level is found to be $y_{\mathrm{k}} = 121$. By interpolating this value in the lower hyperbola we find that the confidence value for the detection limit is $c'_{\mathrm{d}} = 0.011\%$ W. The agreement between this confidence value for the detection limit and the upper limit of the confidence interval for the detection limit, as evaluated in example 7.2, is statistically justified.

7.5.4 Estimation of the detection limit from the standard deviation of analysis errors

We shall again use the model (7.5), so

$$\hat{c}_{\mathrm{d}} = r_{\mathrm{d}}\frac{\sigma_y}{b} \tag{7.68}$$

In this equation, the ratio between the true standard deviation for signal fluctuations σ_y and the true value of b is equal to the true value of the standard deviation of analysis errors, that is:

$$\frac{\sigma_y}{b} = \sigma_c \tag{7.69}$$

Hence, Eq. (7.68) can be written as:

$$\hat{c}_{\mathrm{d}} = r_{\mathrm{d}}\,\sigma_c \tag{7.70}$$

Thus, within the limits of the model used, the detection limit is the product of the signal-to-noise ratio and the standard deviation of analysis errors. By using an estimation value s_c for this standard deviation, we obtain an estimation value for the detection limit,

$$\bar{c}_{\mathrm{d}} = r_{\mathrm{d}}s_c \tag{7.71}$$

Obviously, the higher the number of analytical determinations used to obtain s_c, the closer this is to the true value σ_c, and, consequently, the closer the estimation $\bar{c}_d$ with respect to the true value $\hat{c}_d$ of the detection limit.

In evaluating a confidence interval for the true value of the detection limit estimated as shown above, we may take into account the following theoretical considerations.

The sample standard deviation s_c for the analysis errors, evaluated from n determinations (where n is a large number) will be normally distributed, with mean σ_c (the true value) and a variance $\sigma_{s_c}^2$, that is:

$$s_c \Rightarrow N(\sigma_c; \sigma_{s_c}^2) \tag{7.72}$$

where

$$\sigma_{s_c}^2 = \frac{\sigma_c^2}{2(n-1)} \tag{7.73}$$

Hence, we can accept that $\bar{c}_d$ is normally distributed, with mean equal to the true value $\hat{c}_d$ of the detection limit, and variance $\sigma_{\bar{c}_d}^2$:

$$\bar{c}_d \Rightarrow N(c_d; \sigma_{c_d}^2) \tag{7.74}$$

where

$$\sigma_{\bar{c}_d}^2 = r_d^2\,\sigma_{s_c}^2 = r_d^2\,\frac{\sigma_c^2}{2(n-1)} \tag{7.75}$$

Hence, we can assume that the variable

$$\frac{\hat{c}_d - \bar{c}_d}{s_{\bar{c}_d}} = t \tag{7.76}$$

has a Student distribution with $n-1$ degrees of freedom. Here $s_{\bar{c}_d}$ is the estimation value for the standard deviation of the variable $\bar{c}_d$, obtained from (7.75) by using an estimation value s_c instead of σ_c.

Hence, the confidence interval for the detection limit, corresponding to a confidence probability P, will be given by:

$$\bar{c}_d - t_{(P;n-1)}s_{\bar{c}_d} < \hat{c}_d < \bar{c}_d + t_{(P;n-1)}s_{\bar{c}_d} \tag{7.77}$$

Example 7.4. Under the same analytical conditions as used for obtaining the calibration data given in Table 7.1, a standard steel sample containing 0.007% tungsten is analysed 30 times. The standard deviation of the results obtained is $s_c = 0.0017\%$. Accepting the statistical conditions $[(P_{10})_o]_{y_k} = 0.001$ and $[(P_{11})_d]_{y_k} = 0.998$, i.e. a signal-to-noise ratio $r_d = 6$, the estimation value found from (7.71) for the detection limit is

$$\bar{c}_d = 0.0102\%$$

From (7.75), the standard deviation for $\bar{c}_d$ is:

$$s_{\bar{c}_d} = 0.0013\%$$

For the confidence probability $P = 0.98$, i.e. $t_{(0.98;29)} = 2.47$, the confidence interval for the true value of the detection limit is, from (7.77):

$$0.007\% < \hat{c}_d < 0.0134\%$$

7.5.5 Evaluation of the detection limit when the standard deviation of the signals varies with amount

The case of the standard deviation of the signals varying with the amount is represented in Fig. 7.11 along with a linear calibration curve. This situation is frequently found in chemical analysis.

Near $c = 0$ there is a region, specified as $c < \hat{c}_s$ and designated by I in the figure, where the standard deviation of the signal is practically independent of c. In other words, for this range of concentrations we can accept the hypothesis H_0 that $(\sigma_y^2)_o = (\sigma_y^2)_d$, or that $(s_y^2)_o = (s_y^2)_d$.

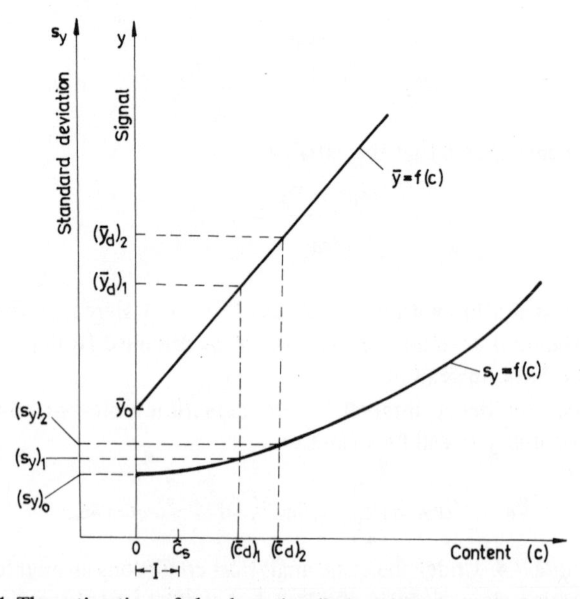

Fig. 7.11 The estimation of the detection limit, when the standard deviation is a function of amount.

Obviously, the previously given method for estimating the detection limit can be rigorously applied only if both the calibration data and the detection limit are in region I.

For the general case represented in Fig. 7.11, the detection limit is evaluated as follows. As many analytical measurements as possible are performed on standard samples containing suitable amounts of the component. From the results we evaluate the calibration function $\bar{y} = f(c)$ and the function $s_y = f(c)$, which is the **characteristic of reproducibility** (precision) of the method. The standard deviation for background signal fluctuations, $(s_y)_0$, is evaluated from the intercept of the reproducibility characteristic on the ordinate. A first evaluation of the detection limit is then made from

$$(\bar{y}_d)_1 = \bar{y}_0 + r_d(s_y)_0 \tag{7.78}$$

Next, $(\bar{y}_d)_1$ is interpolated in the calibration function to give the first approximation $(\bar{c}_d)_1$ for the detection limit. Obviously, if $(\bar{c}_d)_1 < \hat{c}_s$, that is, if $(\bar{c}_d)_1$ is contained in region I, it is already a correct estimation for the detection limit. If, however, $(\bar{c}_d)_1 > \hat{c}_s$, which is the case shown in Fig. 7.11, the signal standard deviation $(s_y)_1$ corresponding to $(\bar{c}_d)_1$ must be evaluated and used to derive, according to (7.43), a new value for the analytical signal corresponding to the detection limit:

$$(\bar{y}_d)_2 = \bar{y}_0 + z_k(s_y)_0 - z_d(s_y)_1 \tag{7.79}$$

where $(s_y)_0$ is the standard deviation for the background signals and $(s_y)_1$ that for an amount equal to $(\bar{c}_d)_1$.

The new value $(\bar{y}_d)_2$ is then interpolated into the calibration function to obtain a new and better approximation $(\bar{c}_d)_2$ for the detection limit.

The procedure is repeated until the difference between two successive approximations is smaller than a preselected critical value.

7.5.6 The frequentometric method for estimating the detection limit

In this procedure [46-50] we choose a definition value for the probability of false detection, $[(P_{10})_0]_{y_k}$, from which, with enough analytical measurements, we evaluate the decision threshold. The detection characteristic (see Section 6.9) is then evaluated with respect to the decision threshold. Writing the decision characteristic as

$$\hat{c} = f(P_{11}) \tag{7.80}$$

(i.e. in a form in which this characteristic is explicit with respect to the amount), the detection limit will be obtained by interpolating the probability of correct detection $[(P_{11})_d]_{y_k}$ in the detection characteristic:

$$\hat{c}_d = f\left([(P_{11})_d]_{y_k}\right) \tag{7.81}$$

We shall now give two methods for estimating the detection limit on this principle.

Method I. From as large a set of calibration data as possible, we evaluate the parameters of the signal–amount relation, and hence the detection characteristic corresponding to a certain value of the decision threshold and, finally, the detection limit corresponding to a certain probability of true detection.

Example 7.5. The following model has been established from the calibration data given in Table 7.1:

$$y \Rightarrow N(\bar{y} = 113 + 1539\,c;\, \sigma_y = 2.395)$$

From this model, evaluate the detection limit corresponding to the statistical conditions $[(P_{10})_0]_{y_k} = 0.001$ and $[(P_{11})_d]_{y_k} = 0.998$.

The method for evaluating the detection characteristic was given in Section 6.9. The results obtained by this method for our model are given in Table 7.3.

Table 7.3 Estimation of the detection characteristic

	$[(P_{10})_0]_{y_k} = 0.001$		$z_k = 3.1$	
c %	$y = 113 + 1539\,c$	$r = \dfrac{\bar{y} - \bar{y}_0}{\sigma_y}$	$z_d = z_k - r$	P_{11}
0.001	114.5	0.63	+2.47	0.007
0.002	116.0	1.27	+1.83	0.034
0.003	117.6	1.90	+1.20	0.115
0.004	119.1	2.54	+0.56	0.287
0.005	120.7	3.17	−0.07	0.528
0.006	122.2	3.80	−0.70	0.758
0.007	123.8	4.44	−1.34	0.909
0.008	125.6	5.07	−1.97	0.975
0.009	126.8	5.71	−2.61	0.995
0.010	128.4	6.34	−3.24	0.999

By using the pairs of data amount (c_i) – correct detection probability (P_{11}), given in the table we obtain the detection characteristic shown in Fig. 7.12. By interpolating the probability of correct detection $[(P_{11})_d]_{y_k} = 0.998$ in the detection characteristic the value of the detection limit is estimated at $\bar{c}_d = 0.0095\%$.

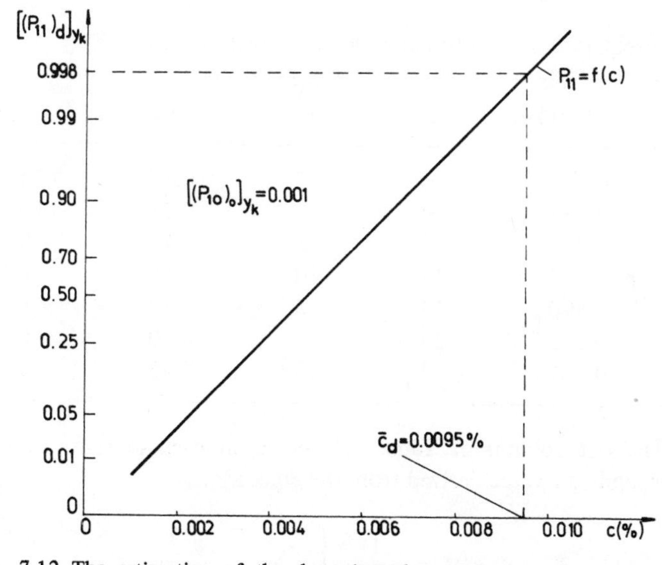

Fig. 7.12 The estimation of the detection characteristic, and detection limit (method I).

Method II. First, from background signal measurements we evaluate the decision threshold corresponding to a probability of false detection $[(P_{10})_0]_{y_k}$. Then repeated analytical measurements are made on a set of m standard samples, containing the component to be detected in amounts equal to $\hat{c}_1, \hat{c}_2, \ldots, \hat{c}_m$, uniformly distributed over the range close to zero.

If the standard sample containing amount $\hat{c}_i$ is analysed N times and there are n_i cases of results greater than the decision threshold, the probability of correct detection at the amount $\hat{c}_i$ will be estimated by the ratio n_i/N. Applying this procedure for each of the m standard samples, we obtain the estimation of the probability of correct detection at each amount in the set $\hat{c}_1, \hat{c}_2, \ldots, \hat{c}_m$.

Next, the detection characteristic is drawn for the set of pairs of data $(\hat{c}_i, n_i/N)$. The estimation value of the detection limit is now obtained by interpolating the preselected probability of correct detection $[(P_{11})_d]_{y_k}$ in the curve of detection characteristic.

Example 7.6. We shall use the example of infrared detection of *N*-methyl-2-pyrrolidone in styrene [47], for the statistical defining conditions $[(P_{10})_0]_{y_k} = 0.025$ and $[(P_{11})_d]_{y_k} = 0.975$.

From 20 measurements of the background signals, the mean and the standard deviation were estimated to be $\bar{y}_0 = 282$ digits and $(s_y)_0 = 6.3$ digits. For the probability of false detection $[(P_{10})_0]_{y_k} = 0.025$, i.e. $z_k = 1.96$, the value of the decision threshold is found to be $y_k = 294$ digits.

The results obtained from 20 analyses of each of 7 standard samples are given in Table 7.4.

Table 7.4 Frequentometric estimation of the detection limit [47]

m	c (mg/100 g)	N	n	n/N	z
1	0.150	20	2	0.10	+1.285
2	0.231	20	5	0.25	+0.675
3	0.320	20	8	0.40	+0.255
4	0.377	20	10	0.50	0.000
5	0.490	20	14	0.70	−0.525
6	0.535	20	16	0.80	−0.842
7	0.701	20	19	0.95	−1.650

The last column of Table 7.4 shows, at each level of amount, the corresponding z value derived from the equation:

$$P_{11} = n/N = \frac{1}{\sqrt{2\pi}} \int_z^\infty \exp\left[-\frac{z^2}{2}\right] dz \qquad (7.82)$$

The simplest way to obtain the value of z corresponding to a specified value of the ratio n/N is to use a table of the normal distribution function and an interpolation procedure.

We give in Fig. 7.13 the estimated characteristic of detection in co-ordinates 'n/N vs. c'. The detection limit corresponding to a probability $[(P_{11})_d]_{y_k} = 0.975$ is obtained by interpolating this probability in the characteristic of detection, and the result is $\bar{c}_d = 0.76$ mg.

Obviously, the linearity of the detection characteristic $P_{11} = f(c)$ on normal probability paper, or of the detection characteristic $z = f(c)$ on an ordinary graph (Fig. 7.14) suggests that the experimental data belong to model (7.5).

The least-squares method applied to the pairs of values (z, c) leads to the following calibration regression:

$$\bar{z} = 1.96 - 5.18c,$$

whence for $[(P_{11})_d]_{y_k} = 0.975$, i.e. for $z_d = -1.96$, the detection limit is found to be $\bar{c}_d = 0.76$ mg.

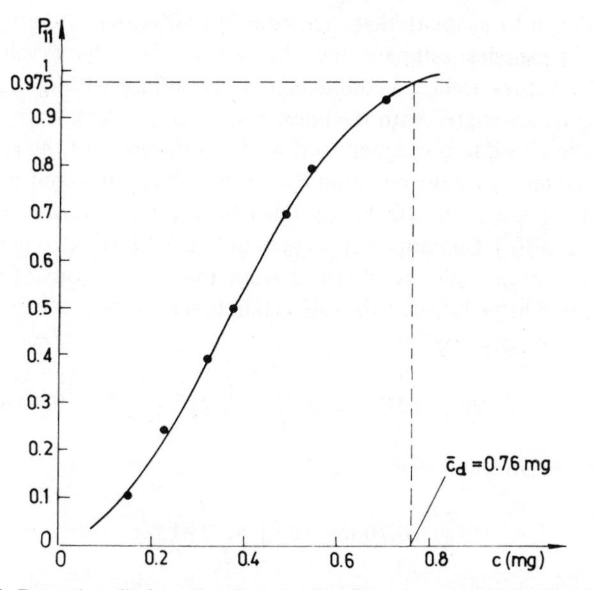

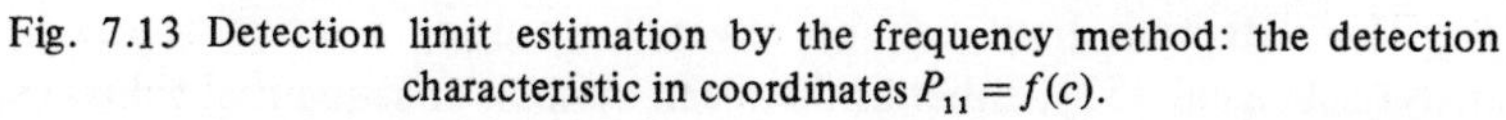

Fig. 7.13 Detection limit estimation by the frequency method: the detection characteristic in coordinates $P_{11} = f(c)$.

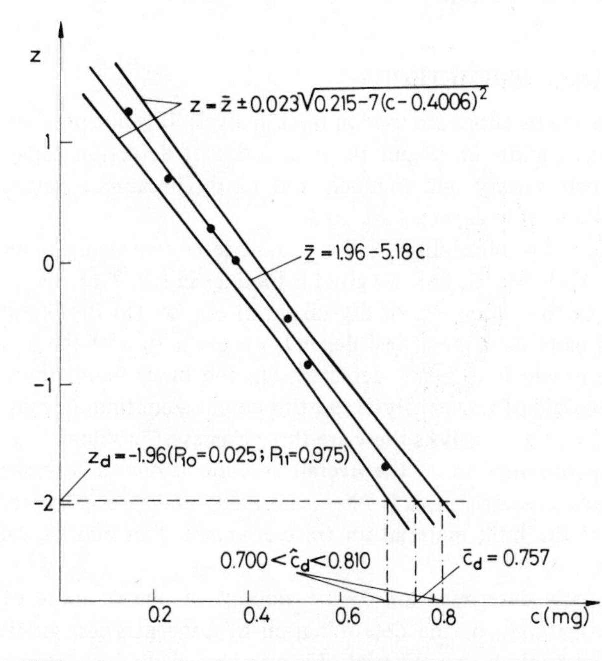

Fig. 7.14 The estimation by the frequency method of the detection limit and of the confidence interval.

Taking into account that, according to Glivenko's theory, the experimental frequencies estimate the theoretical ones, the z values are also estimation values. Hence, a confidence band confined within two hyperbolic arcs must be associated with the linear regression $z = f(c)$.

As the z_d value is uniquely defined from the value of the probability of correct detection by the equation $P_{11} = 0.5 + \Phi(z_d)$, the confidence interval of the detection limit will be obtained from the uncertainty of the linear function $z = f(c)$. Consequently, as seen in Fig. 7.14, the two extreme values of the detection limit are obtained from the intersections of the value z_d with the two hyperbolae of the calibration function [50], which, for $P = 0.95$ and $\nu = 5$, are given by:

$$z = (1.96 - 5.18c) \pm 0.023\sqrt{0.215 + 7(c - 0.4006)^2}$$

For $z = -1.96$ we obtain

$$0.70 \text{ mg} < \hat{c}_d < 0.81 \text{ mg}$$

The detection limit can be lowered by using a set of values of the analytical signal [52], either as the mean or sum, or sequential values (signal sum or frequency sum).

7.6 TRACE ANALYSIS METHODS

A great many methods are used in trace analysis. It is not possible to make a complete classification or to put them in order of detection capacity, partly because of their variety and number, and partly because so many different definitions of detection capacity are used.

The range of applicability of the commonest trace analysis methods has been assessed by Tölg [53, 54]; we give his findings in Fig. 7.15.
These limits are not to be too rigidly taken, of course. On the one hand, for a given class of methods a great deal depends on the nature of the sample and of the component which is being detected. On the other hand, there is always progress in the field of trace analysis and this entails a continuous reassessment.

Generally, in trace analysis there are three classes of problem.

(a) The determination of the overall amount of one or several trace components in a given material: this is what is normally understood by trace analysis, i.e. analysis of the bulk material for trace amounts. This may be called macro trace analysis.

(b) The local determination of the amount of one or more of the components of a material, or the determination of concentration gradients, along one or several directions in a material: this may be called micro trace analysis.

(c) Surface or thin layer analysis, which in some respects is similar to (b).

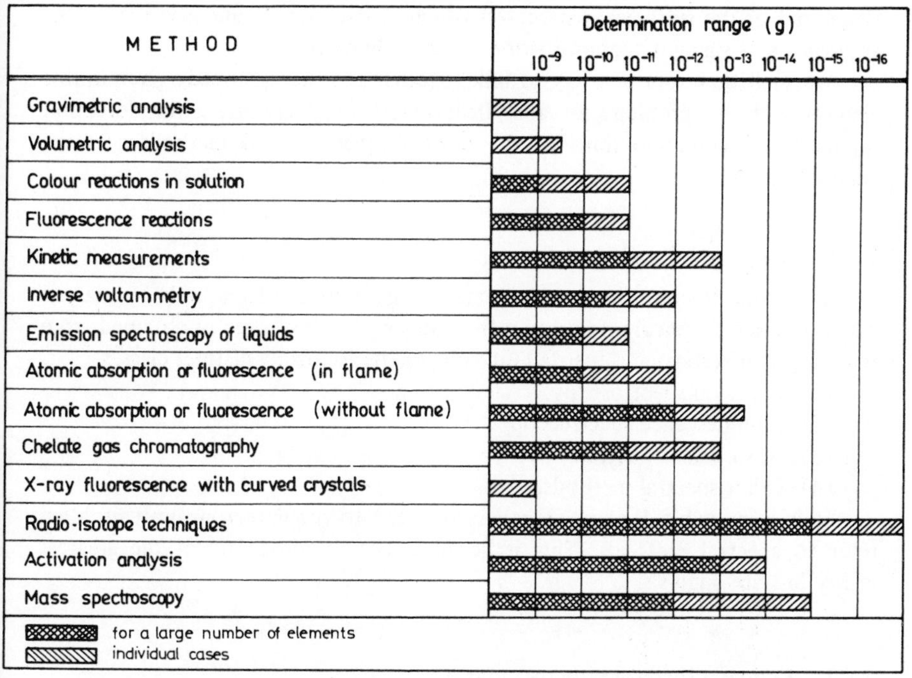

METHOD	Determination range (g)
	10^{-9} 10^{-10} 10^{-11} 10^{-12} 10^{-13} 10^{-14} 10^{-15} 10^{-16}
Gravimetric analysis	
Volumetric analysis	
Colour reactions in solution	
Fluorescence reactions	
Kinetic measurements	
Inverse voltammetry	
Emission spectroscopy of liquids	
Atomic absorption or fluorescence (in flame)	
Atomic absorption or fluorescence (without flame)	
Chelate gas chromatography	
X-ray fluorescence with curved crystals	
Radio-isotope techniques	
Activation analysis	
Mass spectroscopy	

for a large number of elements
individual cases

Fig. 7.15 Limits of applicability of the most important trace analysis methods. (From [54] by permission of Pergamon Press)

These three classes cannot be entirely resolved, and they all pertain to the field of trace analysis.

For dealing with the last two classes special techniques have been developed, for example, laser emission spectroscopy, electron- and ion-microprobe analysis for local analysis problems, and X-ray photoelectron spectroscopy or Auger electron spectroscopy for surface and thin layer analysis, to give but a few examples.

For all methods of analysis, the detection limit is an objective criterion of the detection capacity in the trace range and also for analytical performance.

The detection limit is evaluated in a similar way for all methods, by application of the detection limit theory to the specific features of each analytical process. No matter what the method of analysis, any analytical measurement corresponding to a given amount of substance has a random behaviour. Hence the response function $y = f(c)$ will also have random character. Thus each method of analysis will cease to work if the amount to be detected is below a certain limit.

A rigorous evaluation of this limit requires a proper understanding of the peculiarities of the analytical process of measurement, of the concept of analytical signal and of the statistical parameters which characterize it, and also the proper application of signal detection theory to the evaluation.

We shall therefore now try to link together the theory already given in this chapter and the problems of analytical practice, by discussing the detection capacity and detection limit of the most frequently used methods of trace analysis.

7.6.1 Spectral methods

The name 'spectral methods' covers a large number of methods of analysis which extract chemical information from various kinds of spectra obtained from the samples investigated. Most fall into one of the following distinct classes:

(1) Emission spectroscopy.
(2) Luminescence spectroscopy.
(3) Absorption spectroscopy.
(4) Other spectral methods.

It is remarkable that most papers devoted to the detection limit problem refer to spectral methods. This is because these methods are of the greatest utility in trace analysis.

7.6.1.1 *Emission spectroscopy methods*

These include the arc, spark, flame, or laser atomic emission methods, and also the molecular emission methods.

An analytical system of measurement based on emission spectroscopy consists of the excitation source where the sample for analysis is introduced, the optics which collect and resolve the radiation, and the recording and measuring system.

(a) *Arc and spark emission spectroscopy* In atomic emission spectroscopy, the atomic spectrum is superimposed on the continuous radiation of the background spectrum. The basic equation in emission spectroscopy is the Scheibe-Lamakin equation; it expresses the concentration (or amount) of an atomic species as a function of the intensity of a spectral line characteristic of this species:

$$\hat{c} = aI_{\lambda}^{\eta} \tag{7.83}$$

where the exponent η takes into account self-absorption effects.

Generally, whether the purpose of the experiment is detection or determination, the spectral line intensity I_{λ} is not measured in absolute terms, but with respect to a reference intensity I_r. The reference intensity may be either the intensity of the continuous radiation from a certain spectral region, or the intensity of a line of a different atomic species, suitably chosen, which is used as an internal standard. When relative intensity measurements are used, the fundamental equation becomes

$$\hat{c} = a\left(\frac{I_\lambda}{I_r}\right)^{\eta} \tag{7.84}$$

If the analytical signal corresponding to the detection threshold is $(I_\lambda/I_r)_d$, the detection limit will be given by:

$$\hat{c}_d = a[(I_\lambda/I_r)_d]^{\eta} \tag{7.85}$$

Obviously, Eq. (7.84) and the analytical signal corresponding to the detection limit cannot be ascertained theoretically, and must be evaluated experimentally.

There is a large literature devoted to optimization of the detection capacity, i.e. to the choice of analytical conditions that will lower the detection limit as much as possible.

Thus, optimization of the excitation has been investigated [55-59]; many methods have been established for increasing the intensity of the spectral lines of the atomic species to be detected. These methods are effective in improving the signal-to-noise ratio and thus lowering the detection limit. However, as a general rule, it cannot be ensured that conditions will be simultaneously optimal for the excitation of a large number of atomic species. For this reason, in analysis for a large number of traces, if the species concerned have different excitation behaviour, several samples have to be used, each being excited under the optimum conditions required by suitably chosen smaller groups of atomic species, or even by individual atomic species.

Other research has been concerned with optimization of the optics or the recording and measurement [3, 60-64], etc. The purpose is again to lower the detection limit. There have been major improvements in the optical devices, partly as a result of using diffraction gratings of high resolving power. Further progress has stemmed from use of the photoelectric effect, and especially, from use of photomultipliers.

For a long time, the arc emission method with photographic recording (the spectrographic method) was the most used method in trace analysis. Even now, in spite of the development of other analytical techniques, which have a higher detection capacity, the spectrographic method is still often applied in trace analysis. The problem of the evaluation of the detection limit in emission spectrography has been the subject of a large number of works [12, 14, 25-29, 31-36, 65-69].

In the upper part of Fig. 7.16 we give a photographically recorded spectrum, and below it, the same spectrum recorded on a microphotometer.

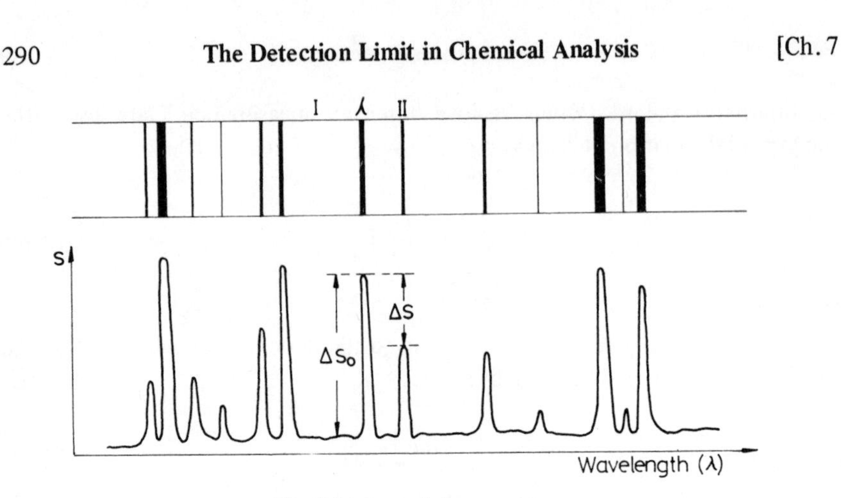

Fig. 7.16 An emission spectrum.

For detection or determination purposes, we measure the plate blackness in the region λ where we should find the spectral line of the atomic species which is to be detected or determined, and also the blackness in a reference region which will be designated either by the symbol I if the background signal is taken as reference, and the symbol II if the reference is a line of another atomic species (an internal standard).

Let us consider the background signal as reference, which is frequently the case in trace analysis. In the region λ, the blackness of the plate, S_λ, will be due to the spectral line intensity I_λ, and also to the intensity of the continuous radiation from this region, and the blackness S_r in the region I is due to the intensity of the continuous radiation from that region. The blackness difference between the regions λ and I is

$$\Delta S_0 = S_\lambda - S_r \tag{7.86}$$

When the component sought is absent, then if, but only if, the continuous radiation intensities in the two regions are equal, the expected value of the blackness difference will be zero, $\Delta \hat{S}_0 = 0$. In many cases, the continuous radiation has different intensities in the two regions, so there is always the possibility of an expected value $\Delta \hat{S}_0 \neq 0$. In order to estimate the mean $\overline{\Delta S_0}$ and the standard deviation $\sigma_{\Delta S_0}$ for the blackness difference of the two regions when the component is absent, we record the spectra of several blank samples (samples without the component of interest), find ΔS_0 for each, and then calculate the mean and standard deviation. We expect that there will be a certain dependence between the blacknesses S_λ and S_r on the same recording; the usually accepted hypothesis is that the variance $\sigma^2_{\Delta S_0}$ of the blackness difference is the sum of the blackness variances of the two regions, so

$$\sigma_{\Delta S_0} = \sqrt{2}\sigma_b \tag{7.87}$$

if the simplifying assumption is made that these blackness variances are equal, the common value being σ_b^2. Equation (7.87) is not rigorously valid, of course.

The type of functional relation between the standard deviation $\sigma_{\Delta S}$ for the blackness difference and the amount $\hat{c}$ was illustrated in Fig. 7.11 [70–73].

For the set of values ΔS_0 for the blackness difference obtained for the blank samples, we shall write the mean and the standard deviation as $\overline{\Delta S}_0$ and $s_{\Delta S_0}$, respectively. Then the detection threshold will be given by:

$$(\Delta S)_k = \overline{\Delta S}_0 + z_k s_{\Delta S_0} \tag{7.88}$$

Let us make the H_0 hypothesis, that the standard deviation $\sigma_{\Delta S_d}$ for the blackness difference at an amount equal to the detection limit is not significantly distinct from the standard deviation $\sigma_{\Delta S_0}$ for the case that the component is absent. On this hypothesis, an estimation value for the analytical signal corresponding to the detection limit will be:

$$(\overline{\Delta S})_d = \overline{\Delta S}_0 + r_d \bar{s}_{\Delta S_0} \tag{7.89}$$

where $\bar{s}_{\Delta S_0}$ is calculated by using Eq. (2.56).

Now, from the relative intensities, the analytical signal corresponding to the detection limit will be:

$$(\overline{I_\lambda / I_r})_d = 10^{-(\Delta S)_d/\gamma} - 1 \tag{7.90}$$

where γ is the contrast factor.

The detection limit is estimated by interpolating the analytical signal corresponding to the detection limit (7.90) in the calibration function written as $\log (I_\lambda / I_r) = f(\log c)$.

It has been proven by Ehrlich [65] that for small amounts, when $I_\lambda / I_r \ll 1$, the relation between $\log [(I_\lambda + I_r)/I_r] = \Delta S/\gamma$ and the amount c is quite linear. Hence, taking the calibration function as $\log [(I_\lambda + I_r)/I_r] = f(c)$, the detection limit will be obtained by interpolating the signal:

$$\log \left(\overline{\frac{I_\lambda + I_r}{I_r}} \right)_d = \frac{(\overline{\Delta S})_d}{\gamma} \tag{7.91}$$

in this calibration function.

Generally speaking, evaluation of the detection limit is much simpler if photoelectric recording is used, mainly because the data required can be obtained much more quickly.

The procedure for evaluating the detection limit in emission spectrography was explained in Section 7.5.2.

(b) *Flame emission spectrometry* Excitation in a flame has many applications in emission spectrometry. In trace analysis the high-temperature oxy-acetylene flame [74–76] and the nitrous oxide-acetylene flame [76–84] are used. The oxygen–hydrogen flame is also important in trace analysis.

In order to improve the signal-to-noise ratio, i.e. to lower the detection limit, many techniques have been developed in flame emission spectroscopy, among which we may mention aerosol modulation [85–87], the aerosol modulation method combined with a phase-lock amplifier for signal measurement [88] and pulsed vaporization [89].

Values for the detection limit in flame emission spectroscopy are given in many works [77–81, 84, 90]. Among them we mention a study by Pickett and Koirtyohann [84] which gives the detection limit for 62 elements, for both emission and absorption flame spectroscopy. We find that among them there are 24 elements for which the detection limit is lower in the flame emission method, 21 for which it is lower in the flame absorption method, and 17 for which it is the same with both methods.

A similar study by Christian and Feldman [81] compares the detection limit for 68 elements in emission and absorption flame spectroscopy: 27 elements have the same detection limit in emission and absorption, 15 have the lower detection limit in the flame emission method, and 26 have the lower detection limit in the absorption method.

Flame emission spectroscopy is suitable for simultaneous analysis for several elements, but optimal detection conditions cannot be realized for a large number of elements simultaneously, and a compromise has to be used [91].

(c) *Laser emission spectroscopy* In trace detection and determination the laser is used as an excitation source for various electrically conductive or dielectric solid materials, such as biological materials, ceramics, minerals, semiconductors, metals, alloys, etc.

There are two ways of using the laser in emission spectroscopy. In one of them [92–98] the excitation is made with the laser alone. The other method, which is applied especially in microanalysis, uses the laser and also an auxiliary excitation [99]. In the latter method, a very small amount of the sample is vaporized by a laser pulse, and the vapour passes into an auxiliary excitation space. By this we obtain an increase of the emission radiation intensity, so the detection limit is correspondingly lowered.

7.6.1.2 *Luminescence spectroscopy methods*

In the luminescence methods, the atomic species is excited by a flux of radiation, electrons, or ions. These methods include X-ray atomic fluorescence spectroscopy, molecular fluorescence spectroscopy, electron- or ion-beam microanalysis, Auger electron spectroscopy, and X-ray photoelectron spectroscopy. All these methods have many applications in trace analysis.

(a) *X-Ray atomic fluorescence spectroscopy* The sample is irradiated with X-rays; the excited atomic species will then give their specific X-ray fluorescence.

Two methods are used for analysing the fluorescence radiation: wavelength dispersive X-ray spectroscopy, and X-ray energy spectroscopy. The fluorescence radiation is analysed according to wavelength in the first, and the photon energy in the second.

Wavelength dispersive X-ray spectroscopy has developed considerably as a result of the remarkable progress in crystal growth technology. The X-ray fluorescence from the sample is diffracted by a suitable crystal. From the resulting spectrum of the fluorescence rays, the region of interest is measured by conversion into electrical pulses with a scintillation or proportional counter.

The X-ray fluorescence method was initially applied to determination of heavy elements in large amounts. The subsequent technical improvements have extended its range to light elements and towards smaller and smaller amounts. At present, the method has many applications for trace analysis on various materials – whether electrical conductors or not. X-Ray fluorescence spectrometers have been developed which can determine more than 40 elements simultaneously.

The relation between X-ray fluorescence intensity and small amounts is linear. The expected (i.e. true) relation will be written as:

$$\hat{I} = \hat{I}_0 + b\hat{c} \qquad (7.92)$$

where $\hat{I}_0$ is the expected value of the background fluorescence intensity.

For a flux of fluorescence of high intensity, the fluctuation of the number of photons per unit time is normally distributed (or, at least, to a very good approximation). The theoretical standard deviation will thus have the expected value:

$$\sigma = \sqrt{\hat{I}} \qquad (7.93)$$

Generally, in X-ray fluorescence spectroscopy, Kaiser's definition of the detection limit is used (see p. 256). Thus, the equations for the analytical signal corresponding to the detection limit, and for the detection limit itself, are

$$\hat{I}_d = \hat{I}_0 + 3\,\sigma_0 \qquad (7.94)$$

and

$$\bar{c}_d = 3\sqrt{\hat{I}_0}/b \qquad (7.95)$$

where the standard deviation for the background fluctuations is:

$$\sigma_0 = \sqrt{\hat{I}_0} \qquad (7.96)$$

This standard deviation varies, as a function of amount. Hence, to evaluate the detection limit in the sense of the minimum amount detectable with certainty we must apply the procedure of Section 7.5.5. The calibration function (7.92), the background signal mean $\hat{I}_0$, the standard deviation for background signals σ_0, and the standard deviation for signal fluctuations corresponding to the detection limit $\sigma_{\bar{c}_d}$ cannot be obtained theoretically, but must be evaluated from suitable experimental measurements.

A statistically based study of X-ray fluorescence spectroscopy applied to detection problems has been given by Pantony and Hurley [100]. Among other things, these authors have evaluated the detection limit for 12 elements (Na, Mg, Al, Cl, Cr, Co, Cu, Mo, Ag, I, W, and Pb) having atomic number between 12 and 82. These elements were embedded in two matrices: a lanthanum trioxide matrix, which has a high absorption coefficient, and a cellulose acetate matrix, with a small absorption coefficient. The values of the detection limit, evaluated by these authors with the equation $\bar{c}_d = 0.89\ s_{Nb}/m$, were between 10^2 and 1 ppm for the lanthanum trioxide matrix and between 10 and 10^2 ppm for the cellulose acetate matrix. The meanings of the symbols used by these authors are: s_{Nb} = the standard deviation for background signal fluctuations, m = the slope of the calibration function.

In some cases, the values of the detection limit for fluorescence analysis are so high that the analysis cannot be performed directly on the samples in their original form. In such cases it is necessary to begin with an enrichment of the trace elements by a suitable procedure, for example by using ion-exchangers [101].

(b) *Fluorescence flame spectroscopy.* As in all flame methods the role of the flame is to release free atoms and ions from the sample. Both can produce fluorescence. Because the fluorescence intensity is a function of the number of fluorescent centres, the flame used must release the highest possible number of free atoms and, at the same time, preferably have a low radiance itself. The hydrogen-oxygen, nitrous oxide-acetylene, or acetylene-air flames are generally used. Excitation was first done by use of electrodeless discharge tubes. Later, hollow-cathode lamps were adapted for atomic fluorescence. In order to increase the radiant energy of these lamps (since the fluorescence intensity is a function of the intensity of the exciting radiation), they are operated in pulses. Lamps have been designed for the determination of one element at a time, and for the simultaneous determination of several elements. Lasers are also good excitation sources.

The relation between amount and signal obtained by atomic flame fluorescence is generally linear, and the distribution of signal fluctuations is either normal, or close to normal. Hence, the detection limit can be estimated by the procedure given in Section 7.5. Many detection limit values for flame atomic fluorescence spectroscopy can be found in the literature [102–106].

To give an overall view of the detection limits in flame atomic fluorescence spectroscopy, and to compare these with values for other flame spectroscopic methods we reproduce the summary given by Fraser and Winefordner [106], in Table 7.5, showing the detection limits for laser-excited atomic fluorescence (AF-laser), atomic fluorescence excited with narrow line sources (AFL), atomic fluorescence excited with continuum sources (AFC), flame atomic emission (AE) and flame atomic absorption with narrow line sources (AAL).

Table 7.5 Comparison of detection limits by atomic flame spectrometry [106] (Reproduced from *Analytical Chemistry* by permission. Copyright the American Chemical Society)

Element	Detection limit (μg/ml)				
	AF-laser	AFL	AFC	AE	AAL
Al	0.03	0.1	–	0.005	0.04
Ca	0.01	0.02	0.1	0.0001	0.0005
Cr	0.03	0.05	10.0	0.005	0.005
Fe	0.3	0.008	1.0	0.005	0.005
Ga	0.3	0.01	5.0	0.01	0.07
In	0.01	0.1	2.0	0.005	0.05
Mn	0.3	0.006	–	0.005	0.002
Sr	0.03	0.03	–	0.0002	0.004
Ti	0.1	–	–	0.02	0.1

AF-laser = laser-excited atomic fluorescence
AFL = atomic fluorescence excited with narrow line source
AFC = atomic fluorescence excited with continuum source
AE = atomic flame emission
AAL = atomic absorption with narrow line source

In Table 7.6 we reproduce the values given by Omenetto *et al.* [105] for the detection limit for rare earth elements for atomic absorption (AA), atomic emission (AE), atomic fluorescence (AF), and ionic fluorescence (IF), with laser excitation.

The problem of lowering the detection limit in flame fluorescence spectroscopy – a problem of tremendous practical importance – has been investigated [107-109]. Good results have been obtained by using modulated radiation sources, and signal-measurement techniques which increase the signal-to-noise ratio [107-111].

(c) *Electron beam microprobe analysis* This technique has many applications in the analysis and microanalysis of metals, alloys, geological or biological materials, etc. When the electron beam hits the sample, the atomic species are excited. By resolution and measurement of the emission radiation from the excited atoms we derive information about the composition of the material investigated.

Table 7.6 Comparison of detection limits by atomic flame spectrometry [105] (Reproduced from *Analytical Chemistry* by permission. Copright the American Chemical Society)

Element	Detection limit (μg/ml)			
	AA	AE	AF	IF
Ce	>500	10	>500	0.5
Dy	0.2	0.05	0.6	0.3
Er	0.1	0.04	0.5	2.5
Eu	0.04	0.0005	0.02	0.2
Gd	4	2	5	0.8
Ho	0.1	0.02	0.15	>500
Ln	3	1	3	–
Nd	2	0.2	2	4
Pr	4	0.07	10	1
Sm	0.6	0.2	0.6	0.15
Tb	2	0.4	1.5	0.5
Tm	0.04	0.02	0.1	5.0
Yb	0.02	0.002	0.01	0.03

AA = atomic absorption
AE = atomic emission
AF = atomic fluorescence
IF = ionic fluorescence (laser excitation)

The local analysis of a material is possible because we can obtain very narrow beams of electrons. On the other hand, the penetration depth is much smaller for electrons than for X-rays, so the depth to which the material is investigated is much smaller with electron beams (around 1 μm for an electron beam accelerated by a potential of 10 keV). This analytical method can be used for concentrations down to 50 ppm, and analyses below this limit can only be done under special conditions [112].

Several definitions of the detection limit in electron microprobe analysis have been advanced [112–116].

According to Heidel [112], the detection limit values, calculated with the 3σ criterion, are in the range 170–340 ppm for Zn, Cu, Ni, Co, Mn, Cr, V, and Ti in a complex silicate matrix containing Al, Fe, Mg, Ca, Na, K. In a similar study [116], the same author has found that, for rare earth elements in a calcium aluminium silicate matrix, the detection limit values are 650–3130 ppm.

(d) *Auger electron spectroscopy* As an analytical technique, this method has been developed in the last few years, and it has many possibilities for surface analysis [117, 118]. This method is non-destructive and is sensitive for light elements.

There is much resemblance between Auger electron spectroscopy and electron microprobe analysis, both techniques using an electron beam projected onto a surface. The excited atomic species return to the ground state by an Auger effect in Auger electron spectroscopy, and by X-ray emission in electron microprobe analysis. The essential difference between these techniques is the depth to which the investigated material is probed. As mentioned above, X-ray absorption is relatively small in comparison with that of electrons, so the depth from which the X-rays can originate is equal to the penetration depth of the incident electron beam, while the depth from which the Auger electrons can arise is much smaller, about 10–20 Å.

Auger electron spectroscopy and electron beam microprobe analysis are complementary analytical techniques; the first is more sensitive for light elements, and the second for heavy elements.

(e) *X-Ray photoelectron spectroscopy.* This method is applied in both qualitative and quantitative chemical analyses. An X-ray beam is projected onto the material investigated. The electrons with a binding energy smaller than the energy of the X-ray photons are ejected from the sample, and their kinetic energy is a function of their binding energy and the X-ray photon energy. The energy spectrum of the ejected photoelectrons has several characteristic maxima superimposed on a continuous background. This spectrum can give information concerning the qualitative composition of the surface layer of the sample.

Applications of this technique in chemical analysis are relatively recent. The first data on the intensities of the photoelectron lines were given by Wagner [119], and then by Berthou and Jorgensen [120], and Giauque *et al.* [121]. The theoretical basis of the quantitative analysis is due to Ebel and Ebel [122], who established the equation relating the photoelectron line intensity, the incident X-ray intensity, the instrumental parameters, the parameters of the matrix containing the atomic species, and the concentration of the atomic species.

The Ebel and Ebel equation, which is fundamental for quantitative analysis, has been developed and verified by Janghorban *et al.* [123], who also characterized the sensitivity by means of the 'sensitivity index', which is the intensity of the photoelectron line obtained from a 1-μg/cm^2 level of the atomic species. Other theoretical considerations have been clarified by Bremser [124].

X-Ray photoelectron spectroscopy is mainly a surface analysis technique, as the emitted photoelectrons usually come from an outer layer about 20 Å deep. Its detection limit is in the range 10–100 ppm, and this limitation makes it inadequate for many of the trace analysis problems of interest in geochemistry, biochemistry, and other fields, where in many cases amounts of the order of one part per milliard (ppM) have to be determined. In order to solve such problems with X-ray photoelectron spectroscopy, suitable techniques must be used in order to concentrate the trace components, and to fix them on suitable surfaces in such a way that the thickness of the layer is of the order of the penetration depth of the photoelectrons. X-Ray photoelectron spectroscopy is then applied to these layers. For the fixation of trace elements from water solutions, Hercules *et al.* [125] used glass fibre disks with chelating properties. In this way they obtained detection limits of the order of 10 ppM for elements such as Pb, Ca, Tl and Hg.

7.6.1.3 *Absorption spectroscopy methods*

These methods are based on the absorption of radiation by the sample. Depending on the nature of the absorbant species, we distinguish between atomic absorption spectroscopy and ionic or molecular absorption spectroscopy.

If $(I_0)_\lambda$ denotes the intensity of radiation of wavelength λ transmitted by a blank sample and I_λ is the intensity of radiation transmitted by the sample being analysed, the equation relating them is

$$I_\lambda = (I_0)_\lambda \, 10^{-\epsilon_\lambda \hat{c} x} \tag{7.97}$$

where $\hat{c}$ is the concentration of the absorbant species, and x the thickness of the absorbing layer.

The transmittance T_λ is defined as the ratio

$$T_\lambda = \frac{I_\lambda}{(I_0)_\lambda} \times 100 \tag{7.98}$$

The absorbance is defined by

$$A_\lambda = \log \frac{(I_0)_\lambda}{I_\lambda} \tag{7.99}$$

From (7.97) and (7.99) there is a linear relation between the absorbance and the concentration of the absorbant species:

$$A_\lambda = \epsilon_\lambda \hat{c} x \tag{7.100}$$

where ϵ_λ is the molar absorption coefficient (molar absorptivity).

This linear relation generally holds for low concentrations, and is the basis for quantitative analysis.

The absorbance corresponding to the amount of component being detected or determined is measured as the difference between the absorbances measured for the sample and for a 'blank' of similar composition except that it does not contain the component of interest. That is, the blank sample is used as a background-signal correction. As both samples are processed at the same time, there may be a certain degree of correlation between the fluctuations of the measurements, in which case the variance of the difference of the two results (i.e. the analytical signal variance) would be smaller than the sum of the variances. Obviously, if the component is absent from the material investigated, the expected value of the analytical signal (which in this case is the difference in the measurements for two blank samples) will be equal to zero, $(A_\lambda)_o = 0$. Hence, in the absence of the component to be detected, the standard deviation for the signal fluctuation will be

$$(\sigma_{A_\lambda})_o \leqslant \sqrt{2}\,\sigma_b \tag{7.101}$$

where σ_b is the standard deviation for fluctuations of the measurements on a single blank sample, the equality sign applying if there is no correlation between the results for the two parallel samples.

If $(\sigma_{A_\lambda})_o$ and $(\sigma_{A_\lambda})_d$, respectively, are the standard deviations for the analytical signal in the absence of the component to be detected and in its presence in an amount equal to the detection limit, the decision threshold will be given by:

$$y_k = z_k(\sigma_{A_\lambda})_o \tag{7.102}$$

and the analytical signal corresponding to the detection limit will be:

$$\hat{y}_d = z_k(\sigma_{A_\lambda})_o - z_d(\sigma_{A_\lambda})_d \tag{7.103}$$

The quantities required for evaluating the detection limit [i.e $(\sigma_{A_\lambda})_o$, $(\sigma_{A_\lambda})_d$ and the calibration function], can be obtained only experimentally, and not theoretically.

(a) *Molecular absorption spectroscopy* Molecular absorption spectroscopy has a long tradition in trace analysis. It includes colorimetry, and absorption spectroscopy in the infrared, ultraviolet and visible regions.

Some authors define the minimum detectable sample concentration (which has the significance of a detection limit) as the concentration of a sample which gives an absorbance of 0.002.

Winefordner *et al.* [126–129] have shown that the minimum detectable concentration is limited by the signal-to-noise ratio of the analytical measurement. If this is taken into account, the limiting detectable sample concentration is defined [22] as the concentration which gives a signal equal to $3\sqrt{2}/\sqrt{n}$ times the standard deviation of the signal, the figure $\sqrt{2}$ arising because the noise of the measurement (the difference between the values for the test sample and the blank) is equal to the sum of the noise for the two measurements; n is the number of experiments.

By using the t distribution, the same authors [22] have defined the limiting detectable sample concentration as the concentration at which the signal is equal to $t\sqrt{2}/\sqrt{n}$ times the noise, where t is defined by the chosen significance level and the number of degrees of freedom, $2n - 2$.

Cetorelli and Winefordner [23] have established a theoretical equation connecting the signal-to-noise ratio and the instrumental parameters, valid for molecular absorption spectrometry in a condensed phase. This equation is then used in the derivation of the equation for the minimum detectable theoretical concentration of the sample, as a function of the instrumental parameters (this concentration corresponds to a signal:noise ratio of $3\sqrt{2}/\sqrt{n}$). From this equation the optimal experimental conditions for trace analysis can be obtained.

The minimum detectable concentration in molecular absorption spectrophotometry is similarly treated in other works, e.g. [21, 27, 28].

Generally, in molecular absorption spectrometry the noise is complex in origin, that is, it has several sources. To investigate these, Ingle and Crouch [130] theoretically derived an equation which shows the way in which the relative standard deviation of the absorbance (i.e. the precision measure) depends on the transmittance and the instrument parameters. Three sources of noise were taken into account: those related to the instrument, the reading method, and the technique of current measurement. The problem was also considered by Rothman *et al.* [131], who derived an equation which expresses the relative standard deviation as a function of the transmittance and the experimental variables. These authors took into account the amplifier noise, the read-out quantization noise, the dark-current shot-noise, the source-flicker noise, and uncertainty in the cell position.

By such studies the individual contribution of various noise sources to the total noise can be established, and hence practical methods can be found for reducing the measurement noise, and consequently lowering the detection limit and improving the analytical performance of the system in general.

(b) *Atomic absorption spectroscopy.* In this method the sample is brought into the gaseous state, containing free atoms (this is called atomization). Monochromatic radiation, at a resonance wavelength of the atomic species of interest, is passed through the gas. The degree to which this radiation is absorbed is a measure of the concentration of the atomic species in the sample.

We can classify the methods according to the atomization technique, into flame, cold vapour, and electrothermal methods. The most commonly used flames are the nitrous oxide-acetylene and air-acetylene flames [74-84]. The cold vapour methods are restricted to elements such as mercury that have an appreciable vapour pressure at ordinary temperatures.

In the electrothermal methods various types of atomizer are used [132-134], including carbon rods, tubes or furnaces, and metal strips or cups. By using these techniques it has been possible, in many cases, to decrease the detection limit to about a thousandth of that for flame atomization.

Table 7.7 Detection limits for atomic flame emission and absorption spectrometry

	Range of the detection limit (ppm)							
	10^{-5}–10^{-4}	10^{-4}–10^{-3}	10^{-3}–10^{-2}	10^{-2}–10^{-1}	10^{-1}–10^{0}	10^{0}–10^{1}	10^{1}–10^{2}	10^{2}–10^{3}
Atomic emission [90]	K; Li	Ca; Eu; Na; Sr	Ag; Ba; Cr; In; Mn; Pb; Yb	Al; B; Co; Cu; Dy; Er; Fe; Ga; Gd; La; Mg; Ni; Pd; Pr; Rh; Tb; Tl; Tm; Y	Cd; Cs; Ge; Ho; Ir; Mo; Nd; Pb; Ru; Sb; Sc; Sm; Sn; Ti; V; W	Au; Be; Lu; Nb; Os; Pt; Si; Ta; Te; U; Zr	As; Ce; Hf; Hg; Th; Zn	Bi; Se
Atomic emission [77–81]	Li	Ca; Eu; Sr; In; Mg; Mn	Al; Ba; Cr; Yb	Ag; Co; Cu; Dy; Er; Fe; Ga; Ho; Ni; Pd; Rh; Ru; Sc; Tl; Tm; V; Y	Au; Ge; Mo; Nd; Pb; Re; Sm; Sn; Te; Ti; W	Cd; Gd; La; Lu; Nd; Pr; Pt; Ta; Zr	Ce	
Atomic absorption [90]		Mn; Na	Ag; Be; Ca; Cd; Co; Cr; Cu; Fe; Li; Mg; Ni; Rb; Sr; Zn	As; Au; Ba; Bi; Cs; En; Ga; In; Mo; Pb; Pd; Pt; Rh; Ru; Sb; Sn; Te; Tl; Tm; V; Yb	Al; Dy; Eu; Ge; Hg; Ho; Os; Re; Sc; Se; Sm; Ti; Y	B; Gd; Ir; La; Lu; Nb; Nd; Pr; Ta; Tb; W; Zr	Hf; U	
Atomic absorption [41]		Ca; Mg; Na	Cr; Cu; Mn; Ag; Sr; Zn	Be; Cs; Co; Au; Fe; Pb; Li; Mo; Ni; K; Si; Ti; Y	Al; Ba; Eu; Ga; Ge; In; Pd; Pt; Rh; Sc; Te; Tl	Sb; As; Bi; B; Nd; Re; Ta; Sn; W; Se	Ce; La; Nb; U; Zr	

As a primary radiation source hollow-cathode lamps are generally used; they may be of the single-element or the multi-element type.

There are many technical solutions for improving the signal-to-noise ratio [110, 111, 135–144], and also the performance parameters of the analytical system, namely the stability, detection limit, and precision. Among these we shall mention first modulation of the primary radiation source [136–138, 142], which improves the detection limit, stability, and precision. Other techniques used for improving the signal-to-noise ratio are tuned amplification [143], the phase-sensitive detector (lock-in amplifier) [138] and signal-averaging [111].

As the detection capacity of a method is limited by the noise associated with the measurement process, identifying the noise sources is a very important problem in atomic absorption spectrometry. Once the noise sources are known, the experimental variables can be optimized with a view to enhancing the signal-to-noise ratio. The noise sources have been investigated several times [145–150].

Atomic absorption methods are frequently used in trace analysis. Detection limit values for this method are reported in several works, but once more there is lack of unanimity in defining and evaluating the detection limit. It is generally defined either as the concentration which gives a signal equal to a certain absorbance value, or (and most usually) as the concentration which gives a signal that is a multiple of the standard deviation for the signal. The value of this multiple is usually arbitrarily chosen (2, 3, 4, 5 or 6 times the standard deviation).

In the analytical literature there are many studies on the detection limit for flame atomic absorption, compared with detection limits for flame atomic emission [41, 84, 90, 151]. Table 7.7 gives an idea of the comparison, and by and large there is not a significant difference between the two methods as far as detection limit is concerned.

7.6.1.4 *Other spectral methods*

Besides the spectral methods mentioned above, the following methods are also applied in trace analysis: mass spectroscopy, nuclear magnetic resonance (NMR) spectroscopy, and electron spin resonance (ESR) spectroscopy.

(a) *Mass spectroscopy.* This method is used for investigating the amounts of atomic species in the sample as a function of the mass/charge ratio.

The mass spectrometer consists essentially of a source of ions (in which the sample is placed), an electrostatic system for ion acceleration and focusing, a magnetostatic system which resolves the beam of ions in terms of mass/charge ratio, and the electrical system for measuring the mass spectrum.

The ionization can be achieved by the following methods: electron ionization, field ionization, photo-ionization, or chemical ionization [152, 153].

The mass spectrum is obtained by ion detection on a photographic plate (mass spectrography) [154], or with electronic detectors of the ion-multiplier type (mass spectrometry). Although electronic detection has some obvious advantages, the photographic plate is still used. The main advantage of the photographic plate is the capacity to simultaneously integrate the ionic current over a wide range of the mass spectrum; in this way, the plate stores a large amount of information which can be processed subsequently. However, the photographic plate has disadvantages related to quantitative analysis of the recorded spectrum [155]. Electronic detection has the advantage that it allows a high degree of automation in collecting and processing the information. Thus, if the mass spectrometer is connected to a computer, the measurement results can be immediately processed, and the information can be stored.

The ionization is frequently obtained by use of an electric spark produced in vacuum between two electrodes made from the material to be analysed (spark source mass spectrometry) [156]. The mechanism of spark ionization is rather complex, being based on the physico-chemical processes which take place in a spark.

Whereas in all the other methods of analysis the detection limit can vary considerably from one atomic species to another, in spark-source mass spectrometry the detection limit is almost constant for all atomic species, and is generally of the order of several parts per milliard (ppM).

The detection limit for mass spectrometry is evaluated in the same way as for emission spectrometry. The definition and evaluation of the detection limit (on the basis of the Kaiser theory) has been given, for spark source mass spectroscopy, by Ehrlich and Mai [32]. They evaluated the detection limit for 66 atomic species, obtaining values between 1 and 100 ppM.

A relatively recent technique is secondary-ion mass spectroscopy [157, 158]. This is based on the emission and measurement of secondary ions ejected from the sample by a bombardment with an ionic beam of high energy (several keV). The secondary ions are analysed, detected, and measured with a mass spectrometer. This method has many applications to surface and thin layer analysis. With this technique, by exploring the composition on the surface and at various depths, it is possible to find the spatial distribution of the elements in the sample. The analysis of the material under the surface is achieved by the controlled decomposition (cratering) of the sample under the action of the incident ionic beam [157].

In analysis of organic substances, valuable results have been obtained by coupling the gas chromatograph to the mass spectrometer. This technique is also applied in trace analysis, and the detection sensitivities have reached the order of a picogram.

(b) *Nuclear magnetic resonance spectroscopy* (NMR). This analytical technique derives chemical information from the NMR spectrum of the sample.

The material is placed in a magnetic field, a radio wave is passed through it, and the induced NMR spectrum is measured. The spectrum is obtained either by varying the magnetic field at a constant frequency of the radio wave, or by varying the radio frequency and maintaining a constant magnetic field. The interpretation of the NMR spectrum gives very valuable information concerning the chemical composition and molecular geometry of the sample. For the analyst, the most important nuclei in this technique are ^{1}H, ^{11}F, ^{13}C, ^{15}N and ^{31}P. The identification of the structure of an organic compound is based on the appearance and evaluation of the 'chemical shift' in the spectrum of the investigated sample.

NMR spectroscopy has a high detection capacity. The signal-to-noise ratio, and hence the detection capacity, can be improved by various techniques. In one of these a computer is connected to the spectrometer and used to accumulate the signals from repetition of the spectrum measurement (see Chapter 9). Thus, the total signal intensity will be proportional to the number of runs n, but the random noise will increase by only a factor of $\sqrt{n}$. Thus, the signal-to-noise ratio will increase by the factor $\sqrt{n}$, so it will be progressively improved [159].

Another procedure for improving the signal-to-noise ratio is the use of a pulsed wave and Fourier transform of the spectrum [160, 161], instead of the usual NMR method which uses a continuous wave.

(c) *Electron spin resonance* (ESR). The ESR method has fewer applications in analytical chemistry than the NMR method. In principle, ESR spectrometers are similar to the NMR spectrometers; the constructional details and the working parameters are, however, different.

ESR signal detection shows the paramagnetic centres pertaining to free radicals or transition metal ions. No other method can compete with ESR for speed and certainty of free radical detection. By this method free radicals can be detected even at a concentration smaller than 10^{11} spins/g.

For study of the hyperfine structure and hence an improved detection of free radicals, a unified method which combines the ESR and NMR methods has been developed: this is called the electron nuclear double resonance method [162].

7.6.2 Radiometric and radiochemical methods

These methods have been developed in the last 30 years, and have greatly contributed to the progress of trace analysis.

Radiometric methods of analysis are those which require only a physical treatment of the samples and radioactivity measurements. In these methods we measure either the natural or induced radioactivity of the sample material.

Radiochemical methods are those in which the samples for analysis are treated chemically before measurement of the radioactivity.

7.6.2.1 *Radiometric methods*

The elements which have naturally radioactive isotopes are those with atomic numbers between 82 and 93, so only a few elements can be determined radiometrically on the basis of natural radioactivity. However, radiometric analysis can also be done on samples having induced radioactivity, and for this reason is useful in many problems of trace analysis.

The sample can be activated by neutrons, charged particles (protons, deuterons, tritons, helium nuclei), and photons.

Neutron activation analysis has many applications in trace analysis, and is recognized as the analytical technique with the highest detection capacity, i.e. the lowest detection limit.

The remarkable progress in radioactivation analysis techniques generally, and in neutron activation analysis in particular, is a consequence of the development and application of the Ge(Li) type of radiation detector. By the development of the methods of *instrumental neutron activation analysis* (INAA) it has become possible to determine simultaneously a large number of trace elements (up to 40) in various materials such as aerosols, biological materials, water, minerals, etc. [163–175]. By the INAA method concentrations of the order of 1 ppM or even less can be detected and determined.

The high value of the signal-to-noise ratio in INAA is due not only to the high refinement of the instrumental technique, but also to the new sample treatment procedures and new methods of measurement. In the classical procedure, a standard sample (used as reference) and the 'unknown' sample are simultaneously irradiated under the same conditions. This procedure has been replaced by new procedures, where either the standard is incorporated into the sample, or a component of the matrix is used as standard [176]. By these means, the errors due to variation in sample geometry, heterogeneity of activation flux, and non-identical treatment of the standard and sample, are almost completely removed.

Radioactivation with a beam of charged particles also has many applications in trace analysis [177–181], and by this method concentrations smaller than 1 ppm can be detected and determined. This technique has the advantage that the activating particles and their energy can be chosen in such a way as to control the nuclear reaction in order to obtain certain analysis products which will simplify the analysis.

Instrumental photon activation analysis (IPAA) has a trace-analysis performance similar to that of INAA [182]. One method of photon activation consists in irradiating the sample with high-energy electrons (>15 MeV), which will generate γ-photons with an energy of about 7 MeV. In turn, these will lead to nuclear reactions of the (γ, n) or (γ, p) type, which give the radioactive products that are then measured for detection or determination.

The problem of defining and evaluating the detection limit for radiometric methods has been dealt with in various papers [18, 39, 40, 42]. We shall use Currie's reasoning [39] regarding the detection limit.

The phenomenon of radioactive disintegration can be described by the Poisson distribution. When the number of particles is large enough, it can be accepted that the activity values are normally distributed. Hence, we accept the hypothesis that the signal distribution is normal.

In the absence of the component of interest, the background activity A_b

is thus normally distributed, with mean $\hat{A}_b$ and standard deviation $\sigma_b = \sqrt{\hat{A}_b}$, that is:

$$A_b \Rightarrow N(\hat{A}_b; \sigma_b) \tag{7.104}$$

When the component is present, the sample activity A_{b+c} has a normal distribution with mean $\hat{A}_{b+c}$ and standard deviation $\sigma_{b+c} = \sqrt{\hat{A}_{b+c}}$, that is:

$$A_{b+c} \Rightarrow N(\hat{A}_{b+c}; \sigma_{b+c}) \tag{7.105}$$

When both measurements are made simultaneously, the background-corrected analytical signal will be given by the difference, that is:

$$A_c = A_{b+c} - A_b \tag{7.106}$$

For generality, we shall use (7.106) with a mean value $\bar{A}_b$ obtained from n measurements; that is:

$$A_c = A_{b+c} - \bar{A}_b \tag{7.107}$$

Because of the Poisson distribution, the variance is equal to the number of events, so the variance for A_c is

$$\sigma_c^2 = \hat{A}_b + \hat{A}_c + \frac{\hat{A}_b}{\sqrt{n}} \tag{7.108}$$

In the case of a sample which does not contain the component of interest, i.e. $\hat{A}_c = 0$, the variance resulting from (7.108) will be:

$$\sigma_0^2 = \hat{A}_b + \frac{\hat{A}_b}{\sqrt{n}} \tag{7.109}$$

Hence, the decision level $\hat{A}_k$ is:

$$\hat{A}_k = z_k \sqrt{\hat{A}_b + \frac{\hat{A}_b}{\sqrt{n}}} = z_k \sigma_0 \tag{7.110}$$

From (7.43), the analytical signal corresponding to the detection limit is:

$$\hat{A}_d = \hat{A}_k + z_d \sqrt{\hat{A}_d + \hat{A}_b + \frac{\hat{A}_b}{\sqrt{n}}} = \hat{A}_k + z_d \sqrt{\hat{A}_d + \sigma_0^2} \tag{7.111}$$

From (7.110) and (7.111) the analytical signal corresponding to the detection limit becomes:

$$\hat{A}_{\mathrm{d}} = \hat{A}_{\mathrm{k}} + \frac{z_{\mathrm{d}}^2}{2}\left[1 + \sqrt{1 + \frac{4\hat{A}_{\mathrm{k}}}{z_{\mathrm{d}}^2} + \frac{4\hat{A}_{\mathrm{k}}^2}{z_{\mathrm{d}}^2 z_{\mathrm{k}}^2}}\right] \tag{7.112}$$

This equation is similar to the equation given by Currie [39].

If the true analytical calculation function is written as

$$\hat{c} = f(\hat{A}) \tag{7.113}$$

the detection limit will be:

$$\hat{c}_{\mathrm{d}} = f(\hat{A}_{\mathrm{d}}) \tag{7.114}$$

To evaluate the detection limit we must evaluate from experimental data both the analytical signal (7.112), which corresponds to the detection limit, and also the analytical calculation function (7.113) for each situation.

7.6.2.2 *Radiochemical methods*

This class of methods comprises tracer, isotopic dilution, substoichiometric isotopic dilution, and radioactivation methods.

In the subclass of tracer methods we shall consider the method of precipitation with labelled reagents. In this case the amount of the component to be determined c_{x}, is obtained from:

$$c_{\mathrm{x}} = \frac{A_{\mathrm{x}}}{\hat{A}_{\mathrm{r}}} E_{\mathrm{r}} \tag{7.115}$$

where A_{x} is the specific radioactivity of the separated material component, $\hat{A}_{\mathrm{r}}$ is the specific radioactivity of the radioactive reagent, and E_{r} is the mass ratio of the component and tracer.

The detection limit is evaluated in the same way as for radioactivity methods, the analytical signal corresponding to the detection limit (7.112) being interpolated in the analytical calculation function (7.115).

The tracer method can be used for detecting concentrations even under 1 ppM.

The isotopic dilution method consists, in principle, in adding a radioactive isotope (with known specific activity) of the component of interest to the sample. The sample is made homogeneous, and the component separated. The specific activity of the separated component is then measured. This method has the great advantage that separation need not be quantitative; all that is needed is a pure product.

We will write the specific activity of the added isotope as:

$$a_0 = \frac{A_0}{m_0} \quad (7.116)$$

After the homogeneous dilution the specific activity becomes:

$$a_1 = \frac{A_0}{m_0 + m_x} \quad (7.117)$$

where m represents mass.

From these two equations we obtain:

$$m_x = m_0 \left(\frac{a_0}{a_1} - 1 \right) \quad (7.118)$$

where a_1 is the specific activity of the quantity m_1 separated, and having activity A_1:

$$a_1 = \frac{A_1}{m_1} \quad (7.119)$$

The detection capacity of this method is limited by the errors involved in the measurement of A_0, A_1, m_0 and m_1. Among these, the major limitation of the detection capacity comes from the errors in measurement of the mass, which originate in the weighing or titration operations.

The inconveniences of the isotopic dilution method are mostly removed in the variant called substoichiometric isotopic dilution. A fixed mass M of a radioisotope of the element sought, with known specific activity, is added to the sample containing an amount M_x of the component sought. A known quantity of reagent is added and a quantity m of the element is separated from the mixture; then the radioactivity of this quantity is measured (a'). The name of the method (the term substoichiometric) arises because the quantity of reagent added must be smaller than the amount needed to react quantitatively with the quantity of element added, M. Indeed it must be sufficiently small for it to be certain that it will be practically completely combined with the element sought, so that a constant amount of the latter will be separated. If a similar experiment is performed with a blank sample and the activity is a, then because the amount of substance counted is the same in both cases the activity can be treated as equivalent to the specific activity. Thus the amount M_x will be given by:

$$M_x = M \left(\frac{a}{a'} - 1 \right) \quad (7.120)$$

The substoichiometric method is more advantageous than the normal isotopic dilution method. If solvent extraction is used as the separation method, the system can easily be automated. The method is not only easier and faster than the normal method, but it also has a high detection capacity. It also has very good detection capacity, even better than that of mass spectrometry and γ-spectrometry [183].

The substoichiometric methods have been reviewed by Kudo and Suzuki [183]. They find that there are more than 40 elements which can be detected and determined by these methods when the elements are present as traces in various materials (metals, glass, or semiconductor materials). They also find that in trace analysis the substoichiometric methods are superior to the other methods (such as mass spectrometry, γ-spectrometry, or atomic absorption) in terms of accuracy, sensitivity, precision, and detection power. The major source of error is the statistical error of the activity measurements, and the ratio a/a' should not be too close to unity.

7.6.3 Electrochemical methods

The more important electrochemical methods for trace analysis are the polarographic methods, anodic stripping voltammetry, coulometric methods, and potentiometric methods with ion-sensitive membrane-electrodes.

The detection capacity data reported in the literature have all been evaluated either on arbitrary criteria, or on the basis of simple and subjective observations.

The statistical criteria given before can be used to define and evaluate the detection limit rigorously in the case of electrochemical methods too, and the procedure will be similar to those described in the previous sections (7.1-7.5). For this reason we shall limit ourselves to some comments on the applications of these methods for trace analysis purposes.

Polarographic methods are based on the variation in the current induced by varying the voltage applied to electrodes immersed in an electrolyte solution which contains the species to be determined. It is over 50 years since the method was discovered and for most of this time, the method most applied has been classical d.c. polarography. The quality of performance is limited, and in many cases does not satisfy trace analysis requirements. The resolution of the classical method is also rather poor and it may not be possible to determine a component in a complex mixture. In such cases various chemical operations must be used. Another drawback is that oxygen must generally be removed from the sample before the polarographic measurement. Further, the detection capacity of the classical polarographic technique is not adequate for trace analysis. Generally, in classical polarography the Faraday current due to an electroactive species at a concentration of about $10^{-5}M$ is of the same order of magnitude as the noise. Hence, the detection limit of classical polarography is about $10^{-5}M$.

These problems have been largely overcome by the new polarographic techniques, such as a.c. polarography [184, 185]. The specificity and sensitivity are improved and the removal of oxygen is no longer required.

To increase the sensitivity and remove the capacitive effect, Barker and Gardner [186] developed pulse polarography. There are three variants: with a normal (or integral) pulse; with differential pulse; and with derivative pulse; the last is more recent. In pulsed polarography, the charging current is decreased and at the same time the Faraday current increased so there is an increase in the signal-to-noise ratio, and an improvement in the detection capacity. In normal pulse polarography the charging current is only partly reduced, but the increase in the Faraday current extends the working range down to concentrations of about $10^{-6}M$. In differential pulse polarography [187-191], the charging current is almost completely removed. The signal-to-noise ratio is higher, and the working region extended to concentrations of about $10^{-8}M$. Differential pulse polarography is one of the most sensitive electrochemical methods, and is applied to trace determinations in various materials [192-200].

Anodic stripping voltammetry is related to polarography. The ion (or ions) of interest is (are) deposited on an electrode, under controlled conditions. The polarity is then reversed and the amplitude of the current obtained from the reoxidation of the deposited metal(s) is measured.

There are three main d.c. variants of this method: linear scan stripping voltammetry, pulsed voltammetric stripping [186, 187] and differential pulse voltammetry [201-203].

There are also variants of a.c. stripping voltammetry: sinusoidal wave, square wave, pulse, and oscillographic. These have good sensitivity and many applications.

Anodic stripping voltammetry has the best detection and determination capacity among the electrochemical methods. It can be used for trace determination on various materials at concentrations even smaller than 1 ppM [204-208].

Coulometric methods are based on the phenomenon of electrolysis. The species to be determined is subjected to an electrochemical reaction, and the species concentration is calculated from the amount of electricity required to complete this reaction. This technique also has good detection capacity, and it can be applied for concentrations of around $10^{-10}M$.

Potentiometry with ion-sensitive membrane-electrodes is an analytical technique which has been largely developed in the last few years, mainly as a result of design of new electrodes. This technique can detect ionic concentrations in solution of about $10^{-7}M$. The detection limit for this method is usually treated in various ways without statistical treatment [209-214]. A rigorous treatment of the detection limit has been given by Liteanu *et al.* [215], who also give the statistical procedure for defining and estimating the detection limit.

7.6.4 Chromatographic methods

It can be said that the present progress in analytical chemistry is inconceivable without the contribution of the chromatographic methods, which have good performance in terms of efficiency of separation, precision, accuracy, detection capacity and speed.

The general development of chromatographic methods is due to the following factors: the development of a rigorous theory of chromatographic processes and the availability of a large variety of chromatographic materials of good quality, high-quality systems of detection and measurement, and also high-precision devices for sampling and injection, as well as development of the analytical theory and the accumulation of a vast number of applications of these methods.

In trace analysis, it must be remembered that in the chromatographic separation process there will be a dilution of the components of the sample because they will be spread along the column so that their concentration profiles will correspond practically to a normal distribution. For closed column chromatography (gas and liquid chromatography), if the mass of a component is m, the maximum concentration of this component in its chromatographic peak is [216]:

$$c_{\max} = \frac{m\sqrt{N}}{V_R\sqrt{2\pi}} \tag{7.121}$$

where N is the number of theoretical plates, and V_R the retention volume for the component.

If the sample had an initial volume v_0 and a concentration c of the component,

$$m = cv_0 \tag{7.122}$$

then from (7.121), the minimum dilution on the column is given by the ratio

$$\frac{c_{\max}}{c} = \frac{v_0\sqrt{N}}{V_R\sqrt{2\pi}} \tag{7.123}$$

Obviously, the dilution effect will decrease the detection capacity of the chromatographic system. From (7.123) it would appear that, in order to obtain small dilution for a given column, a large volume of sample should be taken for analysis. However, the larger the volume of sample the smaller the separation power of the column, because the volume injected will occupy a comparatively large part of the column, and also the components will be spread over a longer length of column during the chromatography. Hence, a compromise must be reached between dilution and resolution power.

Thus, because of the dilution effect, the chromatographic column parameters affect the value of the detection limit and must be optimized [217-220].

Optimization of the column processes is not in itself enough, however; high-performance devices must also be used for measuring the components in the effluent. Such detectors are now available for gas and liquid chromatography. Thus in gas chromatography, the thermal conductivity detectors [221-223] have largely been superseded by flame ionization detectors for trace analysis. For example, the detection limit for hydrocarbons is about 1 ppm with thermal conductivity detectors [223], but 1 ppM or lower with flame ionization detectors.

Environmental samples, such as biological material, have a very large number of components, so the chromatograms obtained are very complex. In many cases, the removal of interferences and the correct detection of a component is quite a difficult problem, even when high-performance columns are used. For analysing such samples there are chromatographic detectors which respond selectively to a certain property of the eluted species (atomic emission spectrum, molecular spectrum, mass spectrum, electrochemical activity, and so on). The most common detectors of this kind, as used in gas chromatography, are as follows.

The flame-photometric detector, used especially for the detection of organic sulphur and phosphorus compounds: the detection limits for these elements (S and P) are smaller than 0.1 ng [224].

The microwave plasma detector: for the elements C, F, Cl, Br, I, P, S, in organic compounds, this detector has a detection limit between 2×10^{-7} and 7×10^{-14} g/sec [225].

The microcoulometric detector: the reported detection limits are around 1 μg [226].

The electron capture detector: as an example, the reported value for the detection limit of chlorine in organic compounds is about 1 pg [224].

There are also various types of detector used in liquid chromatography.

The ultraviolet absorption detector: the reported values for the detection limit are between 0.1 and 0.001 ppm [219].

The refractive index detector: the reported detection limits are between 1 and 10 ppm [219].

The flame-photometric detector: in liquid chromatography this detector has much higher detection limits than in gas chromatography. This is because much of the flame energy is used in vaporization of the eluent.

The coulometric detector: the reported detection limits are between 5×10^{-7} and 5×10^{-10} mole of component [227].

The qualitative performance of chromatographic analysis of complex materials has been significantly increased by using mass spectrometers or infrared absorption spectrometers as detectors. The mixed systems 'gas chromatography-mass spectrometry', and 'gas chromatography-infrared spectrometry' are frequently used for analysing materials of complex composition. These systems are also used in liquid chromatography. A further improvement has been obtained by coupling these systems to electronic computers, which has led to an even higher efficiency.

We must point out that it is not only the column and detector that determine the quality of a chromatographic system; another determining factor is the sampling and injection device [228].

Hence, in trace chromatographic analysis we must use columns optimized for the purpose, detectors with detection limit values as small as possible, and high-performance devices for sampling and injection.

We shall now make some comments on the definition and evaluation of the detection limit in chromatography. We shall assume the case of a material of complex composition, in which we are going to detect a component specified as M. From the point of view of detection, the investigated material can be found in two states: state I – when M is present; state II – when M is absent. Two hypothetical chromatograms corresponding to these two states are given in Fig. 7.17.

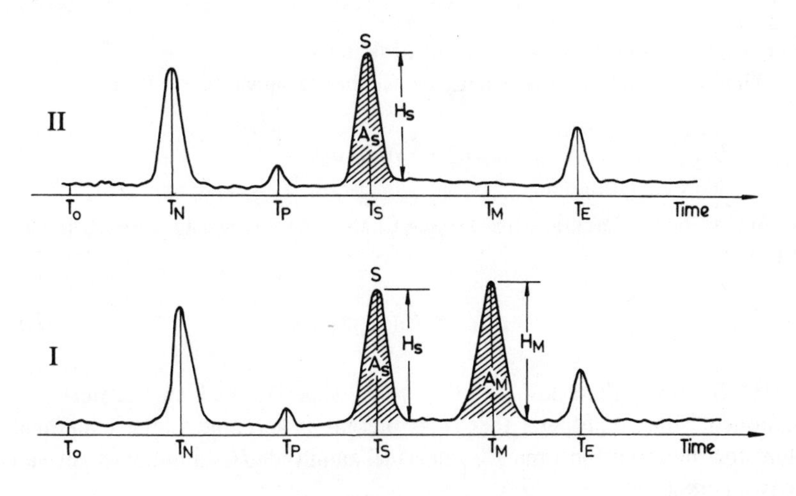

Fig. 7.17 Hypothetical chromatograms.

The detection of M requires evaluation of the retention time T_M, and of the chromatographic signal amplitude (the peak height H_M or peak area A_M). Under real conditions all these variables (T_M, H_M and A_M) have random character. A complete investigation of the sources of errors affecting the retention time T_M is given by Goedert and Guiochon [229].

In a general representation, the retention time T_M is the sum of an expected value $\hat{T}_M$ and a random term ΔT:

$$T_M = \hat{T}_M + \Delta\hat{T} \tag{7.124}$$

Accepting a normal distribution for the retention time we write:

$$T_M \Rightarrow N(\hat{T}_M; \sigma_T) \tag{7.125}$$

where σ_T is the standard deviation.

Hence, the confidence interval for the retention time fluctuations, taken at a confidence level P, will be:

$$\hat{T}_{\mathrm{M}} - |z_{\alpha/2}|\sigma_T < T_{\mathrm{M}} < \hat{T}_{\mathrm{M}} + |z_{\alpha/2}|\sigma_T \tag{7.126}$$

where $\alpha = 1 - P$.

On the other hand, if the fluctuations of the analytical signal amplitude are also normally distributed, at a certain level of component we have:

$$H_{\mathrm{M}} \Rightarrow N(\hat{H}_{\mathrm{M}}; \sigma_{\mathrm{H}}) \tag{7.127}$$

where σ_H is the standard deviation for peak height.

When the component is absent, we accept the signal distribution:

$$H_0 \Rightarrow N(\hat{H}_0; \sigma_0) \tag{7.128}$$

According to detection theory (see Chapter 6) the decision threshold can be written as:

$$H_{\mathrm{k}} = \hat{H}_0 + z_{\mathrm{k}}\,\sigma_0 \tag{7.129}$$

Hence, when there are no other peaks superimposed on the peak of the component, the hypothesis that M is present is accepted if the experimental values for the retention time T_{M} and signal amplitude H_{M} simultaneously satisfy the two conditions:

$$\hat{T}_{\mathrm{M}} - |z_{\alpha/2}|\sigma_T < T_{\mathrm{M}} < \hat{T}_{\mathrm{M}} + |z_{\alpha/2}|\sigma_T \tag{7.130}$$

and

$$H_{\mathrm{M}} > H_{\mathrm{k}} \tag{7.131}$$

This method of formulating the detection decision in chromatography is represented in Fig. 7.18.

According to the theory given at the beginning of this chapter, the analytical signal corresponding to the detection threshold can be written as:

$$\hat{H}_{\mathrm{d}} = \hat{H}_{\mathrm{k}} + z_{\mathrm{d}}\,\sigma_{\mathrm{d}} \tag{7.132}$$

where σ_{d} is the standard deviation for the signal fluctuations at an amount of component equal to the detection limit.

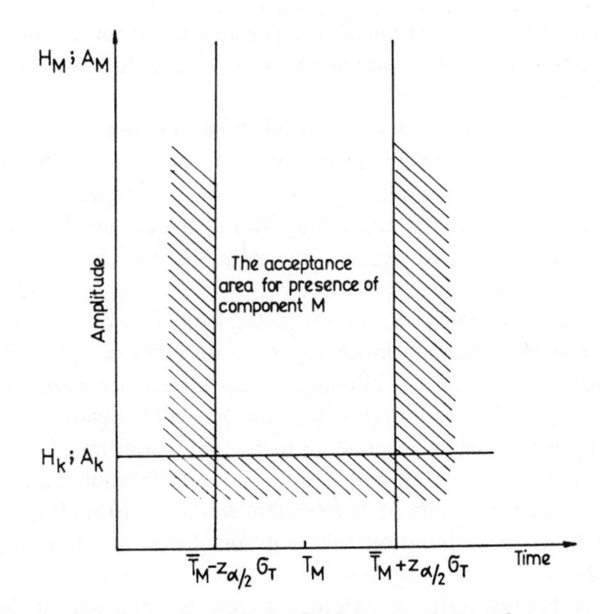

Fig. 7.18 Formulating the chromatographic detection decision for one component.

Let us take into account the expected value of the analytical calculation function, written as amount of component as a function of analytical signal, that is:

$$\hat{m} = f(\hat{H}) \tag{7.133}$$

The detection limit is obtained by using in this equation the analytical signal corresponding to the detection limit, given by (7.132):

$$\hat{m}_{\mathrm{d}} = f(\hat{H}_{\mathrm{d}}) \tag{7.134}$$

We shall accept that the analytical signal is normally distributed, and the mean is linearly dependent on the amount, while the standard deviation is constant, that is:

$$H \Rightarrow N(\hat{H} = \hat{H}_0 + b\hat{m}; \sigma_H) \tag{7.135}$$

Within the framework of this model, the analytical signal corresponding to the detection limit, and the detection limit, will be:

$$\hat{H}_{\mathrm{d}} = \hat{H}_0 + r_{\mathrm{d}}\,\sigma_H \tag{7.136}$$

and

$$\hat{m}_{\mathrm{d}} = r_{\mathrm{d}}\,\frac{\sigma_H}{b} \tag{7.137}$$

where $\hat{H}_0$ is the signal expectation value in the absence of the component, σ_H the standard deviation for signal fluctuations, r_d the signal-to-noise ratio and b the sensitivity of the chromatographic system.

The standard deviation for retention time σ_T and the retention time expectation value $\hat{T}_M$ cannot be accurately predicted theoretically. They can be estimated from a set of experimental data obtained by repeating the chromatographic process on a material containing M and measuring the retention time T_M. From this set of data we evaluate the mean $\bar{T}_M$ and the standard deviation s_T, which estimate the values $\hat{T}_M$ and σ_T.

The decision threshold (7.129) is estimated from a set of experimental signal values, obtained by repeated chromatography on blank samples. The mean $\bar{H}_0$ and the standard deviation s_0 obtained from this set are estimations for the values $\hat{H}_0$ and σ_0 which enter into the decision threshold equation.

The analytical signal corresponding to the detection limit (7.132), and the detection limit (7.137) are estimated on a set of calibration data obtained for samples with known amounts of M (standard samples). From this set we obtain the parameters of the calibration function, and hence the detection limit (see Section 7.5.2).

The signal-to-noise ratio in chromatography is improved by using internal standards. The retention time and signal amplitude for the component investigated are measured with reference to the retention time and signal amplitude of a standard component.

In the hypothetical chromatogram of Fig. 7.17, if the component S is taken as the standard, the relative retention of M with respect to S will be:

$$R_{M,S} = \frac{T_M - T_0}{T_S - T_0} \tag{7.138}$$

Instead of the amplitude of the chromatographic signal (H_M or A_M) a relative value (the ratio H_M/H_S, or A_M/A_S) can be used. In that case, the detection limit will be improved to the same degree as the signal-to-noise ratio. The procedure for detection decision and evaluation of the detection limit remains unchanged.

In order to avoid the confusion and ambiguities which may appear in connection with the detection limit in chromatography, we must distinguish between the following quantities: the detection limit of the chromatographic system as a whole; the detection limit of the detector; the sensitivity of the chromatographic system as a whole; the detector sensitivity.

The detection limit of the chromatographic system must be defined as the minimum amount (expressed in grams) of a component from the sample analysed which can be detected with certainty by a single chromatographic measurement.

The detection limit of the detector must be defined as the minimum concentration of component in the effluent (in g/cm^3) which, when going through the detector, allows the certain detection of the component.

The sensitivity of the chromatographic system must be defined as the variation in chromatographic system response induced by a change of one unit (1 g) of the amount of component present in the sample; hence, if the response is measured in volts, the sensitivity of the system will be given in V/g.

The detector sensitivity must be defined as the change of the detector response when the analysed species concentration changes in the effluent by one unit (1 g/cm^3); hence, if the detector response is in volts, this sensitivity will be in V. cm^3. g^{-1}.

To obtain the relation between the detection limit of the chromatographic system and the detection limit of the detector, we return to Eq. (7.121) which is the expression for the chromatographic dilution effect.

Assuming that all the noise of the chromatographic system is transferred to the detector, and accepting the following model for the dependence between the detector response and the concentration of the component (in the effluent)

$$H \Rightarrow N(\hat{H} = \hat{H}_0 + b_0\hat{c}; \sigma_H) \tag{7.139}$$

the detection limit will be:

$$\hat{c}_d = r_d \frac{\sigma_H}{b_0} \tag{7.140}$$

where $\hat{c}$ is the component concentration (in the effluent), and b_0 the detector sensitivity.

For a component to be detectable, its concentration in the sample must be such that the concentration corresponding to the peak maximum, c_{max}, is greater than or equal to the detection limit of the detector, that is:

$$c_{max} \geqslant \hat{c}_d \tag{7.141}$$

In the limit, from (7.121) it results that

$$\hat{c}_d = \frac{\hat{m}_d\sqrt{N}}{V_R\sqrt{2\pi}} \tag{7.142}$$

Hence, the detection limit of the chromatographic system is

$$\hat{m}_d = \frac{\hat{c}_d V_R\sqrt{2\pi}}{\sqrt{N}} \tag{7.143}$$

From (7.142) and (7.140) we obtain:

$$\hat{m}_{\mathrm{d}} = r_{\mathrm{d}} \cdot \frac{\sigma_H V_{\mathrm{R}} \sqrt{2\pi}}{b_0 \sqrt{N}} \tag{7.144}$$

This final equation gives the dependence between the detection limit of the chromatographic system, the efficiency of the column, the retention volume and the detector sensitivity. This equation can be used for optimizing the column parameters, in order to obtain the smallest possible value for the detection limit.

REFERENCES

[1] Kaiser, H., *Spectrochim. Acta,* **3**, 40 (1947).
[2] Kaiser, H., Massmann, H. and Hagenah, W. D., *Colloq. Spectrosc. Intern. 9th Lyons,* 1961, **3**, 479 (1962).
[3] Kaiser, H., *Optik,* **21**, 309 (1964).
[4] Kaiser, H., *Z. Anal. Chem.,* **209**, 1 (1965).
[5] Kaiser, H., *Z. Anal. Chem.,* **216**, 80 (1966).
[6] Kaiser, H., *Colloq. Spectrosc. Intern., 13th Ottawa,* 1967, 23 (1968).
[7] Kaiser, H., *Anal. Chem.,* **42**, (2), 24A (1970).
[8] Kaiser, H., *Pure Appl. Chem.,* **34**, 35 (1973).
[9] Wilson, A. L., *Analyst,* **86**, 72 (1961).
[10] Kaiser, H. and Specker, H., *Z. Anal. Chem.,* **149**, 46 (1956).
[11] Kaiser, H. and Menzies, A. C., *The Limit of Detection of a Complete Analytical Procedure,* Hilger, London (1968).
[12] Boumans, P. W. J. M. and Maessen, F. J. M. J., *Z. Anal. Chem.,* **220**, 241 (1966).
[13] Boumans, P. W. J. M. and Maessen, F. J. M. J., *Z. Anal. Chem.,* **225**, 98 (1967).
[14] Laqua, K., Hagenah, W. D. and Waechter, H., *Z. Anal. Chem.,* **225**, 142 (1967).
[15] Jolodovsky, P. D., *Spectrosc. Lett.,* **3**, 311 (1970).
[16] Skogerboe, R. K. and Grant, C. L., *Spectrosc. Lett.,* **3**, 215 (1970).
[17] Kuznetsova, A. I. and Raikhbaum, Ya. D., *Zavodsk. Lab.,* **33**, 1076 (1967).
[18] Schulze, W., *Z. Anal. Chem.,* **221**, 85 (1966).
[19] Schulze, W., *Z. Anal. Chem.,* **223**, 1 (1967).
[20] Winefordner, J. D., McCarthy, W. J. and St. John, P. A., *J. Chem. Educ.,* **44**, 80 (1967).
[21] Mandelstam, S. L. and Nedler, V. V., *Spectrochim. Acta,* **17**, 885 (1961).

[22] St. John, P. A., McCarthy, W. J., and Winefordner, J. D., *Anal. Chem.*, **39**, 1495 (1967).
[23] Cetorelli, J. J. and Winefordner, J. D., *Talanta*, **14**, 705 (1967).
[24] Barney, J. E., II, *Talanta*, **14**, 1363 (1967).
[25] Hobbs, D. J. and Smith, D. M., *Canad. Spectry.*, **11**, 5 (1966).
[26] Hobbs, D. J. and Iny, A., *Appl. Spectry.*, **24**, 522 (1970).
[27] Ingle, J. D., Jr., *J. Chem. Educ.*, **51**, 101 (1974).
[28] Winefordner, J. D., *Spectrochemical Methods of Analysis*, p. 516. Wiley, New York (1971).
[29] Broekaert, J. A. C., *Bull. Soc. Chim. Belg.*, **82**, 561 (1973).
[30] Booher, T. R., Elser, R. C. and Winefordner, J. D., *Anal. Chem.*, **42**, 1677 (1970).
[31] Ehrlich, G. and Gerbatsch, R., *Z. Anal. Chem.*, **220**, 260 (1966).
[32] Ehrlich, G. and Mai, H., *Z. Anal. Chem.*, **218**, 1 (1966).
[33] Matherny, M., *Chem. Zvesti*, **24**, 121 (1970).
[34] Ingle, J. D. and Wilson, R. L., *Anal. Chem.*, **48**, 1641 (1976).
[35] Ehrlich, G. and Gerbatsch, R., *Z. Anal. Chem.*, **225**, 90 (1967).
[36] Nalimov, V. V., Nedler, V. V. and Menshova, N. P., *Zavodsk. Lab.*, **27**, 861 (1961).
[37] Ross, J. B., *Analyst*, **87**, 832 (1962).
[38] Ehrlich, G., *Internationales Symposium Reinstoffe in Wissenschaft und Technik*, Dresden 1970, p. 861. Akademie Verlag, Berlin (1972).
[39] Currie, L. A., *Anal. Chem.*, **40**, 586 (1968).
[40] Fisenne, I. M., O'Toole, A. and Cutler, R., *Radiochem. Radioanal. Lett.*, **16**, 5 (1973).
[41] Shifrin, N. and Ramírez-Muñoz, J., *Appl. Spectry.*, **23**, 358 (1969).
[42] Rogers, V. C., *Anal. Chem.*, **42**, 807 (1970).
[43] Hubaux, A. and Vos, G., *Anal. Chem.*, **42**, 849 (1970).
[44] Gabriels, R., *Anal. Chem.*, **42**, 1439 (1970).
[45] Wing, J. and Wahlgren, M. A., *Anal. Chem.*, **39**, 85 (1967).
[46] Liteanu, C. and Florea, I., *Mikrochim. Acta*, 983 (1966).
[47] Liteanu, C. and Panovici, I. I., *Rev. Roumaine Chim.*, **22**, 931 (1977).
[48] Liteanu, C. and Rîcă, I., *Mikrochim. Acta*, 745 (1973).
[49] Liteanu, C. and Rîcă, I., *Pure Appl. Chem.*, **44**, 535 (1975).
[50] Liteanu, C. and Rîcă, I., *Mikrochim. Acta*, 311 (1975 **II**).
[51] Svoboda, V. and Gerbatsch, R., *Z. Anal. Chem.*, **242**, 1 (1968).
[52] Liteanu, C., Rîcă, A. and Hopîtean, E., *Croat. Chim. Acta*, **49**, 869 (1977).
[53] Tölg, G., *Talanta*, **19**, 1489 (1972).
[54] Tölg, G., *Talanta*, **21**, 327 (1974).
[55] Boumans, P. W. J. M., *Theory of Spectrochemical Excitation*, p.203. Hilger-Watts, London (1966).
[56] Clark, L., *The Encyclopaedia of Spectroscopy*, p. 308. Reinhold, New York (1964).

[57] Holdt, G., *Appl. Spectry.*, **16**, 96 (1962).
[58] Schroll, E., *Z. Anal. Chem.*, **198**, 40 (1963).
[59] Vukanović, V. M. and Georgijević, V. M., *Z. Anal. Chem.*, **225**, 137 (1965).
[60] Fratkin, Z. G., *Zavodsk. Lab.*, **26**, 971 (1960).
[61] Schneider, T., *Spectrochim. Acta,* **17B**, 300 (1961).
[62] Haisch, U., *Spectrochim. Acta,* **25B**, 597 (1970).
[63] Haisch, U., *Z. Anal. Chem.*, **259**, 1 (1972).
[64] Haisch, U. and Laqua, K., *Spectrochim. Acta,* **26B**, 651 (1971).
[65] Ehrlich, G., *Z. Anal. Chem.*, **232**, 1 (1967).
[66] Matherny, M., *Spectrosc. Lett.*, **5**, 221 (1972).
[67] Matherny, M., *Z. Anal. Chem.*, **271**, 101 (1974).
[68] Risova, J. and Plsko, E., *Chem. Zvesti,* **27**, 775 (1973).
[69] Bosch, F. M. and Broekaert, J. A. C., *Anal. Chem.*, **47**, 189 (1975).
[70] Zilbershtein, H. I., *Zh. Prikl. Spektrosk.*, **14**, 12 (1971).
[71] Zilbershtein, H. I., and Lageza, S. S., *Zh. Prikl. Spektrosk.*, 8, 6 (1968).
[72] Zilbershtein, H. I., *Zh. Prikl. Spektrosk.*, **8**, 531 (1968).
[73] Zilbershtein, H. I., Polivanova, H.G. and Fratkin, Z.G., *Zh. Prikl. Spektrosk.*, **11**, 204 (1969).
[74] Knisely, R. N., D'Silva, A. P. and Fassel, V. A., *Anal. Chem.*, **35**, 910 (1963).
[75] D'Silva, A. P., Knisely, R. N. and Fassel, V. A., *Anal. Chem.*, **36**, 1287, (1964).
[76] Fassel, V. A. and Golightly, D. W., *Anal. Chem.*, **39**, 466 (1967).
[77] Pickett, E. E. and Koirtyohann, S. R., *Spectrochim. Acta,* **23B**, 235 (1968).
[78] Koirtyohann S. R. and Pickett, E. E., *Spectrochim. Acta,* **23B**, 673 (1968).
[79] Pickett, E. E. and Koirtyohann, S. R., *Spectrochim. Acta,* **24B**, 325 (1969).
[80] Knisely, R. N., Butler, C. C. and Fassel, V. A., *Anal. Chem.*, **41**, 1494 (1969).
[81] Christian, G. D. and Feldman, F. J., *Anal. Chem.*, **43**, 611 (1971).
[82] Kirkbright, G. F., Peters, M. K. and West, T. S., *Talanta,* **14**, 789 (1967).
[83] Winefordner, J. D. and Vickers, T. J., *Anal. Chem.*, **42**, 206 R (1970).
[84] Pickett, E. E. and Koirtyohann, S. R., *Anal. Chem.*, **41**, (14), 28A (1969).
[85] Herrmann, R. and Lang, W., *Z. Anal. Chem.*, **203**, 1 (1969).
[86] Rudiger, K., Gutsche, B., Kirchhof, H. and Herrmann, R., *Analyst,* **94**, 204 (1969).
[87] Antić-Jovanović, A., Bojović, V. and Marinković, M., *Spectrochim. Acta,* **25B**, 405 (1970).
[88] Trampisch, W. and Herrmann, R., *Spectrochim. Acta,* **24B**, 215 (1969).
[89] Prudnikov, E. D., *Zh. Analit. Khim.*, **27**, 2327 (1972).
[90] Christian, G. D. and Feldman, F. J., *Appl. Spectry.*, **25**, 660 (1971).
[91] Busch, K. W. and Morrison, G. H., *Anal. Chem.*, **45**, 712A (1973).
[92] Brokeshoulder, S. F. and Robinson, F. R., *Appl. Spectry.*, **22**, 758 (1968).

[93] Blackburn, W. H., Pelletier, Y. J. A. and Dennen, W. H., *Appl. Spectry.*, 22, 278 (1968).
[94] Runge, E. F., Minck, R. W. and Bryan, F. R., *Spectrochim. Acta,* **20**, 733 (1964).
[95] Runge, E. F., Bonfiglio, S. and Bryan, F. R., *Spectrochim. Acta,* 22, 1678 (1966).
[96] Felske, A., Hagenach, W. D. and Laqua, K., *Z. Anal. Chem.*, **216**, 50 (1966).
[97] Piepmeier, E. H. and Malmstadt, H. V., *Anal. Chem.*, **41**, 700 (1969).
[98] Scott, R. H. and Strasheim, A., *Spectrochim. Acta,* **25B**, 311 (1970).
[99] Rasberry, S. D., Scribner, B. F. and Margoshes, M., *Appl. Opt.*, **6**, 81 (1967).
[100] Pantony, D. A. and Hurley, P. W., *Analyst,* **97**, 497 (1972).
[101] Blount, C. W., Morgan, W. R. and Leyden, D. E., *Anal. Chim. Acta,* **53**, 463 (1971).
[102] Winefordner, J. D., Parsons, M. L., Mansfield, J. M. and McCarthy, W. J., *Anal. Chem.*, **39**, 436 (1967).
[103] Winefordner, J. D., Svoboda, V. and Cline, L. J., *Crit. Rev. Anal. Chem.*, **1**, 233 (1970).
[104] Norris, J. D. and West, T. S., *Anal. Chem.*, **45**, 226 (1973).
[105] Omenetto, N., Hatch, N. N., Fraser, L. M. and Winefordner, J. D., *Anal. Chem.*, **45**, 195 (1973).
[106] Fraser, L. M. and Winefordner, J. D., *Anal. Chem.*, **43**, 1693 (1971).
[107] Hieftje, G. M., Bystroff, R. I. and Lim, R., *Anal. Chem.*, **45**, 253 (1973).
[108] Ingle, J. D., Jr., and Crouch, S. R., *Anal. Chem.*, **44**, 785 (1972).
[109] Murphy, M. K., Clyburn, S. A. and Veillon, C., *Anal. Chem.*, **45**, 1468 (1973).
[110] Hieftje, D. J., *Anal. Chem.*, **44**, No. 6, 81A (1972).
[111] Hieftje, D. J., *Anal. Chem.*, **44**, No. 7, 69A (1972).
[112] Heidel, R. H., *Anal. Chem.*, **43**, 1907 (1971).
[113] Ziebold, T. O., *Anal. Chem.*, **39**, 858 (1967).
[114] Ziebold, T. O. and Ogilvie, R. E., *Anal. Chem.*, **36**, 322 (1964).
[115] Heidel, R. H., *Anal. Chem.*, **44**, 1860 (1972).
[116] Heidel, R. H., *Anal. Chem.*, **46**, 2038 (1974).
[117] Harris, L. A., *Anal. Chem.*, **40**, (14), 24A (1968).
[118] Palmberg, P. W., *Anal. Chem.*, **45**, 549A (1973).
[119] Wagner, C. D., *Anal. Chem.*, **44**, 1050 (1972).
[120] Berthou, H. and Jørgensen, C. K., *Anal. Chem.*, **47**, 482 (1975).
[121] Giauque, R. D., Goulding, F. S., Jaklevic, J. M. and Pehl, R. H., *Anal. Chem.*, **45**, 671 (1973).
[122] Ebel, H. and Ebel, M., *X-Ray Spectry.*, **2**, 19 (1973).
[123] Janghorbani, M., Vulli, M. and Starke, K., *Anal. Chem.*, **47**, 2200 (1975).
[124] Bremser, W., *Z. Anal. Chem.*, **259**, 204 (1972).

[125] Hercules, D. M., Cox, L., Onisick, S., Nichols, G. D. and Carver, J. C., *Anal. Chem.*, **45**, 1973 (1973).

[126] Winefordner, J. D., and Vickers, T. J., *Anal. Chem.*, **36**, 1939 (1964).

[127] Winefordner, J. D., and Vickers, T. J., *Anal. Chem.*, **36**, 1947 (1964).

[128] Winefordner, J. D., McCarthy, W. J. and St. John, P. A., *J. Chem. Educ.*, **44**, 80 (1967).

[129] St. John, P. A., McCarthy, W. J. and Winefordner, J. D., *Anal. Chem.*, **38**, 1828 (1966).

[130] Ingle, J. D., Jr., and Crouch, S. R., *Anal. Chem.*, **44**, 1375 (1972).

[131] Rothman, L. D., Crouch, S. R. and Ingle, J. D., Jr., *Anal. Chem.*, **47**, 1226 J. P., *Anal. Chem.*, **43**, 211 (1971).

[132] West, T. S. and Williams, X. K., *Anal. Chim. Acta*, **45**, 27 (1969).

[133] Amos, M. D., Bennett, P. A., Brodie, K. G., Lung, P. W. Y. and Matousek, J. P., *Anal. Chem.*, **43**, 211 (1971).

[134] Dipierro, S. and Tessari, G., *Talanta*, **18**, 707 (1971).

[135] Omang, S. H., *Anal. Chim. Acta*, **56**, 470 (1971).

[136] Walsh, A., *Spectrochim. Acta*, **7**, 108 (1955).

[137] Parsons, M. L. and Winefordner, J. D., *Appl. Spectry.*, **21**, 368 (1967).

[138] Russell, B. J., Shelton, J. P. and Walsh, A., *Spectrochim. Acta*, **8**, 317 (1957).

[139] Coor, T., *J. Chem. Educ.*, **45A**, 583 (1968).

[140] Snelleman, W., *Spectrochim. Acta*, **23B**, 403 (1968).

[141] Marinković, M. and Vickers, T. J., *Anal. Chem.*, **42**, 1613 (1970).

[142] Hieftje, G. M., Holder, B. E., Maddux, A. S., Jr. and Lim, R., *Anal. Chem.*, **45**, 238 (1973).

[143] Box, C. F. and Walsh, A., *Spectrochim. Acta*, **16**, 255 (1960).

[144] Holder, B. E., Lim, R., Maddux, A. and Hieftje, G. M., *Anal. Chem.*, **44**, 1716 (1972).

[145] Roos, J. T., *Spectrochim. Acta*, **25B**, 539 (1970).

[146] Roos, J. T., *Spectrochim. Acta*, **28B**, 407 (1973).

[147] Parsons, M. L., McCarthy, W. J. and Winefordner, J. D., *J. Chem. Educ.*, **44**, 214 (1967).

[148] Winefordner, J. D. and Veillon, C., *Anal. Chem.*, **37**, 416 (1965).

[149] Ingle, J. D., Jr., *Anal. Chem.*, **46**, 2161 (1974).

[150] Posma, F. D., Smit, H. C. and Rooze, A. F., *Anal. Chem.*, **47**, 2087 (1975).

[151] Campion, J. J., *Anal. Chem.*, **46**, 1145 (1974).

[152] Munson, B., *Anal. Chem.*, **43**, (13) 28A (1971).

[153] Franzen, J. and Schuy, K. D., *Z. Anal. Chem.*, **225**, 295 (1967).

[154] Schuy, K. D. and Franzen, J., *Z. Anal. Chem.*, **225**, 260 (1967).

[155] Frisch, M. A. and Reuter, W., *Anal. Chem.*, **45**, 1889 (1973).

[156] Ahearn, A. J., *Trace Analysis by Mass Spectrometry*, Academic Press, New York (1972).

[157] Simons, D. S., Baker, J. E. and Evans, C. A., Jr., *Anal. Chem.*, **48**, 1341 (1976).
[158] Morabito, J. M. and Lewis, R. K., *Anal. Chem.*, **45**, 869 (1973).
[159] *Jeol News*, **9a**, 16 (1971).
[160] Farrar, T. C., *Jeol News*, **10a**, 2 (1972).
[161] Farrar, T. C. and Becker, E. D., *Pulse and Fourier Transform NMR*, Academic Press, New York (1971).
[162] *Jeol News*, **9a**, 6 (1972).
[163] R. A. Nadkarni and Morrison, G. H., *Anal. Chem.*, **45**, 1957 (1973).
[164] Nadkarni, R. A. and Ehmann, W. D., *J. Radioanal. Chem.*, **3**, 175 (1969).
[165] Nadkarni, R. A., Flieder, D. E. and Ehmann, W. D., *Radiochim. Acta*, **11**, 97 (1969).
[166] Nadkarni, R. A. and Ehmann, W. D., *Radiochem. Radioanal. Lett.*, **2**, 161 (1969).
[167] Nadkarni, R. A. and Ehmann, W. D., *Radiochem. Radioanal. Lett.*, **4**, 325 (1970).
[168] Nadkarni, R. A. and Ehmann, W. D., *Radiochem. Radioanal. Lett.*, **6**, 89 (1971).
[169] Morrison, G. H. and Potter, N. M., *Anal. Chem.*, **44**, 839 (1972).
[170] Steinnes, E., Birkelung, O. R. and Johansen, O., *J. Radioanal. Chem.*, **9**, 269 (1971).
[171] Samsahl, K., Wester, P. O. and Landström, O., *Anal. Chem.*, **40**, 181 (1968).
[172] Biloen, P., Dorrepaal, J. and van der Heijde, H. B., *Anal. Chem.*, **45**, 288 (1973).
[173] Anders, O. U., *Anal. Chem.*, **41**, 428 (1969).
[174] Dams, R., Robbins, J. A., Rahn, K. A., and Winchester, J. W., *Anal. Chem.*, **42**, 861 (1970).
[175] Turkstra, J., Pretorius, P. J. and de Wet, W. J., *Anal. Chem.*, **42**, 835 (1970).
[176] Heurtebise, M. and Lubkowitz, J. A., *Anal. Chem.*, **43**, 1218 (1971).
[177] Rosenstein, A. W. and Nir, A., *Anal. Chem.*, **45**, 1707 (1973).
[178] Debrun, J. L., Riddle, D. C. and Schweikert, E. A., *Anal. Chem.*, **44**, 1386 (1972).
[179] Swindle, D. L. and Schweikert, E. A., *Anal. Chem.*, **45**, 2111 (1973).
[180] Ricci, E., *Anal. Chem.*, **43**, 1866 (1971).
[181] Bankert, S. F., Bloom, S. D. and Sauter, G. D., *Anal. Chem.*, **45**, 692 (1973).
[182] Aras, N. K., Zoller, W. H., Gordon, G. E. and Lutz, G. J., *Anal. Chem.*, **45**, 1481 (1973).
[183] Kudo, K. and Suzuki, N., *J. Radioanal. Chem.*, **26**, 327 (1975).
[184] Bond, A. M., *Anal. Chem.*, **44**, 315 (1972).
[185] Bond, A. M., *Anal. Chem.*, **45**, 2026 (1973).
[186] Barker, G. C. and Gardner, A. W., *Z. Anal. Chem.*, **173**, 79 (1960).
[187] Parry, E. P. and Osteryoung, R. A., *Anal. Chem.*, **37**, 1634 (1965).

[188] Parry, E. P. and Osteryoung, R. A., *Anal. Chem.*, **36**, 1366 (1964).
[189] Devaleriola, M. and Nangniot, P., *Talanta,* **15**, 759 (1968).
[190] Osteryoung, J. G. and Osteryoung, R. A., *Intern. Laboratory,* Sept./Oct., 10 (1972).
[191] Parry, E. P. and Anderson, D. P., *Anal. Chem.*, **45**, 458 (1973).
[192] Gilbert, D. D., *Anal. Chem.*, **37**, 1102 (1965).
[193] Peker, C., Herlem, M. and Badoz-Lambling, J., *Z. Anal. Chem.*, **224**, 302 (1967).
[194] Abdullah, M. I. and Royle, L. G., *Anal. Chim. Acta*, **58**, 283 (1972).
[195] Milner, G. W. C., Wilson, J. D., Barnett, G. A. and Smales, A. A., *J. Electroanal. Chem.*, **2**, 25 (1961).
[196] Parry, E. P. and Oldham, K. B., *Anal. Chem.*, **40**, 1031 (1968).
[197] Wolff, G. and Nürnberg, H. W., *Z. Anal. Chem.*, **216**, 169 (1969).
[198] de Silva, J. A. F. and Hackman, M. R., *Anal. Chem.*, **44**, 1145 (1972).
[199] Garber, R. W. and Wilson, C. E., *Anal. Chem.*, **44**, 1357 (1972).
[200] Myers, D. J. and Osteryoung, J., *Anal. Chem.*, **45**, 267 (1973).
[201] Flato, J. B., *Anal. Chem.*, **44**, (11), 75A (1972).
[202] Christian, G. D., *J. Electroanal. Chem.*, **23**, 1 (1969).
[203] Copeland, T. R., Christie, J. H., Osteryoung, R. A. and Skogerboe, R. K., *Anal. Chem.*, **45**, 2171 (1973).
[204] Kemula, W. and Kublik, Z., *Nature,* **189**, 57 (1961).
[205] Eisner, U. and Ariel, M., *J. Electroanal. Chem.*, **11**, 26 (1966).
[206] Bond, A. M., *Anal. Chem.*, **42**, 1165 (1970).
[207] Levit, D. I., *Anal. Chem.*, **45**, 1291 (1973).
[208] MacLeod, K. E. and Lee, R. E., Jr., *Anal. Chem.*, **45**, 2380 (1973).
[209] Pungor, E., Tóth, K. and Havas, J., *Acta Chim. Acad. Sci. Hung.*, **48**, 17 (1966).
[210] Morf, W. E., Kahr, G. and Simon, W., *Anal. Chem.*, **46**, 1538 (1974).
[211] Woodson, J. H. and Liebhafsky, H. H., *Anal. Chem.*, **41**, 1894 (1969).
[212] Moody, C. J. and Thomas, J. D. R., *Selective Ion-Sensitive Electrodes,* p. 77. Merrow, Watford (1971).
[213] Frant, M. S., *Cronache di Chimica,* No. 44, 3 (1974).
[214] *IUPAC Inform. Bull.*, No. 13, 2, (1975). (Recommendations for Nomenclature of Ion-selective Electrodes).
[215] Liteanu, C., Hopîrtean, E. and Popescu, I. C., *Anal. Chem.*, **48**, 2013 (1976).
[216] Glueckauf, E., *Ion Exchange and its Applications,* p. 34. Society of Chemical Industry, London (1955).
[217] Huber, J. F. K., Hulsman, J. A. R. J. and Meijers, C. A. M., *J. Chromatog.*, **62**, 79 (1971).
[218] Maijers, C. A. M., Hulsman, J. A. R. J. and Huber, J. F. K., *Z. Anal. Chem.*, **261**, 347 (1972).

[219] Kalmanovskii, V. I. and Zhukhovitskii, A. A., *J. Chromatog.*, **18**, 243 (1965).
[220] Karger, B. L., Martin, M. and Guiochon, G., *Anal. Chem.*, **46**, 1640 (1974).
[221] Teranishi, R. and Mon, T. R., *Anal. Chem.*, **36**, 1490 (1964).
[222] Camin, D. L., King, R. W. and Shawhan, S. D., *Anal. Chem.*, **36**, 1175 (1964).
[223] Pecsar, R. E., DeLew, R. B. and Iwao, K. R., *Anal. Chem.*, **36**, 2191 (1973).
[224] Stevens, R. K., Mulik, J. D., O'Keeffe, A. E. and Krost, K. J., *Anal. Chem.*, **43**, 827 (1971).
[225] Natusch, D. F. S. and Thorpe, T., *Anal. Chem.*, **45**, 1184A (1973).
[226] Williams, F. W. and Umstead, M. E., *Anal. Chem.*, **40**, 2232 (1968).
[227] Takata, Y. and Muto, G., *Anal. Chem.*, **45**, 1864 (1973).
[228] Bowen, B. E., Cram, S. P., Leitner, J. E. and Wade, R. L., *Anal. Chem.*, **45**, 2185 (1973).
[229] Goedert, M. and Guiochon, G., *Anal. Chem.*, **45**, 1188 (1973).

Chapter 8

Trace Determination

8.1 INTRODUCTION

An analytical determination is the evaluation of the amount of a component by means of a suitable method of analysis. Under real conditions, the result will always include an error (an uncertainty). The total error, i.e. the deviation of the result from the true value can generally be separated into two components: one due to inhomogeneity in the material analysed, the other generated by perturbations in the determination process (the determination error).

When evaluating the determination performance of an analytical method, we are obviously interested only in the second of these errors. It can be evaluated either by analysing homogeneous materials (so that the inhomogeneity is negligible), or by statistically designing analytical experiments on non-homogeneous materials in such a way as to separate (by processing the results) the determination error from the total error.

Hence, with regard to determination performance, we shall take into account only the determination error.

For an analysis of a homogeneous material, we shall call the result of a determination c_i, and the true value $\hat{c}$, so the absolute and relative errors of determination can be written as:

$$\Delta c_i = c_i - \hat{c} \tag{8.1}$$

$$(\Delta c_i)_r = (c_i - \hat{c})/\hat{c} \tag{8.2}$$

or

$$(\Delta c_i)\% = 100(c_i - \hat{c})/\hat{c} \tag{8.3}$$

In trace analysis, the third method of expressing the error is usually used, because it is simple and there is no risk of confusing it with the absolute error (as there is in the case of major components) since the absolute error will be expressed in ppm or some similar type of unit.

There are several works devoted to the statistical processing and interpretation of analytical errors [1-8]. In this section we shall recall only some procedures used in defining and evaluating the quantities which characterize the performance of the methods of trace determination.

The determination error is usually quite complex; it always contains a random component, sometimes also a gross component, and it may include a systematic component. The random component itself arises as the sum of several elementary random errors arising in the various steps of the process. The processing and interpretation of the random errors is almost always based on the normal distribution model:

$$\Delta c \Rightarrow N(0; \sigma_c) \tag{8.4}$$

i.e. on the assumption that the random errors belong to a normal distribution around a mean equal to zero, with a standard deviation σ_c. Obviously, if the random errors Δ_c for a single result correspond to the model (8.4), the random errors $(\overline{\Delta c})_n$ for results obtained as a mean of n single results will correspond to the model:

$$(\overline{\Delta c})_n \Rightarrow N\left(0; \frac{\sigma_c}{\sqrt{n}}\right) \tag{8.5}$$

Let us assume that $c_1, c_2, \ldots, c_n$ are results obtained by repeated analyses on a material, and that there are no systematic errors; then, the mean result $\bar{c} = \sum_{i=1}^{n} c_i/n$ will tend towards the true value as the number of determinations n is increased.

Investigations of the nature of the random error distribution have shown cases when this distribution is normal, and also cases when there are deviations from the normal distribution [9-11]. Large deviations from the normal distribution are, most usually, a consequence of various deficiencies in the process of analysis.

Investigations of the precision (reproducibility) of analytical results have shown that, generally, the standard deviation varies with the amount of component to be determined [12-22], in a manner similar to the model given in Fig. 8.1.

It is also generally accepted that random errors are not time-correlated. That is, no matter what the time interval between repetitions there is no correlation between the random errors $(\Delta c)_t$ and $(\Delta c)_{t+a}$ associated with determinations repeated at times t and $t + a$, respectively, which is expressed mathematically as:

$$\text{cov.}\ [(\Delta c)_t, (\Delta c)_{t+a}] = 0 \tag{8.6}$$

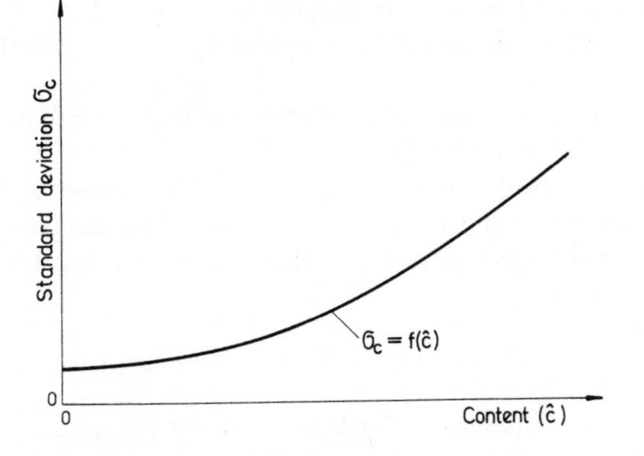

Fig. 8.1 Model of the dependence between the standard deviation and the amount present.

However, we would expect there to be a certain correlation between successive determinations, especially if the time interval is short. This is so because analytical systems work in a stationary covariant regime (see Section 4.2). For example, two samples of the same material are likely to give closer agreement of results if they are analysed simultaneously than if the second is analysed some days after the first, because there is a greater likelihood of exact reproduction of the procedural manipulations.

The performance of a method of analysis can obviously be estimated from the accuracy attained. Owing to the random nature of the error, however, the performance can be rigorously evaluated only on a statistical basis. Two concepts are used in order to characterize it: precision (reproducibility) and accuracy. The precision shows how closely the results of repeated analysis of a homogeneous material are grouped around the mean value, and is measured by the sample standard deviation:

$$s_c = \sqrt{\frac{\sum_{i=1}^{n} (c_i - \bar{c})^2}{n - 1}} \tag{8.7}$$

where $\bar{c} = \sum_{i=1}^{n} c_i/n$ is the experimental mean.

The precision is conveniently expressed as the relative standard deviation (sometimes called the coefficient of variation). The abbreviations rsd and c.v. are often used, and the quantity is expressed in per cent:

$$\text{rsd or c.v.} = 100\, s_c/\bar{c}\% \tag{8.8}$$

The true value of the standard deviation remains unknown, but the estimation value can be obtained from a set of experimental results. Then, by using the χ^2 distribution, we can obtain a confidence interval for the true value of the standard deviation (see Section 2.5.2).

The precisions of two methods of analysis can be compared by applying the F-test to their sample variances (Section 2.6.3.2).

Closely related to the concept of precision is the concept of random error. At a confidence probability P, the t-distribution, within the framework of the random error model (8.4), gives the confidence intervals for the amplitude of the random errors Δc as

$$-s_c|t_{(\alpha/2;n-1)}| < \Delta c < + s_c|t_{(\alpha/2;n-1)}| \qquad (8.9)$$

or

$$-(\text{c.v.})|t_{(\alpha/2;n-1)}| < (\Delta c)\% < + (\text{c.v.})|t_{(\alpha/2;n-1)}| \qquad (8.10)$$

where $\alpha = 1-P$ is the significance level, $t_{(\alpha/2;n-1)}$ the t-parameter corresponding to the significance level α and $n-1$ degrees of freedom, and n the number of results.

To characterize the precision of a method of analysis, Kaiser and Specker [23] introduced the quantity

$$A = \frac{\bar{c}}{s_c} \qquad (8.11)$$

which is the reciprocal of the relative sample standard deviation. Hence, this quantity has the meaning of an estimation value for the signal-to-noise ratio, corresponding to a certain value of the content to be determined.

According to Fujimori and co-workers [24, 25], the appreciation and selection of methods of analysis should be made on the criterion:

$$\eta = \sigma_{\mathrm{M}}^2/\sigma_{\mathrm{E}}^2 \qquad (8.12)$$

which also has the meaning of signal-to-noise ratio. Here the quantities σ_{M}^2 (variance of content) and σ_{E}^2 (variance of error) are obtained from the results of analysis of a set of n samples covering the range of content expected in the materials for which the method is to be used. Each of the n samples is analysed m_i times. The results are then processed by single-factor variance analysis (see Section 2.6.6.1), and the quantity σ_{M}^2 is estimated from the n mean results, and the quantity σ_{E}^2 from all the results.

The precision of a method can also be evaluated by using quantities defined in information theory. Thus, the entropy associated with normally distributed random errors will be

$$H(\Delta c) = \sigma_c\sqrt{2\pi e} \qquad (8.13)$$

or

$$H[(\Delta c)\%] = \text{c.v.}\sqrt{2\pi e} \qquad (8.14)$$

Either (8.13) or (8.14) can be taken as a measure of precision.

The accuracy is concerned with the total error of analysis (the systematic error plus the random error), so it measures the deviation of the results from the true value. It can be studied and evaluated by analysing standard samples, i.e. samples for which the amount of component to be determined is rigorously known. The total error always contains a random component, so accuracy can only be evaluated by statistical treatment of the data.

The significance of a systematic error can be established by using the t-test (see Section 2.6.2.1). Let us assume that a standard sample containing the true amount $\hat{c}$ of the component to be determined has been analysed n times, and that the mean and standard deviation are $\bar{c}$ and s_c. The significance of a systematic error is established by calculating the quantity

$$t_{\text{exp}} = \frac{(\hat{c} - \bar{c})\sqrt{n}}{s_c} \tag{8.15}$$

and the hypothesis that there is a systematic error is accepted if

$$|t_{\text{exp}}| > |t_{(\alpha/2;n-1)}|$$

It is obvious that, if systematic errors are absent, a rigorous measure both for the precision and for the accuracy is given by the standard deviation σ_c or equally well by the coefficient of variation c.v. In the absence of systematic errors, the confidence interval for the true value of the amount $\hat{c}$, corresponding to a confidence probability P, is obtained from the t distribution and is given by:

$$\bar{c} - |t_{(\alpha/2;n-1)}|\frac{s_c}{\sqrt{n}} < \hat{c} < \bar{c} + |t_{(\alpha/2;n-1)}|\frac{s_c}{\sqrt{n}} \tag{8.16}$$

In order to characterize the accuracy of a method (taking into account both the systematic and the random error). McFarren *et al.* [26] introduced the following quantity, designated the total error:

$$\text{total error} = \frac{d + 2\sigma_c}{\hat{c}} \times 100 \tag{8.17}$$

The significance of (8.17) is shown in Fig. 8.2 [26] : $\hat{c}$ is the true value and $\hat{c}_R$ the true value of the mean for the distribution of the results. Hence d has the significance of a systematic error, and $2\sigma_c$ is half the confidence interval for random errors corresponding to a confidence probability $P = 0.954$.

Considering any practical case of evaluation of the total error, it is obvious that the mean and the standard deviation of the results, $\bar{c}_R$ and s_c, are always evaluated from a finite set of results, so they are estimation values. Hence, a result $d = \hat{c} - \bar{c}_R$ different from zero can appear not only because of a systematic error, but also because of random errors. For this reason, according to Eckschlager [27], we should use in (8.17) the experimental value $d = \hat{c} - \bar{c}_R$ only when the t-test leads to the conclusion that the systematic error is significant; if this is not the case, we should take $d = 0$.

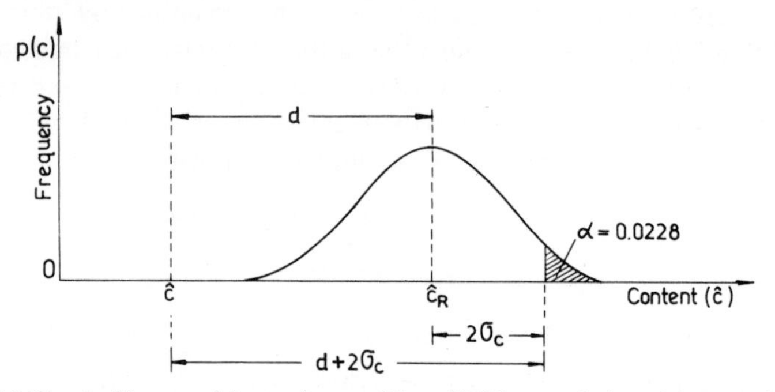

Fig. 8.2 The significance of the total error. (From [26] by permission of the copyright holder, the American Chemical Society)

Later, Eckschlager [28] returned to the total error, and proposed to take the quantity:

$$T_B = \frac{100\,(2.066\,\sigma_c/\sqrt{n} + \theta)}{\hat{c}} \tag{8.18}$$

where $2.066\sigma_c$ represents half of the informational entropy $H(\Delta c) = \sigma_c\sqrt{2\pi e}$ coresponding to a normal distribution, $\theta = d - 2.066\,\sigma_c/\sqrt{n}$, and $d = |\hat{c} - \bar{c}_R|$. For $d < 2.066\,\sigma_c/\sqrt{n}$, we take $\theta = 0$, which gives:

$$T_B = \frac{2.066\sigma_c}{\hat{c}\sqrt{n}} \tag{8.19}$$

For $d > 2.066\,\sigma_c/\sqrt{n}$, we take $\theta = d - 2.066\,\sigma_c/\sqrt{n}$, that is:

$$T_B = 100\,d/\hat{c} \tag{8.20}$$

When the mean $\bar{c}_R$ and the standard deviation s_c are evaluated from a small number of determinations, T_B is calculated in (8.18) by using the t value corresponding to the number of degrees of freedom and the chosen significance level, instead of the value $\sqrt{2\pi e}/2 = 2.066$.

From the discussion above it is clear that the definition of a statistical quantity which would rigorously characterize the accuracy of a method is not a completely solved problem; in the next section we shall consider a new quantity for characterizing the accuracy of a method of analysis.

8.2 THE RELIABILITY OF A METHOD OF ANALYSIS WITH RESPECT TO TWO ERROR LIMITS

By reliability of a method of analysis with respect to two error limits we mean the probability that the results lie in the interval defined by the two chosen limits. It gives a rigorous method for evaluating the correctness of a method of analysis in relation to the two limits for error. The meaning of this concept is illustrated in Fig. 8.3.

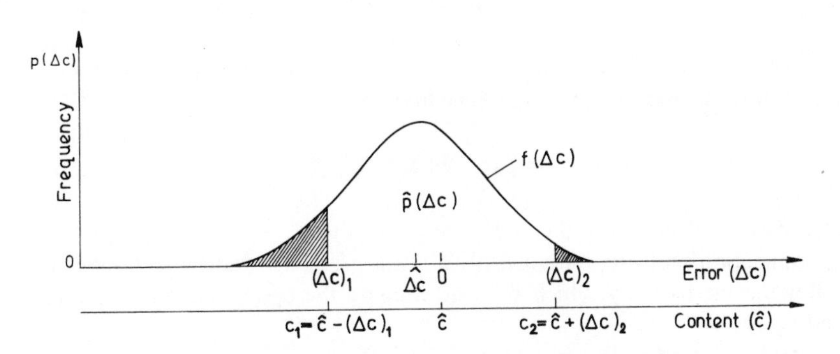

Fig. 8.3 The meaning of the reliability of a method of analysis with respect to two error limits.

The symbols in Fig. 8.3 have the following meanings: $f(\Delta c)$ is the probability density of the errors; $(\Delta c)_1$ and $(\Delta c)_2$ are the error limits with respect to which the reliability is to be evaluated; $(\widehat{\Delta c})$ is the true mean corresponding to the distribution of the errors (and is also the systematic error); $\hat{c}$ is the true amount; $c_1 = \hat{c} - (\Delta c)_1$ and $c_2 = \hat{c} + (\Delta c)_2$ are the result limits with respect to which the reliability is evaluated; $\hat{p}(\Delta c)$ is the true value of the reliability, in the case that the errors are given by their absolute values.

From the definition given above, it follows that the reliability of a method of analysis with respect to two error limits is evaluated by the integral:

$$\hat{p}(\Delta c) = \int_{(\Delta c)_1}^{(\Delta c)_2} f(\Delta c)\mathrm{d}(\Delta c) \tag{8.21}$$

When the analysis errors are expressed in per cent, the reliability will be written as $\hat{p}[(\Delta c)\%]$ and will be evaluated in the corresponding way.

There are several statistical procedures [29-31] by which we can evaluate the probability $\hat{p}(\Delta c)$ which measures the accuracy of a method of analysis with respect to two error limits. All use a set of results obtained for one (or several) standard sample(s), with rigorously known content of the component to be determined. It is assumed that the random errors are normally distributed, and that the method of analysis is in a stationary stability regime.

(a) *The binomial distribution* We shall assume that n analytical determinations (where n is large) have been made on a standard sample with content $\hat{c}$ of the component to be determined, and that m of these results fall within the interval defined by the limits $c_1 = \hat{c} + (\Delta c_1)$ and $c_2 = \hat{c} + (\Delta c)_2$. With respect to these error limits, the reliability of the method is estimated by the ratio:

$$p(\Delta c) = \frac{m}{n} \tag{8.22}$$

The true reliability value $\hat{p}(\Delta c)$ is not known, but we can use the binomial distribution to establish its confidence interval:

$$p_1 < \hat{p}(\Delta c) < p_2 \tag{8.23}$$

at a chosen value P for the confidence probability. As shown in Section 2.5.3 in connection with the binomial distribution, for large values of n there are the following methods by which we can evaluate the confidence interval limits p_1 and p_2:

(i) by operating with the binomial distribution and solving Eqs. (2.38) and (2.39);

(ii) by finding the normal distribution which approximates the binomial distribution and solving Eqs. (2.40) and (2.41);

(iii) by accepting a normal distribution which is only a coarse approximation of the binomial distribution and solving Eqs. (2.42) and (2.43).

(b) *The normal distribution* Let $c_1, c_2, \ldots, c_n$ be the set of results for the standard sample with true content $\hat{c}$. The mean and standard deviation for a large set of n results are:

$$(\bar{c})_n = \frac{\sum_{i=1}^{n} c_i}{n} \tag{8.24}$$

$$s_c = \sqrt{\frac{\sum_{i=1}^{n} [c_i - (\bar{c})_n]^2}{n-1}} \tag{8.25}$$

The normal deviates are

$$z_1 = \frac{[\hat{c} - (\Delta c)_1] - (\bar{c})_n}{s_c} \tag{8.26}$$

$$z_2 = \frac{[\hat{c} + (\Delta c)_2] - (\bar{c})_n}{s_c} \tag{8.27}$$

With respect to the error limits $(\Delta c)_1$ and $(\Delta c)_2$, the reliability is estimated by the following quantity:

$$p(\Delta c) = \frac{1}{\sqrt{2\pi}} \int_{z_1}^{z_2} \exp\left[-\frac{z^2}{2}\right] dz \tag{8.28}$$

which is evaluated by using tables of the normal distribution. The variable $p(\Delta c)$ evaluated in this way from a large number of determinations is normally distributed and has the following variance [30]:

$$\sigma_p^2 = \frac{1}{n} k_1^2 \left(1 + \frac{z_1^2}{2}\right) + k_2^2 \left(1 + \frac{z_2^2}{2}\right) + 2k_1 k_2 \left(1 + \frac{|z_1||z_2|}{2}\right) \tag{8.29}$$

where k_1 and k_2 are the normal (Gaussian) curve ordinates at z_1 and z_2. Hence, for a certain value of the confidence probability P, the limits of the confidence interval will be:

$$p_1 = p(\Delta c) - |z_{\alpha/2}|\sigma_p \tag{8.30}$$

$$p_2 = p(\Delta c) + |z_{\alpha/2}|\sigma_p \tag{8.31}$$

The methods given above can be applied only if the number of determinations is large. However, there are also statistical methods which can be used if the number of analytical determinations is small [30].

Example 8.1. The individual results obtained by repeated determinations of arsenic in a standard sample of steel are given in the histogram in Fig. 8.4 (the determinations were made by X-ray fluorescence spectroscopy). The arsenic concentration in the standard sample of steel was $\hat{c}_{As} = 0.024\%$. From these data we wish to evaluate the reliability of the method with respect to the error limits $(\Delta c)_1 = -0.0025$ and $(\Delta c)_2 = +0.0025$ (absolute errors) or respectively $(\Delta c)_1 = -10\%$ and $(\Delta c)_2 = +10\%$ (relative error) for a confidence probability $P = 0.95$.

There are $n = 35$ results. Of these $m = 28$ are found within the error limits of ± 0.0025 (or ± 10%). Hence, with respect to these limits, the reliability is estimated by:

$$p(\Delta c) = \frac{28}{35} = 0.80$$

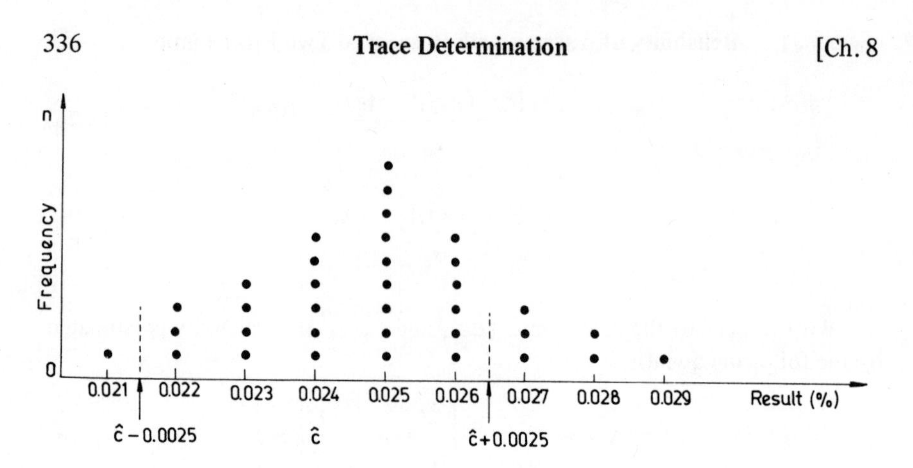

Fig. 8.4 Histogram of results for arsenic in a standard steel sample.

From Eqs. (2.38) and (2.39), taking $n = 35$, $m = 28$, and $P = 0.95$, the two limits of the reliability confidence interval are $p_1 = 0.63$ and $p_2 = 0.91$, so:

$$0.63 < \hat{p}(\Delta c) < 0.91$$

For the same data, by using Eqs. (2.40) and (2.41), we find

$$0.63 < \hat{p}(\Delta c) < 0.92$$

and with Eqs. (2.42) and (2.43):

$$0.67 < \hat{p}(\Delta c) < 0.93$$

By the method just shown, based on the normal distribution, we find:

$$(\bar{c})_n = 0.0249; s_c = 0.00171; z_1 = -1.96; z_2 = 0.96;$$

$$k_1 = 0.0584; k_2 = 0.252; s_p = 0.0675$$

From (8.28) the estimation value of the reliability is:

$$p(\Delta c) = 0.81$$

At a confidence probability $P = 0.95$, i.e. $z = 1.96$, from (8.30) and (8.31) the confidence interval of the true value of the reliability is:

$$0.68 < \hat{p}(\Delta c) < 0.94$$

8.3 THE DETERMINATION CHARACTERISTIC OF A METHOD

The parameters by which we evaluate the determination performance of a method (i.e. the absolute standard deviation, variation coefficient, entropy, signal-to-noise ratio, total error, reliability with respect to two error limits) vary with the content to be determined. Hence, it is not enough to establish these parameters (or some of them) at a single level of content, and their dependence on content of component must also be determined. We shall designate this dependence the **determination characteristic**.

The model for the determination characteristic was given in Fig. 8.1. For convenience, and because it is the method most frequently used, we shall express the determination characteristic in terms of relative (percentile) errors.

Obviously, when there is no systematic error, the determination characteristic expressed as c.v. vs. content, or $H[(\Delta c)\%]$ vs. content will give a complete evaluation for the determination performance of a method of analysis (for one component).

It is clear that in trace analysis the determination characteristic of a method must be evaluated for contents in the immediate region of zero.

We shall assume that the dependence between the absolute standard deviation and the content is well described by the model of Fig. 8.1. In the region of amount immediately greater than zero, the standard deviation can practically be regarded as constant, so the determination characteristic expressed as (c.v.) vs. content is approximated by a hyperbola, according to the equation:

$$\text{c.v.} = \frac{\sigma_c}{\hat{c}} \times 100\% \tag{8.32}$$

When the results used are the mean of n determinations, the determination characteristic will be given by:

$$(\text{c.v.})_n = \frac{\sigma_c}{\hat{c}\sqrt{n}} \times 100\% \tag{8.33}$$

or

$$(\text{c.v.})_n = \frac{\text{c.v.}}{\sqrt{n}} \tag{8.34}$$

Hence, when means of results are used instead of single results, the detection characteristic is improved because the random error fluctuations are smaller. This change in the determination characteristic is shown in Fig. 8.5.

By taking logarithms in Eq. (8.33) we obtain

$$\log (\text{c.v.})_n = \left(\log \frac{\sigma_c}{\sqrt{n}} \times 100\right) - \log \hat{c} \tag{8.35}$$

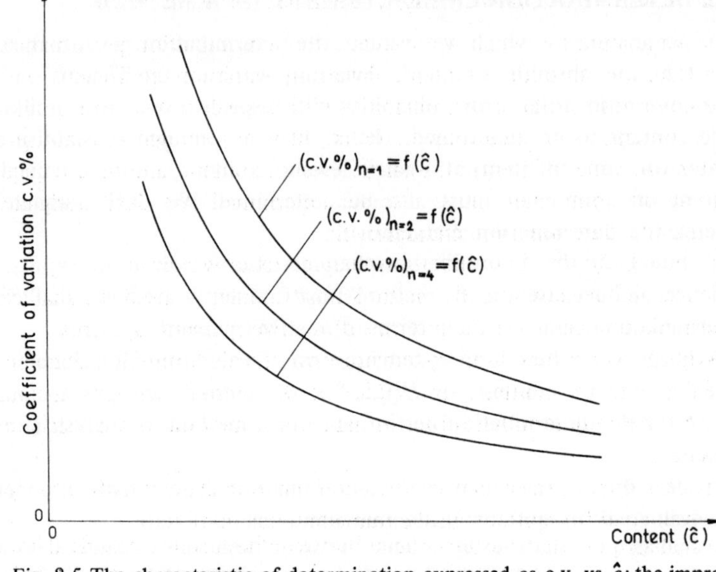

Fig. 8.5 The characteristic of determination expressed as c.v. vs. $\hat{c}$; the improvement obtained by using mean results.

Hence, within a region for which the standard deviation σ_c can be regarded as independent of the content there is a linear relation between the logarithm of the coefficient of variation and the logarithm of the content. This is shown graphically in Fig. 8.6. Once again, the determination characteristic is improved by using means of results instead of single results.

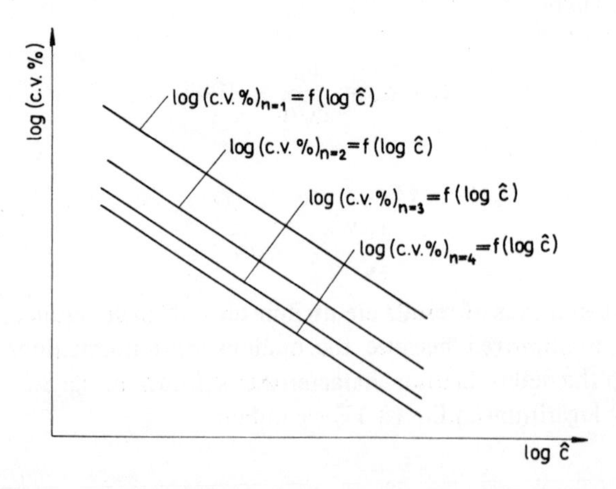

Fig. 8.6 The characteristic of determination expressed as log(c.v.%) vs. log $\hat{c}$; the improvement obtained by using mean results.

If the errors are expressed as percentage, for a normal distribution the random error entropy is:

$$H[(\Delta c)\%] = \left(\log \frac{100\sigma_c}{\hat{c}\sqrt{n}} \times \sqrt{2\pi e}\right) \tag{8.36}$$

From this, for a region where σ_c can be taken as constant, there is a linear relation between the entropy of the errors and the logarithm of the content:

$$H[(\Delta c)\%] = \left(\log \frac{100\sigma_c}{\sqrt{n}} \times \sqrt{2\pi e}\right) - \log c \tag{8.37}$$

This relation will be called the **entropy characteristic of determination for a method of analysis**. The model for this is shown in Fig. 8.7.

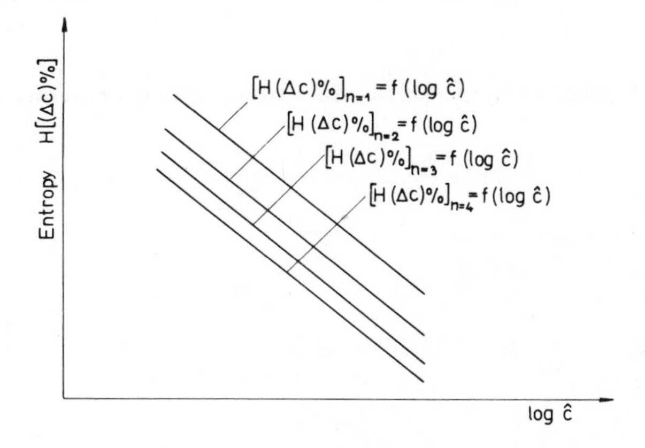

Fig. 8.7 The entropy characteristic of determination and the change induced by using mean results.

The model for the distribution of the random errors (percentile relative values) as a function of content is given in Fig. 8.8. When systematic errors are absent, the reliability of the method with respect to two error limits $(\Delta c)_1\%$ and $(\Delta c)_2\%$ is represented by the hatched areas.

The dependence on content ($\hat{c}$) of the reliability $\hat{p}[(\Delta c)\%]$, with respect to two error limits, $(\Delta c)_1\%$ and $(\Delta c)_2\%$, is shown in Fig. 8.9.

In the absence of systematic errors and when the random errors are normally distributed, the determination characteristic expressed as $p[(\Delta c)\%]$ vs. content can be obtained in a very simple way from the determination characteristic expressed as c.v. or $H[(\Delta c)\%]$ vs. content, as shown in the next section.

This discussion of the concept of the determination characteristic shows the following.

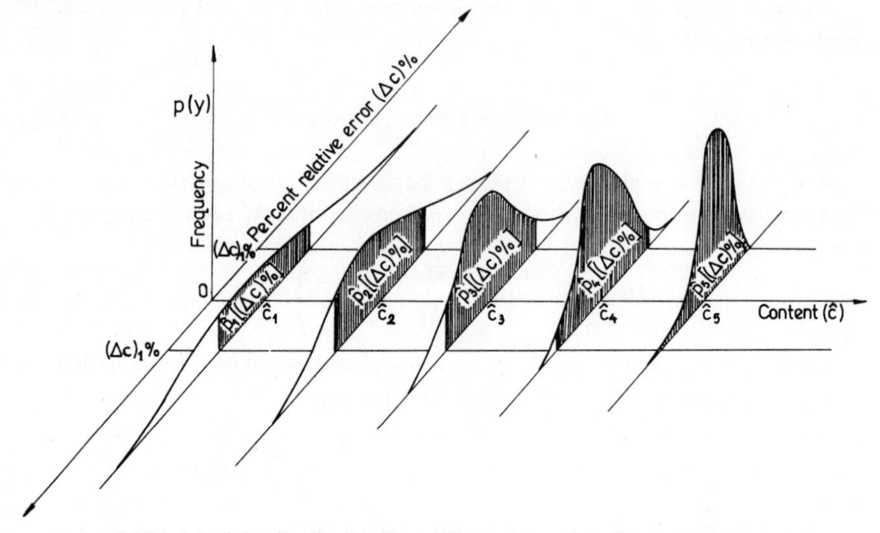

Fig. 8.8 Model of the distribution for relative error as a function of amount present.

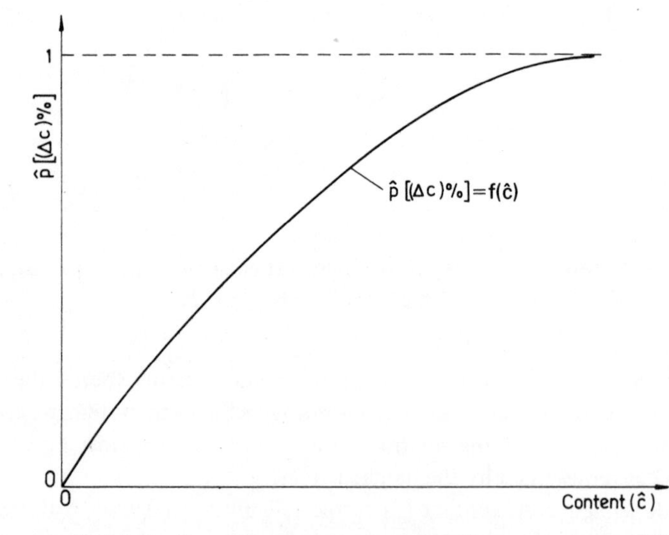

Fig. 8.9 Model for the determination characteristic expressed as reliability vs amount present.

(a) A determination characteristic is peculiar to each method of analysis and to the particular conditions under which the determinations are done. Hence, practically, for each method and set of working conditions, the corresponding characteristic must be evaluated.

(b) The requirement to have a constant standard deviation σ_c for the analysis errors (i.e. that σ_c should be independent of the content of component) is not in general satisfied. For this reason the determination characteristic must usually be evaluated from results obtained on several standard samples covering the entire range of contents likely to be encountered.

(c) For any method of analysis, the determination characteristic deteriorates as the concentration decreases. Hence, for each method of analysis, the capacity of determination has a certain limit at low concentrations.

8.4 THE DETERMINATION LIMIT

8.4.1 Introduction

For any method of analysis, the determination performance becomes progressively poorer with decrease in the content of component to be determined. Hence, for any method of analysis, there is a specific level for this content, below which the method will cease to give satisfactory evaluation of the content. This value will be called the **determination limit**.

Unlike the detection limit, which has been extensively discussed in the analytical literature, the determination limit has seldom been considered.

The determination limit for radiochemical methods was defined by Currie [32] as the limit 'at which a given analytical procedure will be sufficiently precise to give a satisfactory quantitative estimate'. Hence, according to this author, the determination limit is the minimum content at which the coefficient of variation reaches a preselected limiting value. Currie chose a limiting value of 10%. An arbitrary value of c.v. = 10% for the evaluation of the determination limit has also been suggested by other authors [33,34].

The value of the determination limit suggested by Matherny [35] for spectrographic methods of analysis is the amount $\hat{c}_D$ of component to be determined, at which the slope of a plot of c.v. vs. $\hat{c}$ is −1 (Fig. 8.10).

More generally, if the determination characteristic is used in an explicit form with respect to the content of component to be determined:

$$\hat{c} = f(\mathrm{M}) \tag{8.38}$$

where M is the parameter by which the performance of the method of determination is measured (that is, c.v., $H[(\Delta c)\%]$ or $p[(\Delta c)\%]$), the determination limit will be:

$$\hat{c}_D = f(\mathrm{M}_D) \tag{8.39}$$

where M_D is the accepted limiting value for the performance parameter.

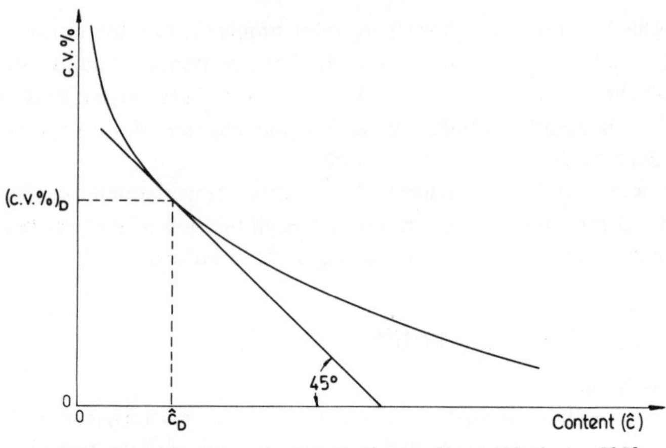

Fig. 8.10 The determination limit according to Matherny [35].

8.4.2 The determination limit in terms of the coefficient of variation

In the absence of systematic errors, the percentile coefficient of variation is a measure of both the precision and the accuracy. The determination limit is defined by the value $\hat{c}_D$ for the component to be determined which corresponds to the accepted limiting value $(c.v.)_D$ of the coefficient of variation. This last quantity is also called the **acceptability threshold**.

In other words, when the component to be determined is present at a level below the determination limit so defined, the coefficient of variation is greater than the threshold of acceptability.

The definition and evaluation of the determination limit by this method are shown in Fig. 8.11. The determination limit is evaluated by interpolating the threshold value $(c.v.)_D$, i.e. the preselected maximum value of the coefficient of variation, in the determination characteristic. It can be seen in the same figure that the determination characteristic can be improved by using means of determination results instead of individual results. Hence, the determination limit can be shifted to lower levels of the content by increasing the number of results which contribute to the means.

We shall assume that for the region of interest the standard deviation σ_c can be regarded as constant. For single results, the determination limit corresponding to an acceptability threshold $(c.v.)_D$ is given by:

$$\hat{c}_D = \frac{100\ \sigma_c}{(c.v.)_D} \tag{8.40}$$

For means of results obtained from n single determinations, the determination limit will be:

$$(\hat{c}_D)_n = \frac{100\, \sigma_c}{(\text{c.v.})_D \sqrt{n}} \tag{8.41}$$

With the notation

$$r_D = \frac{100}{(\text{c.v.})_D} \tag{8.42}$$

we have

$$\hat{c}_D = r_D \sigma_c \tag{8.43}$$

and

$$(\hat{c}_D)_n = r_D \frac{\sigma_c}{\sqrt{n}} \tag{8.44}$$

The quantity r_D will be called the **signal-to-noise ratio corresponding to the determination limit**, because it is equal to the ratio of the determination limit $\hat{c}_D$ to its standard deviation:

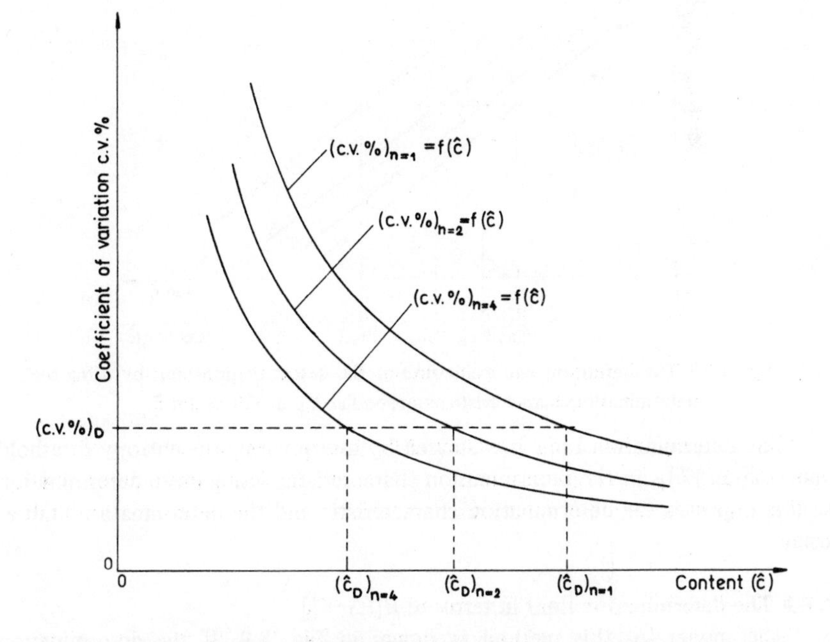

Fig. 8.11 The definition and evaluation of the determination limit by using the determination characteristic expressed as c.v. vs. amount present.

$$r_{\mathrm{D}} = \frac{\hat{c}_{\mathrm{D}}}{\sigma_c} \tag{8.45}$$

Thus, for a threshold value $(\mathrm{c.v.})_{\mathrm{D}} = 10\%$, from (8.42) $r_{\mathrm{D}} = 10$, so the determination limit will be

$$\hat{c}_{\mathrm{D}} = 10\sigma_c \tag{8.46}$$

Generally, however, σ_c is not independent of c and the determination characteristic must be evaluated from a set of standard trace samples, as mentioned on p. 341.

8.4.3 The determination limit in terms of $H[(\Delta c)\%]$

When the entropy determination characteristic is used, the determination limit can be defined as the content $\hat{c}_{\mathrm{D}}$ which corresponds to the entropy acceptability threshold $H[(\Delta c)\%]_{\mathrm{D}}$. Thus when the component is below the determination limit $\hat{c}_{\mathrm{D}}$ defined in this way, the entropy of the error will be higher than the threshold. The definition and evaluation of the determination limit in this way is shown in Fig. 8.12.

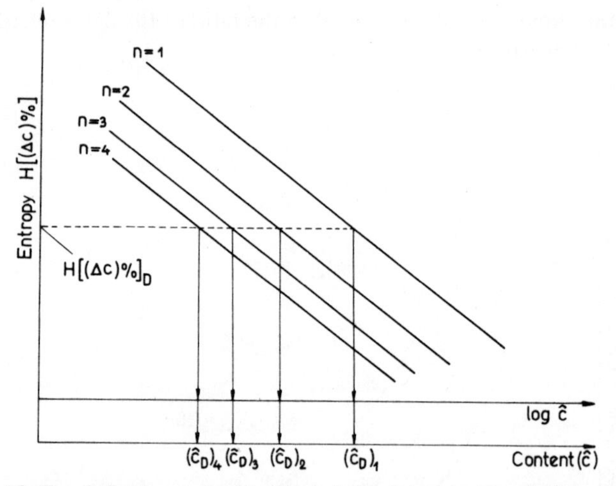

Fig. 8.12 The definition and evaluation of the determination limit by using the determination characteristic expressed as $H[(\Delta c)\%]$ vs. $\log \hat{c}$.

The determination limit is evaluated by interpolating the entropy threshold value $H[(\Delta c)\%]_{\mathrm{D}}$ in the determination characteristic. Using mean determination results improves the determination characteristic and the determination limit as usual.

8.4.4 The determination limit in terms of $p[(\Delta c)\%]$

The model for this method is shown in Fig. 8.9. If the determination performance parameter is the reliability with respect to two percentile error limits this parameter must be higher than some statistically selected threshold value $p[(\Delta c)\%]_{\mathrm{D}}$; it is obvious that, for contents below a certain value $\hat{c}_{\mathrm{D}}$, the method will cease to satisfy this performance requirement.

Hence, on the basis of this determination characteristic, the determination limit is defined as the content $\hat{c}_{\mathrm{D}}$ of the component to be determined at which the reliability is equal to the statistical threshold of acceptability $p[(\Delta c)\%]_{\mathrm{D}}$.

We shall consider the determination characteristic in the form in which it is obtained with respect to the content:

$$\hat{c} = f\left(p[(\Delta c)\%]\right) \tag{8.47}$$

The determination limit is obtained from (8.47) by interpolating the value of the threshold of acceptability. Hence:

$$\hat{c}_{\mathrm{D}} = f\left(p[(\Delta c)\%]_{\mathrm{D}}\right) \tag{8.48}$$

Obviously, when the amount of the component present is smaller than the determination limit c_{D} defined in this way, the reliability $p[(\Delta c)\%]$ will be smaller than the statistical threshold of acceptability $p[(\Delta c)\%]_{\mathrm{D}}$.

We have considered three methods for defining and evaluating the determination limit. There is a correspondence between these methods, so that we can go from one to another, as will be shown in the following example.

Example 8.2. The three methods given above for estimating the determination limit will be exemplified with the data in Table 8.1, which gives

Table 8.1 Data for evaluation of determination characteristics

Standard sample no.	1	2	3	4	5
Content: $\hat{c}_{\mathrm{W}}$ %	0.001	0.006	0.012	0.022	0.053
Mean of determinations: $\bar{c}_{\mathrm{W}}$ %	0.001	0.005	0.012	0.024	0.050
Standard deviation: s_c	0.0013	0.0014	0.0018	0.0022	0.0027
Coefficient of variation: %	130	23	15	10	5.1
Entropy of relative error: $H[(\Delta c)\%]$ dits	2.75	1.97	1.78	1.61	1.32
Reliability: $p[(\Delta c)\%]$ with respect to the error limits ± 20%	0.124	0.497	0.816	0.962	0.999

the tungsten contents of five standard steels, and, for each sample, the mean and standard deviation obtained from 15 determination results. From these data, for each standard sample the coefficient of variation, the entropy of the percentile error, and the reliability with respect to error limits of +20% and −20% have been evaluated.

Figure 8.13 gives the determination characteristic determined from the coefficient of variation. This characteristic is also represented in logarithmic coordinates in Fig. 8.14.

For the coefficient of variation, we shall assume the threshold of acceptability $(c.v.)_D = 10\%$. By interpolating this value into the determination characteristic, the determination limit is estimated at $\bar{c}_D = 0.022\%$. Hence, if tungsten content $\hat{c}_W$ is smaller than 0.022%, the coefficient of variation for the results of determination will be greater than the threshold value $(c.v.)_D = 10\%$.

From the data for $H[(\Delta c)\%]$ in Table 8.1 we obtain the estimated determination characteristic shown in Fig. 8.15. For the coefficient of variation of 10%, the corresponding value of the error entropy is $H[(\Delta c)\%]_D = 1.6$ dits. By interpolating this value into the determination characteristic, the estimation value of the determination limit is again found to be $\bar{c}_D = 0.022\%$. The dit (or Hartley) is the base-10 analogue of the bit.

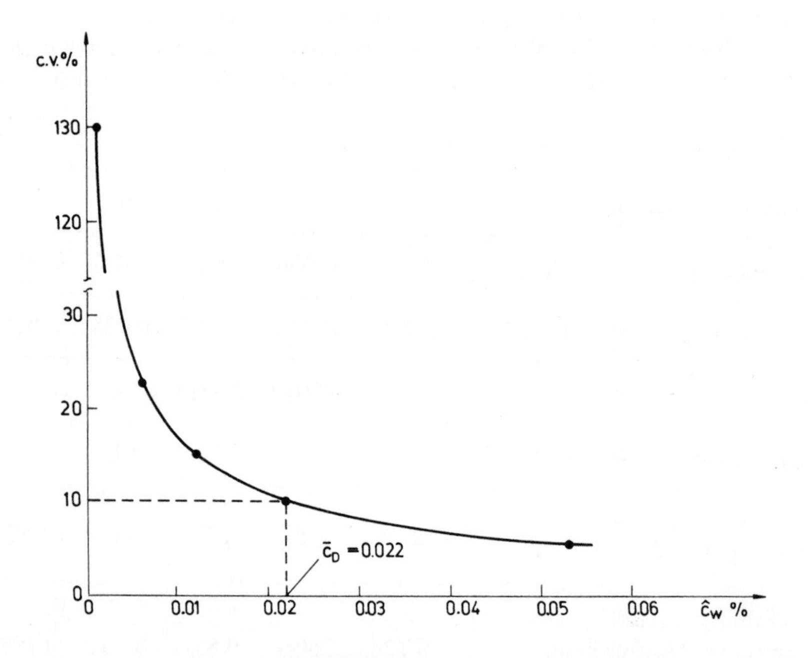

Fig. 8.13 The estimation of the determination characteristic expressed as c.v. vs. amount present.

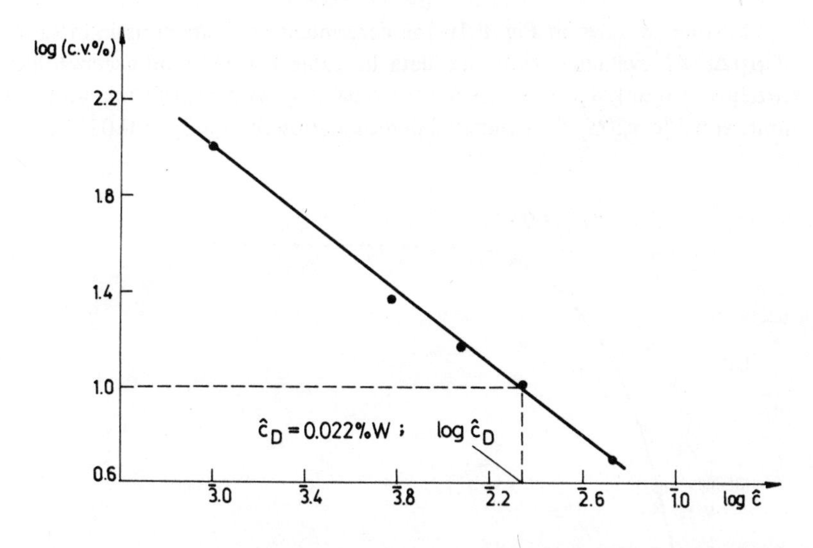

Fig. 8.14 The estimation of the determination characteristic expressed as log(c.v.) vs. log $\hat{c}$.

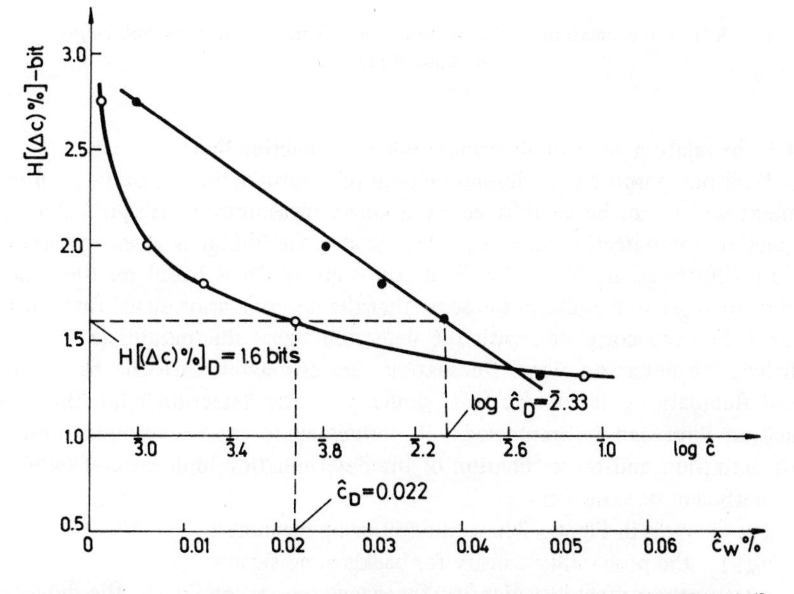

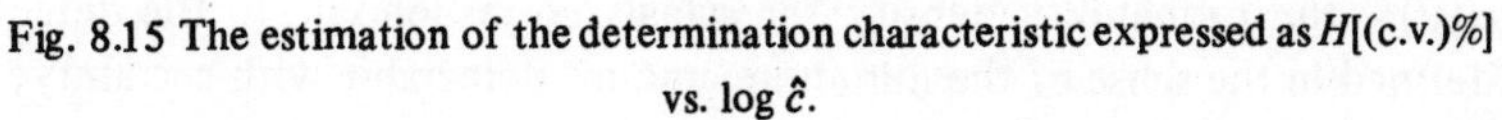

Fig. 8.15 The estimation of the determination characteristic expressed as H[(c.v.)%] vs. log $\hat{c}$.

Further, we give in Fig. 8.16 the determination characteristic in terms of $p[(\Delta c)\%]$ evaluated from the data in Table 8.1. With an acceptability threshold $p[(\Delta c)\%]_D = 0.975$ for the reliability with respect to two error limits equal to ±20%, the estimated determination limit is $\bar{c}_D = 0.023\%$.

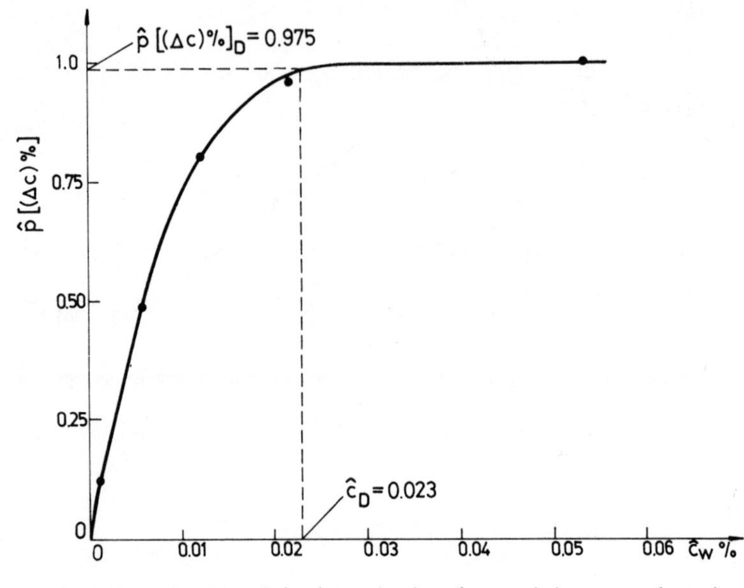

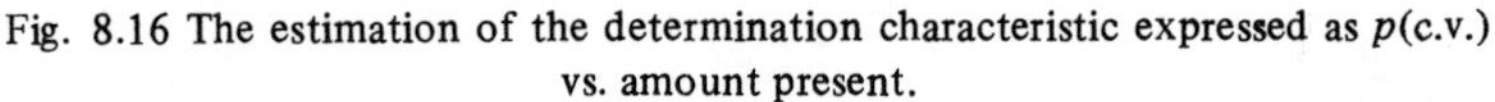
Fig. 8.16 The estimation of the determination characteristic expressed as p(c.v.) vs. amount present.

8.4.5 The relation between determination and detection limits

Here our purpose is to define the limit of determination $\hat{c}_D$ as the minimum content which can be established by a single measurement as significant with respect to the detection limit $\hat{c}_d$ [36]. Such a definition is given by the procedure illustrated in Fig. 8.17. This four-step model is based on the general theory of signal detection, in the sense that the determination signal fluctuations ($I > 1$ bit) are correlated with the detection signal fluctuations ($I = 1$ bit). Further, the detection signal fluctuations are correlated with the background signal fluctuations. It follows that, similarly to the detection limit, the determination limit can be expressed with reference to the background noise $\hat{y}_0$. This definition and the estimation of the determination limit are not based on the coefficient of variation.

The symbols in Fig. 8.17 have the following meanings:

$p_0(y)$ – the probability density for background signals;

$p_d(y)$ – the probability density for signals corresponding to the detection limit defined in the sense of the minimum amount detectable with certainty;

$p_D(y)$ – the probability density for signals corresponding to the amount $\hat{c}_D$ (the determination limit) which can be established by a single analytical experiment as significant with respect to the detection limit;

y_k – the decision level between the background signals and signals corresponding to the detection limit;

y'_k – the decision level between the signals corresponding to the detection limit and signals corresponding to the determination limit.

It is assumed that the standard deviation is independent of the content of component to be determined.

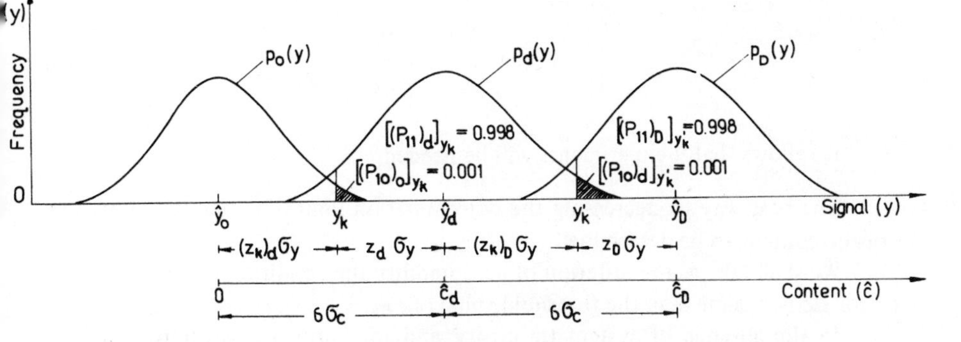

Fig. 8.17 The minimum amount which can be established by a single analytical measurement as significant with respect to the detection limit.

For the definition of the detection limit $\hat{c}_d$ the values $[(P_{10})_o]_{y_k} = 0.001$ and $[(P_{11})_d]_{y_k} = 0.998$ are used, and to define $\hat{c}_D$, the values $[(P_{10})_d]_{y'_k} = 0.001$ and $[(P_{11})_D]_{y'_k} = 0.998$.

The signal-to-noise ratio corresponding to the determination limit defined in this way (see Fig. 8.17) is $r_D = (\hat{y}_D - \hat{y}_b)/\sigma_c = (z_k)_d + z_d + (z_k)_D + z_D = 12$, so

$$\hat{c}_D = 12\sigma_c,$$

and

$$\hat{y}_D = \hat{y}_o + r_D\sigma_y$$

When the component to be determined is present in amount $\hat{c}_D$ equal to the determination limit so defined, the corresponding value of the coefficient of variation is:

$$(\text{c.v.})_D = 8.3\%$$

and the corresponding entropy is:

$$H[(\Delta c)\%]_D = 1.5 \text{ dits}$$

8.4.6 Lowering the determination limit by repeated determinations

The decrease in the determination limit when a set of results is used (expressed as the grand mean) can be seen in Figs. 8.5, 8.6, and 8.7, by examining the corresponding change in the determination characteristic.

Example 8.3. Suppose that when the content of component to be determined is $\hat{c} = 0.005\%$, the corresponding standard deviation for individual results is $\sigma_c = 0.0015\%$, the coefficient of variation will be c.v. = 30%. We wish to establish the number of experiments n required to reduce the coefficient of variation for the mean result of these experiments to 10%, i.e. $(\text{c.v.})_n = 10\%$. From the equation:

$$(\text{c.v.})_n = \frac{100\sigma_c}{\hat{c}\sqrt{n}}\ \% \tag{8.49}$$

it follows that 9 experiments will be needed.

The best way of decreasing the determination limit is sequential estimation of the content to be determined.

We shall take as the criterion of acceptability the condition that the absolute error Δc is smaller than the threshold value $(\Delta c)_D$.

In the absence of systematic errors, and for normally distributed random errors, the confidence interval for the true value of the content to be determined is given by:

$$(\bar{c})_n - |t_{(\alpha/2;n-1)}| \times \frac{s_c}{\sqrt{n}} < \hat{c} < (\bar{c})_n + |t_{(\alpha/2;n-1)}| \times \frac{s_c}{\sqrt{n}} \tag{8.50}$$

where $(\bar{c})_n$ and s_c are respectively the mean and the sample standard deviation for n analytical determinations, $t_{(\alpha/2;n-1)}$ is Student's t, and α is the significance level.

Table 8.2 Data for sequential estimation of the tungsten content in a steel sample

n	Result of determination	Mean of results $(\bar{c})_n$	$\lvert t_{(\alpha/2;n-1)}\rvert \cdot \frac{s_c}{\sqrt{n}}$	n	Result of determination	Mean of results $(\bar{c})_n$	$\lvert t_{(\alpha/2;n-1)}\rvert \cdot \frac{s_c}{\sqrt{n}}$
1	0.0039	–	–	7	0.0045	0.0049	0.00130
2	0.0071	0.0055	0.02034	8	0.0045	0.0048	0.00109
3	0.0032	0.0047	0.00517	9	0.0045	0.0048	0.00094
4	0.0045	0.0047	0.00270	10	0.0032	0.0046	0.00090
5	0.0045	0.0046	0.00183	11	0.0022	0.0044	0.00094
6	0.0065	0.0050	0.00160	12	0.0044	0.0044	0.00086

Consequently, in order to guarantee at a significance level α that $\Delta c < (\Delta c)_D$, as many determinations must be made as required to obtain:

$$|t_{(\alpha/2;n-1)}| \times \frac{s_c}{\sqrt{n}} < (\Delta c)_D \tag{8.51}$$

and the result is expressed as the mean value.

Example 8.4. This sequential method will be exemplified with the data in Table 8.2. The significance level is $\alpha = 0.05$ and the threshold of acceptability for absolute errors is $(\Delta c)_D = 0.001$.

From Table 8.2 it follows, that from the 9th determination onwards, the condition $(\Delta c)_n < (\Delta c)_D = 0.001$ is fulfilled and experimentation can be stopped.

To lower the determination limit we can use the frequentometric method [37], in either the $P_{11} = f(c)$ form or the $z = f(c)$ form (see Section 7.5.6). Therefore the dependent variable in the transfer function is not physical in character but statistical, whether in non-linear form $P_{11} = f(c)$, in which case the probability of positive runs P_{11}:$y > y_0$ [38–40], or $y > y_k$ [41-43] is considered, or in linear form, $z = f(c)$, in which case the normal deviate, z, for $P_{11} > 0.5$, $\Phi(z) = P_{11} - 0.5$ or for $P_{11} < 0.5$, $\Phi(z) = 0.5 - P_{11}$, is considered.

8.4.7 Conclusions concerning the determination limit

(1) Because of perturbations, any analytical determination result includes an error (uncertainty).

(2) The performance of a method of determination, which is measured by the coefficient of variation, percentile error entropy or reliability with respect to two percentile error limits, is a function of the content to be determined.

(3) The *determination characteristic* of a method of analysis is the dependence between one of the performance parameters and the value of the content to be determined.

(4) The accuracy of estimation of the content to be determined is estimated by the coefficient of variation, the entropy of the relative error, or the reliability with respect to two limits for the relative error. The smaller the content to be determined, the poorer the accuracy. Hence, for contents below a certain value the accuracy becomes unacceptable for practical purposes.

(5) The determination limit is a measure of the capacity of a method to determine small contents. In general, if we use the determination characteristic of a method in the form

$$\hat{c} = f(M) \tag{8.52}$$

where M is the estimation accuracy expressed as (c.v.), $H[(\Delta c)\%]$, $p[(\Delta c)\%]$ or

$\hat{y}_D$, the determination limit will be:

$$\hat{c}_D = f(M_D) \tag{8.53}$$

where M_D is the threshold of acceptability for the parameter by which the estimation accuracy is evaluated.

(6) The detection limit $\hat{c}_d$ and the determination limit $\hat{c}_D$ are two different performance parameters for an analytical system.

The detection limit measures the capacity of a method to detect small contents. The definition of this quantity is based on a false and a true detection probability, $[(P_{10})_0]_{y_k}$ and $[(P_{11})_d]_{y_k}$ respectively. The detection limit defines the domain of contents $\hat{c} > \hat{c}_d$ for which the analytical system can detect the component investigated with a true detection probability higher than the threshold value $[(P_{11})_d]_{y_k}$.

The determination limit measures the capacity of a method of analysis to determine small contents. It defines the minimum content of component to be determined for which the method can be used with an accuracy greater than the threshold of acceptibility M_D. That is, the accuracy of the estimations is acceptable only when the content of component to be determined is $\hat{c} > \hat{c}_D$.

If the detection limit is defined as the minimum content detectable with certainty [36], it corresponds to a signal-to-noise ratio $r_d = 6$ (see Section 7.2). The coefficient of variation corresponding to a content equal to this detection limit is c.v. = 17%. This value is usually too large to be accepted. Hence, for practical purposes, the determination limit is always larger than the detection limit, i.e. $\hat{c}_D > \hat{c}_d$. The relative positions of the values of these two parameters is shown graphically in Fig. 8.18.

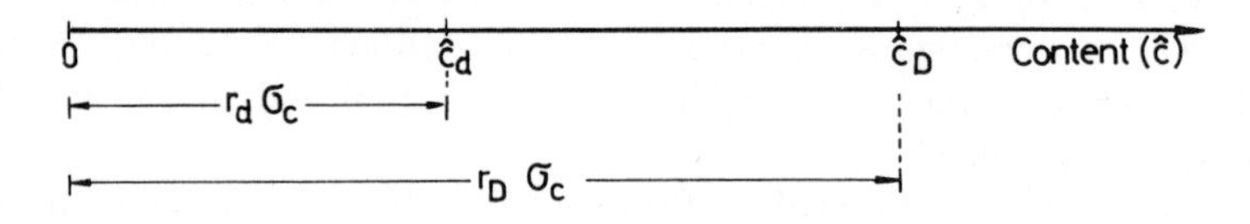

Fig. 8.18 The limit of detection and the limit of determination.

(7) The minimum content which can be established by a single analytical measurement, as significant with respect to the detection limit, is a measure of the determination capacity. When the absolute standard deviation is constant, the signal-to-noise ratio for this content is $r_D = 12$, the coefficient of variation is $(\text{c.v.})_D = 8.3\%$, and the entropy of the relative error is $H[(\Delta c)\%]_D = 1.5$ dits.

(8) The determination capacity of an analytical system can be lowered by using either the mean of several analytical results, or a sequential estimation from them.

8.5 ANALYSIS OF NON-HOMOGENEOUS MATERIALS

In the case of a non-homogeneous material, we may be required to determine the mean content of one or several of the material components, or to evaluate the degree of non-homogeneity (the variability) of the material, or to solve both these problems.

Unlike the case of homogeneous materials, where there are no sampling difficulties, taking the necessary samples for non-homogeneous materials becomes a complex problem, both theoretically and practically. Needless to say, the difficulties are much greater when the analytical determination is made for trace components.

8.5.1 The determination of the mean content of a component in a non-homogeneous material

The classical procedure for this is to take samples from as many points as possible throughout the material; each of these samples is then analysed, and the mean value of the results is obtained. This procedure requires a large number of analytical determinations, and hence a high expenditure of time and materials.

The alternative is to use a statistically planned sampling strategy. This usually takes the form of the following three stages.

In the first stage, several samples are extracted from the bulk of the material, and a 'primary mean sample' is made up from them. The number of points from which the samples are extracted, the disposition of these points within the material, and the mass extracted at each point, are all established according to the physical and chemical nature of the material, the type of inhomogeneity, and the total mass of the material investigated.

The second stage consists of homogenization of the primary mean sample, and then reduction of the sample mass in several steps. The last step of this stage ends with the extraction of the 'laboratory sample'.

In the third stage, a further homogenization treatment is applied to the laboratory sample. In the end, one or several samples for analysis are extracted – these are the samples on which the analytical determinations will actually be performed.

Obviously, when the purpose is the determination of the mean content of a constituent, the concentration c_i in the samples for analysis must be close to the true mean value $\hat{c}$; the closer these values, the more accurate the sampling. Let us write $c_1, c_2, \ldots, c_n$ for the concentrations in samples for analysis obtained by repeating completely the sampling process. If the sampling is accurate, the quantity $\sum_{i=1}^{n} (c_i - \hat{c})$ tends asymptotically towards zero as n increases. The variance σ_R^2 of the results of analysis of a non-homogeneous material consists of two components: one of them, σ_S^2, is due to the non-homogeneity, and the other, σ_A^2, is due to analytical determination errors. We have:

$$\sigma_R^2 = \sigma_S^2 + \sigma_A^2 \tag{8.54}$$

When we have to determine the mean content of a constituent, the sampling process must be planned and conducted in such a way as to minimize the sampling variance σ_S^2.

In the following, we shall consider the case of a one-stage sampling process. That is, the samples for analysis are extracted directly from the given material. The sampling variance σ_S^2 then depends on the degree of inhomogeneity of the material, the dimensions of the constituent particles of the material, the relative density of the constituents, and the mass of the samples for analysis.

An important problem is to find the sample mass required in order to make the variance σ_S^2 smaller than a certain threshold value. It is very difficult to derive theoretically a relation between the variance σ_S^2 and the mass of the sample, assuming the physico-chemical and statistical properties of the material to be known. Such a relation can be derived only for simplified material models.

Harris and Kratochvil [44] take into account a simple material model, for a one-stage sampling process. It is assumed that the material consists of two types of spherical particles – A and B. All the particles have the same diameter, and they are randomly distributed in the material. The component to be determined is present only in the particles of type A, but is completely absent in the particles of type B. From the binomial distribution, for samples containing n particles, the standard deviation of the number of type A particles is given by:

$$\sigma_S = \sqrt{np(1-p)} \tag{8.55}$$

where p is the number fraction of the type A particles in the material.

The coefficient of variation for the number of A-type particles in samples containing n particles will be:

$$(\text{c.v.})_S = 100\sqrt{(1-p)/np} \tag{8.56}$$

Benedetti-Pichler [45] also considered the case of a model material consisting of two kinds of particles, 1 and 2. The component is present in both types of particles, but in different concentrations. This author finds the following equation which relates the number of particles of a sample, n, to the coefficient of variation for the content of component to be determined:

$$n = p(1-p)\left(\frac{d_1 d_2}{d^2}\right)^2 \left[\frac{(P_1 - P_2)}{P_{av}\,(\text{c.v.})_S} \times 100\right]^2 \tag{8.57}$$

where d_1 and d_2 are the particle densities, d is the average density of the material, P_1 and P_2 are the percentages of component to be determined, in the two kinds of particles, P_{av} is the average percentage of the component in the material.

The problem of analysis error arising from material inhomogeneity has been treated in several works [46–55], in connection with the determination of trace components in solid geological materials. Wilson [46] has considered a system of two mineral species of particles which are binomially distributed. He has established theoretically the following two equations

$$(\text{c.v.})_S = \sqrt{K'/n} \tag{8.58}$$

$$(\text{c.v.})_S = \sqrt{K/m} \tag{8.59}$$

where n is the number of particles in the sample, m is the sample mass and K' and K are constants with values depending on the concentrations and densities of the two mineral species.

Ingamells and Switzer [47–49] call constant K the **sampling constant**. From (8.59), K is equal to the sample mass required to obtain a coefficient of variation $(\text{c.v.})_S$ of 1% for the component determined. If K is known, the sample mass required to obtain any desired value of the coefficient of variation can be established from (8.59).

The simple model used by Wilson [46] is not generally valid, however. For example, it disregards segregation. Visman *et al.* [50–52] have developed a theory which also takes segregation into account. They obtain the following equation, in which the coefficient of variation is related to the sample mass m:

$$(\text{c.v.}\%)_S = \sqrt{\frac{K}{m} + K''} \tag{8.60}$$

where K has the same meaning as Wilson's constant, and K'' is the component of the variance due to the segregation phenomenon.

Visman's work has been examined by Duncan [54], who has shown that Eq. (8.60) does not have general validity; it still shows a degree of approximation for many materials.

In the case of geochemical materials, the problem of the sample for analysis has also been examined by Ingamells and Switzer [47–49]. For a material with two mineral constituents which appear in the form of uniform particles, they have derived the following sampling constant:

$$K_S = \left(\frac{Bd_B - Hd_H}{K}\right)^2 \times \frac{p_W q_W u^3 W}{d_B W_H + d_H W_B} \times 10^{-8}\,\text{g} \tag{8.61}$$

where B and H are the content of the component to be determined, in the two kinds of particle, K is the mean content of the component in the material, W is the mass of the sample, i.e. W_B g of one mineral species and W_H g of the other, p_W and q_W are the weight fractions of the two minerals, u is the linear dimension of the particles (assumed to be cubic), and K_S is the sampling constant for the component, i.e. the mass of the sample required for analysis in order to have $(c.v.)_S = 1\%$. It has been pointed out that if $Bd_B = Hd_H$, then $K_S = 0$, but this point has been further considered by Ingamells [55a].

A practical study of the effect of the material inhomogeneity on the error of determination for trace constituents in mineral rocks has been made by Jaffrezic [53]. He found that the sample constant K obtained from (8.60) fails to characterize completely the inhomogeneity of powdered materials. He also showed that correct use of mineral rock standards can only be made when the function $(c.v.)_S = f(m)$ is known, i.e. the relation between the coefficient of variation and the mass of sample analysed. This function is usually evaluated experimentally.

8.5.2 Experimental design for evaluating the compositional inhomogeneity of a material

The planning of an analytical experiment for the determination of the compositional inhomogeneity (variability) of a material can be seen Fig. 8.19.

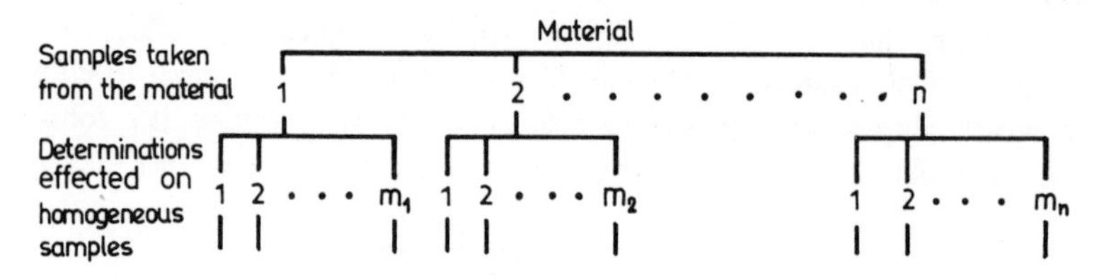

Fig. 8.19 Design for determination of variability of composition.

According to this plan, samples with a certain mass are extracted from n points of the material. Each of these samples is homogenized so that it will have finally a negligible compositional variability. Repeated analyses are then performed on each sample.

Let us assume the general case when the number of repeated analyses varies from one sample to another. The n samples will be labelled by an index i ($i = 1, 2, \ldots, n$), and the number of determinations on sample i will be written as m_i. The order number of each of these determinations will be shown by an index j ($j = 1, 2, \ldots, m_i$). The analysis results will be written according to the model:

$$c_{ij} = \hat{c} + (\Delta c)_i + (\Delta c)_{ij} \qquad (8.62)$$

where $\hat{c}$ is the true value of the mean content, $(\Delta c)_i$ is the error due to the inhomogeneity of the material (in other words, the deviation of the content in the sample i from the true value $\hat{c}$ of the mean content for the overall material), and $(\Delta c)_{ij}$ the error introduced by the process of analysis, i.e. the determination error made in determination j of the content of sample i.

We shall assume that $(\Delta c)_i$ – the material inhomogeneity – is a random variable with a mean equal to zero and a variance σ_S^2; further, for all i and j, the analytical determination error $(\Delta c)_{ij}$ is assumed to be a random variable with zero mean value and variance σ_A^2. The variance of the determination results, σ_R^2, will be the sum of the two separate variances:

$$\sigma_R^2 = \sigma_S^2 + \sigma_A^2 \tag{8.63}$$

The degree of inhomogeneity of the material is measured by the variance σ_S^2. In order to estimate this variance, the experimental results c_{ij} must be processed in such a way as to separate the term σ_S^2. The set of analytical determination results c_{ij}, obtained as shown, has the structure given in Table 8.3. The calculations which must be made in order to evaluate the variances σ_S^2 and σ_A^2 are given in Table 8.4 [56, 57].

Table 8.3 Set of analytical results obtained for evaluating variability of material composition

Sample	Result of analytical determinations
1	c_{11} c_{12} c_{13} $\cdots$ c_{1m_1}
2	c_{21} c_{22} c_{23} $\cdots$ c_{2m_2}
$\vdots$	$\vdots$ $\vdots$ $\vdots$
i	c_{i1} c_{i2} c_{i3} $\cdots$ c_{ij} $\cdots$ c_{im_i}
$\vdots$	$\vdots$ $\vdots$ $\vdots$
n	c_{n1} c_{n2} c_{n3} $\cdots$ c_{nm_n}

In Table 8.4, the meaning of the symbols is as follows:

$\bar{c}_{i.} = \dfrac{\sum_j c_{ij}}{m_i}$, the mean of the analysis results pertaining to one sample;

$\bar{c}_{..} = \dfrac{\sum_i m_i \bar{c}_{i.}}{\sum_i m_i}$, the general mean, obtained with all the analysis results:

$N = \sum_i m_i$, the total number of analysis results;

$$k = \left(N - \frac{\sum_i m_i^2}{N} \right) / (n-1).$$

From Table 8.4, it follows that the variance σ_S^2, i.e. the measure of the variability of the composition of the material, is estimated by the quantity:

$$s_S^2 = \frac{MS_S - MS_A}{k} \tag{8.64}$$

Table 8.4 Analysis of variance

Source of variance	Degrees of freedom	Sum of squares	Mean squares	Expected mean square
Between samples	$n-1$	$SS_S = \sum_i m_i(c_{i.} - \bar{c}_{..})^2$	$MS_S = \frac{SS_S}{n-1}$	$\sigma_A^2 + k\sigma_S^2$
Within sample	$N-n$	$SS_A = \sum_i \sum_j (c_{ij} - \bar{c}_{i.})^2$	$MS_A = \frac{SS_A}{N-n}$	σ_A^2
Total	$N-1$	$SS_T = \sum_i \sum_j (c_{ij} - \bar{c}_{..})^2$	$MS_T = \frac{SS_T}{N-1}$	σ_R^2

The true mean value of the content $\hat{c}$ is estimated by the mean $\bar{c}_{..}$, and the variance corresponding to this variable is:

$$\sigma_{\bar{c}..}^2 = \frac{\sigma_S^2}{N} + \frac{\sum_i m_i}{N^2} \sigma_A^2 \tag{8.65}$$

which itself is estimated by the quantity:

$$s_{\bar{c}..}^2 = \frac{MS_T}{N} \tag{8.66}$$

8.5.3 The criterion for accepting the purity of a material

In many cases, purity is an important characteristic of materials to be used in analysis. In terms of quality, a material is accepted if the content of one or several of the components is smaller than the corresponding tolerance limit prescribed.

Owing to the variability of the analysis results, which is due to material inhomogeneity and analytical determination errors, there are two classes of error which may appear in classifying the purity of a material with respect to a tolerance limit c_L.

(1) If we accept the hypothesis $\hat{c} < c_L$ (i.e. that the content of impurity is smaller than the tolerance limit), when actually the true hypothesis is $\hat{c} > c_L$, the material is erroneously accepted as pure.

(2) If we accept the hypothesis $\hat{c} > c_L$ when actually the true hypothesis is $\hat{c} < c_L$ we erroneously reject the material as impure.

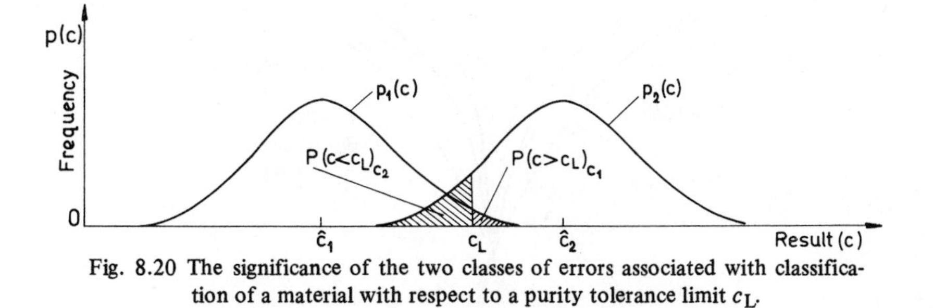

Fig. 8.20 The significance of the two classes of errors associated with classification of a material with respect to a purity tolerance limit c_L.

The meaning of these two classes of error is shown in Fig. 8.20, where we represent the probability densities $p_1(c)$ and $p_2(c)$ for results of analysis corresponding respectively to two values, $\hat{c}_1$ and $\hat{c}_2$, of the true contents, situated on opposite sides of the tolerance limit c_L. We also show $P(c > c_L)_{\hat{c}_1}$, the probability for a result $c < c_L$, when the true content is $\hat{c}_1$, and $P(c < C_L)\hat{c}_2$, the probability for a result $c < c_L$, when the true content is $\hat{c}_2$. The values of these two probabilities of classification with respect to a tolerance limit of purity are functions of the true content $\hat{c}$ and the standard deviation σ_R associated with the fluctuations of the results. This is shown in Fig. 8.21, where, for three discrete values of the standard deviation σ_R, we represent the probabilities corresponding to the two types of classifications, as a function of the true value of the content. We have assumed a normal distribution for the analysis results.

Separately for each case, we represent by the segments I, II, and III the width of the indifference regions calculated on the 3σ criterion. A correct classification of a material, i.e. a classification which is practically free from error, can be achieved on the basis of the result of a single determination on a sample extracted from an arbitrary point of the material, only if the true value of the composition of the material is outside the indifference region. The closer the true value of the content $\hat{c}$ to the tolerance limit c_L, the higher the probability for an erroneous classification.

In order to make sure that we do not erroneously accept a material on the basis of a single determination – as far as purity is concerned – from Fig. 8.21 it follows that:

(i) for $\sigma_R = 1$ we must reject all materials for which the analysis result is $c > 10$ units;

(ii) for $\sigma_R = 0.5$ we must reject any material for which the analysis result is $c > 11.5$ units;

(iii) for $\sigma_R = 0.3$ we must reject any material for which the analysis result is $c > 12.1$ units.

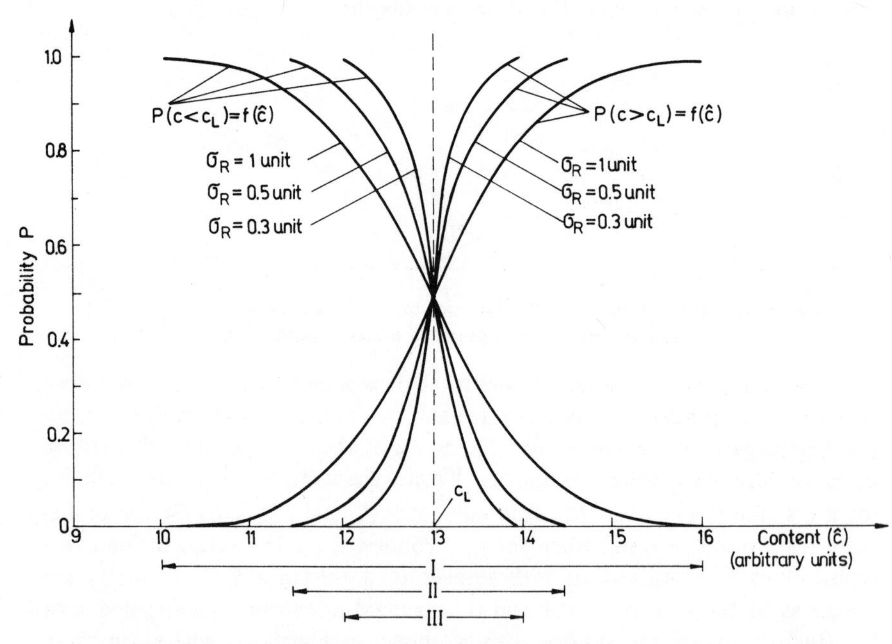

Fig. 8.21 The dependence between the true value of the amount and the probability of classification of a material with respect to a tolerance limit, at three discrete values of the standard deviation.

By proper planning of the experiments, the width of the indifference region can be diminished, i.e. we can reduce the risk of 'wrong' classification of materials. To exemplify this,. several methods of planning experiments with a view to evaluating the content and classifying the material will now be examined. We shall consider the particular case for which the standard deviations for inhomogeneity, and for errors of analysis, have respectively the values $\sigma_S = 1$ and $\sigma_A = 0.2$.

1. A single determination is made on a sample from an arbitrary point of the material. In this situation, the individual result variance is:

$$\sigma_R^2 = \sigma_S^2 + \sigma_A^2 \tag{8.67}$$

or, with the values assumed above,

$$\sigma_R^2 = 1^2 + 0.2^2 = 1.04$$

The dependence between the content and the two probabilities of classification with respect to the tolerance limit $c_L = 13$ units is shown in Fig. 8.22 for a variance $\sigma_R^2 = 1.04$ (see the curves labelled a).

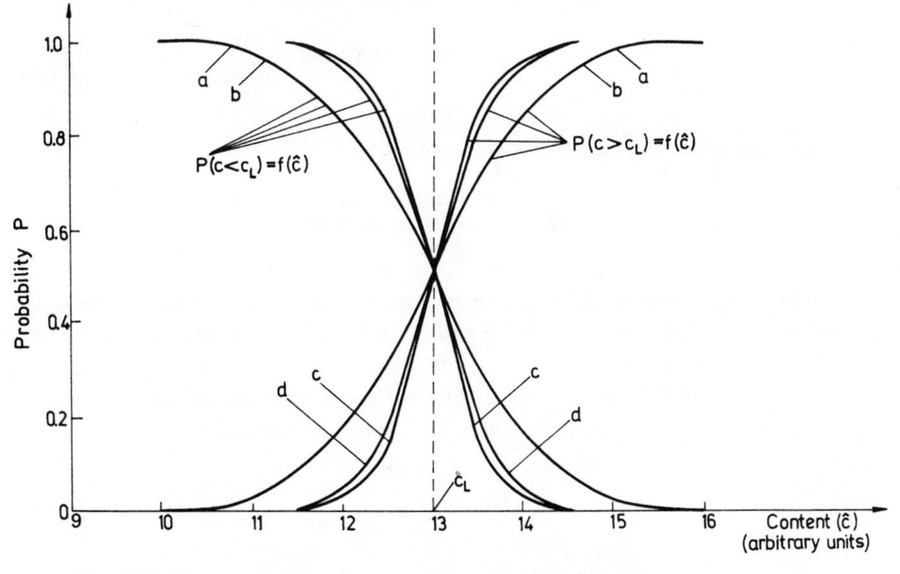

Fig. 8.22 The importance of experimental planning when classifying a material with respect to a tolerance limit for purity.

For purity control with respect to the tolerance limit $c_L = 13$, by means of a single determination on a sample extracted from an arbitrary point of the material, the indifference region corresponding to the 3σ criterion is between 9.94 and 16.06 units.

2. A sample extracted from a single point of the material is homogenized, then analysed 4 times. The mean result is calculated. The mean result variance will be given by:

$$\sigma_R^2 = \sigma_S^2 + \frac{\sigma_A^2}{4} \tag{8.68}$$

or, in our particular case:

$$\sigma_R^2 = 1 + \frac{(0.2)^2}{4} = 1.01$$

As $\sigma_S^2 \gg \sigma_A^2$, the analytical determination results have practically the same variance as case 1. Hence, such a planning of the experiment does not appreciably diminish the indifference region (see the curves labelled b in Fig. 8.22) relative to that for the case of a single determination; the indifference region has practically the same value (9.98–16.02 units).

3. Four samples are extracted from different points of the material, and each sample is analysed once. The result is expressed as the average of the four determinations. For this experiment the result variance is

$$\sigma_R^2 = \frac{\sigma_S^2 + \sigma_A^2}{4} \quad (8.69)$$

For the particular case considered, we obtain:

$$\sigma_R^2 = \frac{1 + 0.04}{4} = 0.26$$

Hence, in this case the indifference region corresponding to the 3σ criterion is considerably narrowed, to the interval 11.47–14.53 units, shown in Fig. 8.22 by the curves labelled c.

4. Four samples are extracted from different material points, mixed and homogenized, and then analysed once. The result variance will be:

$$\sigma_R^2 = \frac{\sigma_S^2}{4} + \sigma_A^2 \quad (8.70)$$

or, in our case:

$$\sigma_R^2 = \frac{1}{4} + 0.04 = 0.29$$

This plan also results in a substantial decrease in the variance. The indifference region is reduced to the interval 11.38–14.62 units, shown in Fig. 8.22 by the curves labelled d.

This last plan proves to be the most advantageous, because it requires only a single analytical determination, and the classification errors are practically the same as in the case 3, where four analytical determinations were made.

The conclusion is that planning analytical experiments on statistical principles is important in studying the degree of purity of materials.

REFERENCES

[1] Youden, W. J., *Statistical Methods for Chemists,* Wiley, New York (1962).
[2] Bennett, C. A. and Franklin, N. L., *Statistical Analysis in Chemistry and the Chemical Industry,* Wiley, New York (1954).
[3] Gottschalk, G., *Statistic in der quantitativen chemischen Analyse,* Enke Verlag, Stuttgart (1962).
[4] Nalimov, V. V., *The Application of Mathematical Statistics to Chemical Analysis,* Pergamon, London (1963).
[5] Echschlager, K., *Fehler bei der chemischen Analyse,* Akad. Verlag, Leipzig (1965); *Errors and Measurement in Chemical Analysis,* Van Nostrand, London (1969).
[6] Doerffel, K., *Statistic in der analytischen Chemie.* VEB Deutscher Verlag für Grundstoffindustrie, Leipzig (1966).
[7] Ceaușescu, D., *Tratarea statistică a datelor chimico-analitice,* Ed. Tehnică, București (1973).
[8] Lacroix, Y., *Analyse chimique. Interpretation des resultats par la calcul statistique,* Masson, Paris (1973).
[9] Nalimov, V. V., *The Application of Mathematical Statistics to Chemical Analysis,* p. 74, Pergamon, London (1963).
[10] Behrends, K., *Z. Anal. Chem.,* **235**, 391 (1968).
[11] Doerffel, K., *Z. Anal. Chem.,* **185**, 20 (1962).
[12] Doerffel, K., *Z. Anal. Chem.,* **185**, 91 (1962).
[13] Püschel, R., *Mikrochim. Acta,* 783 (1968).
[14] Koch, W. and Sauer, K. H., *Arch. Eisenhütten.,* **35**, 861 (1964).
[15] Lassner, E., *Z. Anal. Chem.,* **222**, 170 (1966).
[16] Schwarz-Bergkampf, E., *Z. Anal. Chem.,* **221**, 143 (1966).
[17] Nalimov, V. V., Nedler, V. V. and Menshova, N. P., *Zavodsk. Lab.,* 27, 861 (1961).
[18] Miskaryants, V. G., Kaplan, B. Ya. and Nedler, V. V., *Zavodsk. Lab.,* **37**, 170 (1971).
[19] Zilbershtein, H. I. and Legeza, S. S., *Zh. Prikl. Spektroskopi,* 8, 6 (1968).
[20] Zilbershtein, H. I. and Legeza, S. S., *Zh. Prikl. Spektroskopi,* **8**, 531 (1968).
[21] Zilbershtein, H. I., Polivanova, N. G. and Fratkin, Z. G., *Zh. Prikl. Spektroskopi,* **11**, 204 (1969).
[22] Zilbershtein, H. I., *Zh. Prikl. Spektroskopi,* **14**, 12 (1971).
[23] Kaiser, H. and Specker, H., *Z. Anal. Chem.,* **149**, 46 (1956).
[24] Fujimori, T., and Ishikawa, K., *Fuel,* **51**, 247 (1972).
[25] Fujimori, T., Miyazu, T. and Ishikawa, K., *Microchem. J.,* **19**, 74 (1974).
[26] McFarren, E. A., Lishka, R. J. and Parker, J. H., *Anal. Chem.,* **42**, 358 (1970).
[27] Eckschlager, K., *Anal. Chem.,* **44**, 879 (1972).
[28] Eckschlager, K., *Collection Czech. Chem. Commun.,* **39**, 1426 (1974).

[29] Larson, H. R., *Ind. Qual. Control,* **23**, 270 (1966).
[30] Drnas, T., *Ind. Qual. Control,* **23**, 118 (1966).
[31] Owe, D. B., *Technometrics,* **6**, 377 (1964).
[32] Currie, L. A., *Anal. Chem.,* **40**, 586 (1968).
[33] Kaiser, H., *Spectrochim. Acta,* **3**, 40 (1947); *Z. Anal. Chem.,* **209**, 15 (1965).
[34] Adams, P. B., Passmore, W. C. and Campbell, D., Paper No. 13, *Symposium on Trace Characterization, Chemical and Physical,* National Bureau of Standards (Oct. 1966).
[35] Matherny, M., *Spectrosc. Letters,* **5**, 221 (1972).
[36] Liteanu, C., Rîcă, I., and Hopîrtean, E., *Rev. Roumaine Chem.,* in the press (1980).
[37] Liteanu, C., *Mikrochim. Acta,* 715 (1970).
[38] Liteanu, C. and Florea, I., *Mikrochim. Acta,* 983 (1966).
[39] Liteanu, C. and Alexandru, R., *Mikrochim. Acta,* 639 (1968).
[40] Liteanu, C. and Mihálka, St., *Rev. Roumaine Chim.,* **16**, 275 (1971).
[41] Liteanu, C. and Rîcă, I., *Mikrochim. Acta,* 745 (1973).
[42] Liteanu, C. and Rîcă, I., *Mikrochim. Acta,* 311 (1975 **II**).
[43] Liteanu, C. and Rîcă, I., *Pure Appl. Chem.,* **44**, 535 (1975).
[44] Harris, W. E. and Kratochvil, B., *Anal. Chem.,* **46**, 313 (1974).
[45] Benedetti-Pichler, A. A., *Physical Methods of Chemical Analysis,* W. M. Berl, ed., Vol. 3, p. 183, Academic Press, New York (1956).
[46] Wilson, A. D., *Analyst,* **89**, 18 (1964).
[47] Ingamells, C. O. and Switzer, P., *Talanta,* **20**, 547 (1973).
[48] Ingamells, C. O., *Talanta,* **21**, 141 (1974).
[49] Ingamells, C. O., *Talanta,* **23**, 263 (1976).
[50] Visman, J., *Mat. Res. Stds.,* **9**, (11), 8 (1969).
[51] Visman, J., Duncan, A. J., and Lerner, M., *Mat. Res. Stds.,* **11**, (8), 32 (1971).
[52] Visman, J., *J. Materials,* 7, 345 (1972).
[53] Jaffrezic, H., *Talanta* **23**, 497 (1976).
[54] Duncan, A. J., *Mat. Res. Stds.,* **11**, (1), 25 (1971).
[55] Seille, G. J. and Morrison, G. H., *Anal. Chem.,* **49**, 1529 (1977).
[55a] Ingamels, C. O., *Talanta,* **25**, 732 (1978).
[56] Bennett, C. A. and Franklin, N. L., *Statistical Analysis in Chemistry and the Chemical Industry,* p. 319, Wiley, New York (1954).
[57] Wernimont, G., *Mat. Res. Stds.,* **9**, (10), 8 (1969).

Chapter 9

Increasing the Signal-to-Noise Ratio in Analytical Chemistry

It has been shown that the performance of detection and determination for a given method of analysis is limited by the noise inherent in any analytical process. In order to improve the performance we must find methods for increasing the signal-to-noise ratio. The practical procedures used for doing this in analytical chemistry i.e. for obtaining analytical systems able to yield the largest possible amount of information, can be grouped into three classes, as follows.

(1) *The optimization of analytical conditions.* This involves finding the analytical conditions under which the signal-to-noise ratio is maximal. The most reliable way of doing this is by using statistical methods.

(2) *The improvement of analytical instrumentation.* The production of analytical instruments able to give the highest possible signal-to-noise ratio for an analysis process stems from the general progress in science and technology and also from the application of the theory of signal transmission and measurement to the problems of measurement of analytical signals.

(3) *The statistical-mathematical processing of analytical measurements.* Analytical measurements have two basic characteristics: they always include a certain noise, and in many cases have a complex structure and composition. Owing to these circumstances, it often happens that suitable mathematical and statistical treatments must be applied to the analytical measurements, with a view to increasing the signal-to-noise ratio, because this is the only way to extract all the information contained. Remarkable progress has been made in this field by using techniques based on electronic computers.

We shall now give a short account of these three lines of action for increasing the signal-to-noise ratio.

9.1 THE OPTIMIZATION OF ANALYTICAL CONDITIONS

9.1.1 Introduction

The basic purpose of research in analytical chemistry is the design of optimum methods of analysis. In order to have a clear formulation for a problem of optimization, the concept of 'optimum' must always be defined by a certain **objective function** (or **purpose function**). In analytical chemistry, the objective function can be the error of analysis, the time required by a determination, the cost of one determination, the simultaneous determination of several components in the material analysed, the capacity of detection and determination of trace components, and so on.

In optimization of the detection and determination capacity of a method, the objective function will be the signal-to-noise ratio. In this case, among all the possible analytical conditions, we must choose those which ensure a maximum signal-to-noise ratio.

Let us now discuss this particular case, and consider the analytical system in the form given in Fig. 9.1. The material investigated is placed at the input of the system, and the measurement result (from which we obtain the analysis result) is at the output. The analytical conditions are specified by $x_1, x_2, \ldots, x_n$ (for instance, the pH, temperature, time, pressure, amount of reagent, voltage, etc.), and we are looking for their optimum values.

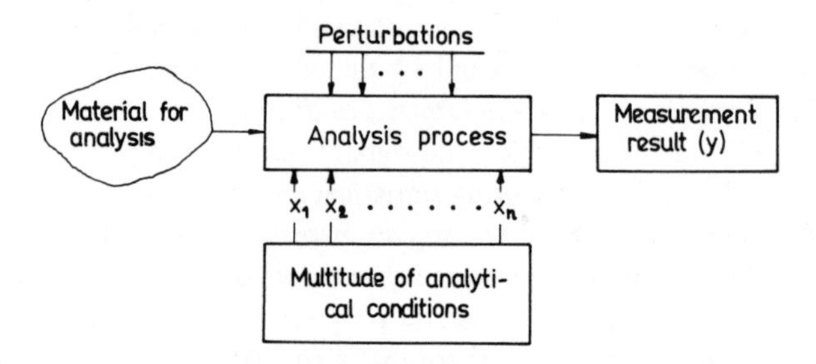

Fig. 9.1 Analytical system.

The analytical conditions will be treated as input variables, and the measurement result as the output variable. Thus, we shall have a system with several inputs and one output. The material analysed is usually kept constant while controlled variations are made in the variables $x_1, x_2, \ldots, x_n$, the corresponding system response being evaluated each time. The relation between the analytical signal (the measurement result) and the values of the variables $x_1, x_2, \ldots, x_n$ is given by the **response function (objective function)**, which is written as:

$$\hat{y} = f(x_1, x_2, \ldots, x_n) \tag{9.1}$$

In practice, various restrictions are imposed on the analytical conditions (i.e. on the variables $x_1, x_2, \ldots, x_n$). That is, the values of these variables must be confined to certain ranges. Hence, the problem of optimization contains the following elements: the objective function, the variables to be optimized, and the restrictions imposed on these variables.

In what follows, we shall specify by x_i ($i = 1,2, \ldots, n$) the ith analytical condition, and represent the set of n analytical conditions by an n-dimensional vector X; in this n-dimensional space, a co-ordinate axis will be associated with each analytical condition. Hence, the response function (9.1) can be written as:

$$\hat{y} = f(X) \tag{9.2}$$

The problem of optimizing the signal-to-noise ratio, in such a way as to satisfy the restrictions for the variables $x_1, x_2, \ldots, x_n$, consists in finding the *position vector*:

$$X_0 = (x_1, x_2, \ldots, x_n) \tag{9.3}$$

for which the response has a maximum signal-to-noise ratio. If the response is accompanied by a Gaussian noise with a mean equal to zero and a standard deviation σ_y, the optimization of the signal-to-noise ratio consists in solving the problem

$$\hat{y}/\sigma_y \rightarrow \text{maximum} \tag{9.4}$$

Obviously, the ratio (9.4) can be improved by increasing the signal amplitude $\hat{y}$, by diminishing the noise σ_y, or by these two methods together. Generally speaking, greater efforts are required in order to diminish the noise (i.e. the standard deviation σ_y), so the signal-to-noise ratio can be much more efficiently improved by increasing the signal amplitude.

If we accept that, for the values which can be taken by the analytical conditions, the noise can be regarded as constant, the response optimization will be reduced to the problem of finding the particular values of the analytical conditions which ensure the maximum value for the signal amplitude (i.e. for the response), that is:

$$\hat{y} = f(X) \rightarrow \text{maximum} \tag{9.5}$$

Ususally, in analytical chemistry, the problem of response optimization is restricted to solving this problem maximizing the response.

In order to state this problem as clearly as possible we shall consider the hypothetical relation (shown in Fig. 9.2) between the value of the response amplitude $\hat{y}$ and the value of one variable to be optimized, x_1.

As said before, the signal amplitude may be the absorbance of a solution, the intensity of a frequency line, the height of a chromatographic peak, etc., while the value to be optimized may be the pH, temperature, time, amount of reagent, voltage, surface area, etc.

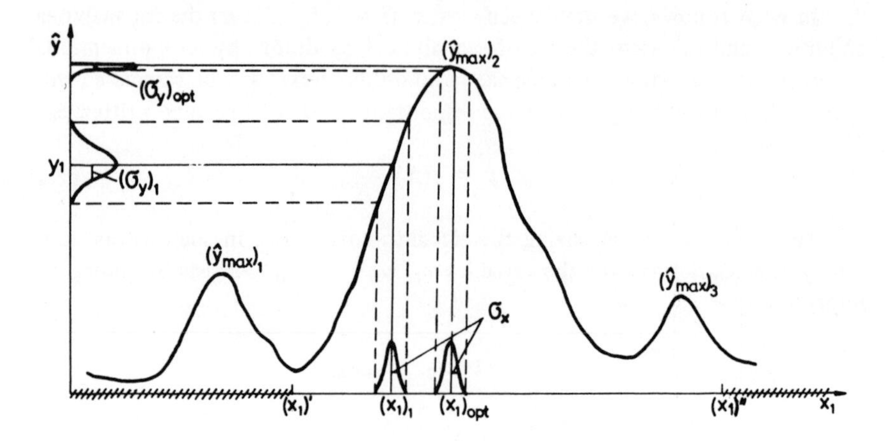

Fig. 9.2 Optimization for one analytical variable; error transfer from the variable to the response.

As shown in the figure, there are two restrictions $(x_1)'$ and $(x_1)''$ imposed upon the variable x_1, such that:

$$(x_1)' \leqslant x_1 \leqslant (x_1)'' \tag{9.6}$$

In the hypothetical situation considered, the response function has three significant maxima: $(\hat{y}_{max})_1$, $(\hat{y}_{max})_2$ and $(\hat{y}_{max})_3$, but the restriction (9.6) is satisfied only by the last two of them. Obviously, the optimum value for the variable x_1 is $(x_1)_{opt}$, because at this value of x_1 we find the highest maximum signal $(\hat{y}_{max})_2$; the other maxima, $(\hat{y}_{max})_1$ and $(\hat{y}_{max})_3$ are less important maxima.

Further, the way in which errors associated with input quantities are unavoidably transformed into response errors is also shown in Fig. (9.2). In order to clarify this question of error transmission from input quantities to the response, we shall consider two different values of the variable x_1, namely $(x_1)_1$ and $(x_1)_{opt}$. Obviously, no matter which of them is chosen as the working condition, its exact value cannot be exactly reproduced in a repeated process of analysis. We can reasonably accept that these values are reproduced with a Gaussian error with a mean equal to zero, and a standard deviation σ_x. From the point of view of the analysis errors, it is far from immaterial which of the two values is prescribed as the working condition, because the errors made in reproducing the working conditions are differently transmitted in the two cases. As can be seen in Fig. 9.2, the condition $(x_1)_1$ is much more conducive to errors than the condition $(x_1)_{opt}$. The standard deviation $(\sigma_y)_1$ of the response obtained under working conditions set to the value $(x_1)_1$ is much larger than the standard deviation $(\sigma_y)_{opt}$ corresponding to working conditions set to the value $(x_1)_{opt}$.

Hence, the value $(x_1)_{opt}$ is indeed the best, both because the signal amplitude is maximal, and also because there is minimum error transmission. This value of the optimized variable will ensure the maximum signal-to-noise ratio for the response signal.

Thus, the optimum analytical conditions are usually those at which the response reaches the maximum value. Further, to increase the signal-to-noise ratio, instead of choosing the working conditions in such a way as to reduce the signal fluctuations σ_x, it is much more efficient to choose them so as to obtain the maximum response amplitude.

As can be seen in Fig. 9.3, in the case of two independent variables x_1 and x_2, the response function can be represented in three-dimensional space by a surface, generally described by an equation of the form:

$$\hat{y} = f(x_1, x_2) \tag{9.7}$$

For this case, the optimal values of the variables $(x_1)_{opt}$ and $(x_2)_{opt}$ are those corresponding to the highest maximum $\hat{y}_{max}$.

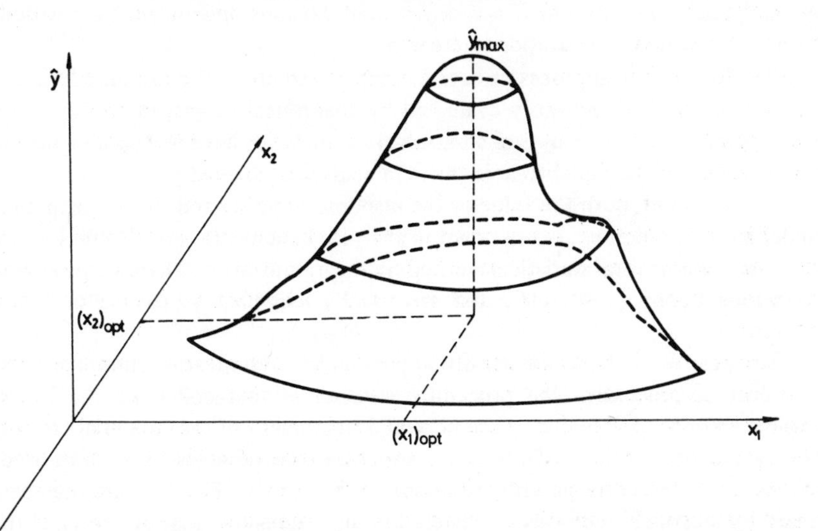

Fig. 9.3 Response surface for two independent variables.

When we have to optimize n independent variables (where $n > 2$), the response function is represented by a hypersurface in $(n + 1)$-dimensional space. The hypersurfaces $f(X)$ = constant are called 'constant level hypersurfaces' of the response function. In the case of a function of two variables $\hat{y} = f(x_1, x_2)$, the hypersurfaces $f(X)$ = constant are described by contours, as represented in Fig. 9.4.

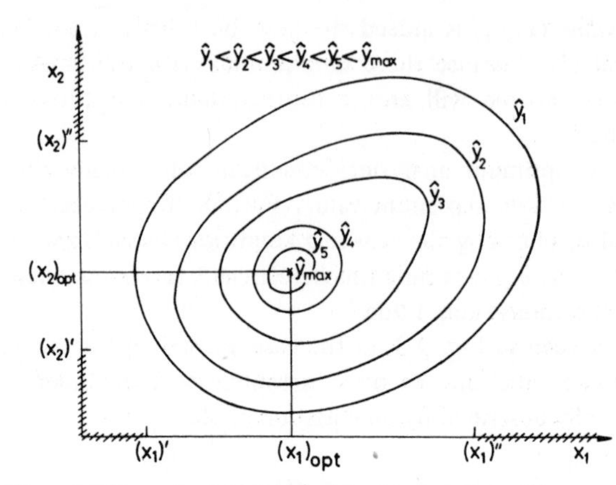

Fig. 9.4 Outline of $f(X)$ = constant for an objective function with two variables.

When the response function $\hat{y} = f(X)$ (the objective function) is rigorously known, the optimum values of the variables to be optimized can be established by analytical methods which use differential calculus and numerical methods described in works on operational research.

Because analytical processes have a complex structure, the response function $\hat{y} = f(X)$ cannot be rigorously evaluated by theoretical means, except in a very few cases. Hence, the analytical and numerical methods have few applications in maximization of the signal-to-noise ratio in analytical chemistry.

In most cases, both the value of the response function and the mathematical model for this function are unknown in analytical chemistry. For this reason, we must use experimental statistical methods in the optimization. These are described in various books [1-4], etc., and are usually classified as passive or active methods.

The passive methods are usually applied in the optimization (improvement) of industrial processes. The procedure consists in first collecting the largest possible amount of data during the normal development of the industrial process. The data contain values of the input variables (the variables to be optimized) and values of the corresponding response of the system. The data are then processed by statistical methods of correlation and regression analysis, to yield the estimation value of the response function. From this estimation value the influence of various factors (variables) on the response of the system can be evaluated. This information is then used for optimizing the experimental conditions, or, more precisely, improving the process, i.e. shifting it towards the optimal.

The applications of passive methods to industrial processes are based on the fact that the input variables usually have large fluctuations in such processes. Consequently, by processing the input and output data, we can ascertain the influence of the input variables on the response of the process.

In contrast to this, in the case of analytical processes the analytical conditions are kept under fairly strict control at the values prescribed in the working instructions. For this reason, the data collected during a normal run of the analytical process will not contain any information concerning the relationship between the variables to be optimized and the system response.

The active method is therefore based on an experimental pattern which is set from the beginning. For this reason, we say that we use an experimental design. Provisions are usually made for changing the experimental design in the course of the experiments.

The investigation of a process by active methods to improve process performance is based on the cybernetic principle of the black-box. Let us assume that we wish to find how the system response is affected by the variables x_1, $x_2, \ldots, x_n$. For this, rigorously controlled changes are induced in these variables according to the design of the experiment, and the corresponding system response is evaluated for each case. The experimental data will include the values of the input variables and the corresponding system response. Statistical methods of data processing are then used to derive the response function, the next direction of development of the experiment (if neccessary), and, finally, the optimum conditions for development of the process. The active methods have many and various applications in investigations of chemical and analytical processes, with the purpose of optimization of the signal-to-noise ratio.

An overall view of the evolution and level attained by the methods of response optimization in analytical chemistry can be obtained by examining the main reviews concerning applications of statistical methods in chemistry.

From Hill and Brow's review [5], published in 1966, it can be seen that after 1951 a large number of papers and books on active experiment methods appeared, devoted to theoretical problems and also to applications in process investigations. Experimental design methods have already many applications in investigations concerning the improvement of performance of industrial processes. However, this review was limited to active experiment methods applied in research and optimization of analytical chemistry processes.

The more recent review of 1968 by the same authors [6] has shown an increase in the number of works on theoretical and practical problems connected with process investigation and optimization by using active experiment statistical methods. The **response surface method**, given by Box and Wilson [7] in 1951, as well as its variant known as the **steepest ascent method** [7, 8], and the **evolutionary operation method** (EVOP) [9] are frequently applied to improve the performance of industrial processes. From this review it appears that up to that time these statistical experimental methods had not been applied to the improvement or optimization of analytical chemistry processes.

The review in 1972 by Currie [10] *et al.* also shows a large increase in the number of works on statistical methods of experiment design and process optimization. The authors found that, in spite of the existence of very advanced statistical theories in this field, these theories had very few applications to practical problems. Only a few works were mentioned in connection with experiment design for the optimization of analytical processes. By that time, besides the response surface methodology and the steepest ascent method, the **simplex method**, first given by Spendley *et al.* [11] and subsequently modified by Nedler and Mead [12], had also been applied to problems of response optimization in analytical chemistry.

The most recent review is by Shoenfeld and DeVoe [13]. Although only a small number of publications were reviewed, for the period 1971–1976, they mentioned a large number of works on the statistical experimental optimization of the response in analytical chemistry. Besides response surface methodology and the steepest ascent method, the simplex method had become a recognized and widely used optimization technique in analytical chemistry.

From this discussion, and also from examination of the recent literature, it seems that these three methods are generally accepted as efficient procedures of investigation and optimization of the response in analytical chemistry. It is to be expected that the applications will be considerably extended in future.

We shall now make some comments on these methods.

9.1.2 The response surface method – RSM

This is a strategy for investigating the response of the system, first developed and described by Box and Wilson [7]. Its theoretical basis is described in several papers [7, 8, 14–17], and books [1, 2].

It is successfully applied to the investigation of industrial processes for the improvement of their performance, and is equally applicable to analytical chemistry for optimization of the response.

The response function of an analytical process can be generally written as:

$$\hat{y} = f(X) \tag{9.8}$$

but as a result of the complexity of the analytical processes, both the value and the model of this function are usually unknown. In this situation, the response surface method is an efficient method of investigation of analytical processes, with a view to maximizing the signal-to-noise ratio. In Fig. 9.5 we give schematically the basic strategy of this method.

In connection with the response function, given generally by:

$$\hat{y} = f(x_1, x_2, \ldots, x_n) = f(X) \tag{9.9}$$

the following concepts are used: the factors, i.e. the independent variables $x_1, x_2, \ldots, x_n$; the factor space, i.e. the n-dimensional space associated with the n factors; the response surface, i.e. the response function represented in the factor space.

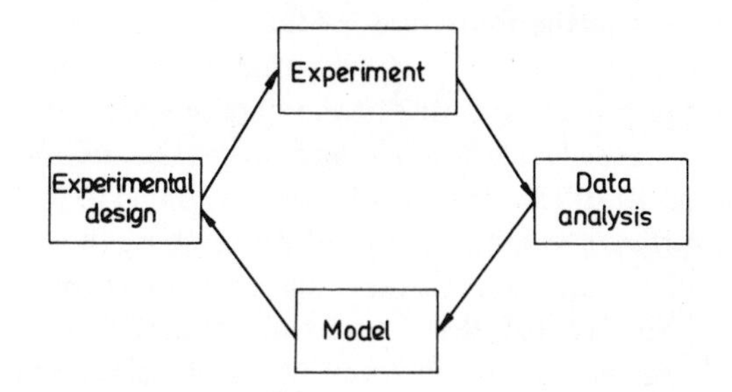

Fig. 9.5 Strategy in the response surface method.

In the response surface method, the unknown model of the response function is approximated within a certain region of the factor space by a polynomial of the first degree, that is:

$$\hat{y} = \beta_0 + \sum_{i=1}^{n} \beta_i x_i \tag{9.10}$$

or by a second-degree polynomial, that is:

$$\hat{y} = \beta_0 + \sum_{i=1}^{n} \beta_i x_i + \sum_{\substack{i,j=1 \\ i \neq j}}^{n} \beta_{ij} x_i x_j + \sum_{i=1}^{n} \beta_{ii} x_i^2 \tag{9.11}$$

or, in some cases, by a polynomial of higher degree.

The polynomial coefficients in (9.11) are designated as follows: β_0 – the constant term; β_i – linear effect of the ith factor; β_{ij} – the interaction effect of the factors; β_{ii} – the squared effect of the ith factor.

Of course, neither (9.10) nor (9.11) will give an exact representation of the true response function (9.9), which remains unknown. The polynomials (9.10) and (9.11) are only approximations to the real model of the response function within a restricted region of the factor space.

The experiment design is made on the basis of the model accepted for the response function. The experiment design principles are described in several works [7, 18, etc.].

The data analysis is usually done with the following purposes: the evaluation of the coefficients of the accepted model, the determination of their statistical significance, and the evaluation of the degree of agreement between the evaluated polynomial and the response function. The procedure used is described in detail in the book by Davies [2].

In Fig. 9.5 we give the four stages of the response function methodology. First, a model is postulated, then the experiment design is mapped out, the experiments are performed, and finally the data are analysed. However, these four stages are not usually carried out only once. On the results obtained by analysing the initial set of experimental data, it is possible to decide to perform a new cycle starting from a modified model of the response function.

When we are looking for response maximization (optimization) by the steepest ascent method [7, 19], the advance towards the optimum is accomplished by steps on the response surface; the cycle of Fig. 9.5 is performed at least once for each step. When the conditions are far from the optimum region, a first degree polynomial is usually accepted as the model, but in the optimum region a polynomial of higher degree is used.

In the following, in connection with experiment design problems, we shall discuss the concept of the **orthogonal factorial experiment**.

We shall consider a linear model of the first degree (9.10), for which we have to estimate the model coefficients. The design matrix of the independent variables is given by

$$X = \begin{bmatrix} x_{01}, & x_{11}, & \ldots, & x_{n1} \\ x_{02}, & x_{12}, & \ldots, & x_{n2} \\ x_{03}, & x_{13}, & \ldots, & x_{n3} \\ \cdots & \cdots & \cdots & \cdots \\ \cdots & \cdots & \cdots & \cdots \\ x_{0N}, & x_{1N}, & \ldots, & x_{nN} \end{bmatrix} \tag{9.12}$$

where n is the number of independent variables, and N the number of experiments. The elements from a row of the matrix represent the values given to the input variables in the corresponding experiment. The elements of the first column in the matrix (9.12) are always taken as equal to 1.

The response values corresponding to every row of the matrix (9.12) are arranged into a column matrix as follows:

$$Y = \begin{bmatrix} y_1 \\ y_2 \\ \cdot \\ \cdot \\ y_N \end{bmatrix} \tag{9.13}$$

Similarly, for the polynomial coefficients we write the column matrix:

$$B = \begin{bmatrix} b_0 \\ b_1 \\ \cdot \\ \cdot \\ b_n \end{bmatrix} \tag{9.14}$$

In (9.14), the coefficients $\beta_0, \beta_1, \ldots, \beta_n$ have been replaced by $b_0, b_1, \ldots, b_n$, in order to underline the fact that the values obtained after the experiment are only estimation values for the polynomial coefficients in (9.10).

The least-squares method leads to the following system of normal equations for the determination of the coefficients $b_0, b_1, \ldots, b_n$:

$$\begin{aligned}
b_0 \sum_{k=1}^{N} x_{0k}^2 + b_1 \sum_{k=1}^{N} x_{0k}x_{1k} + \ldots + b_n \sum_{k=1}^{N} x_{0k}x_{nk} &= \sum_{k=1}^{N} x_{0k}y_k \\
b_0 \sum_{k=1}^{N} x_{1k}x_{0k} + b_1 \sum_{k=1}^{N} x_{1k}^2 + \ldots + b_n \sum_{k=1}^{N} x_{1k}x_{nk} &= \sum_{k=1}^{N} x_{1k}y_k \\
\cdots\cdots\cdots\cdots\cdots\cdots\cdots\cdots\cdots\cdots\cdots\cdots & \\
\cdots\cdots\cdots\cdots\cdots\cdots\cdots\cdots\cdots\cdots\cdots\cdots & \\
b_0 \sum_{k=1}^{N} x_{nk}x_{0k} + b_1 \sum_{k=1}^{N} x_{nk}x_{1k} + \ldots + b_n \sum_{k=1}^{N} x_{nk}^2 &= \sum_{k=1}^{N} x_{nk}y_k
\end{aligned} \tag{9.15}$$

This system can be written as the following matrix equation:

$$X^{\mathrm{T}}XB = X^{\mathrm{T}}Y \tag{9.16}$$

whence

$$B = X^{\mathrm{T}}Y/X^{\mathrm{T}}X \tag{9.17}$$

The superscript T indicates the transposed matrix, so from (9.12) we have:

$$X^{\mathrm{T}} = \begin{bmatrix} x_{01}, x_{02}, x_{03}, \ldots, x_{0N} \\ x_{11}, x_{12}, x_{13}, \ldots, x_{1N} \\ \cdots\cdots\cdots\cdots \\ \cdots\cdots\cdots\cdots \\ x_{n1}, x_{n2}, x_{n3}, \ldots, x_{nN} \end{bmatrix} \tag{9.18}$$

Further, the matrices $X^{\mathrm{T}}X$ and $X^{\mathrm{T}}Y$ are defined by:

$$X^{\mathrm{T}}X = \begin{bmatrix} \sum_{k=1}^{N} x_{0k}^2, & \sum_{k=1}^{N} x_{0k}x_{1k}, \ldots, & \sum_{k=1}^{N} x_{0k}x_{nk} \\ \sum_{k=1}^{N} x_{1k}x_{0k}, & \sum_{k=1}^{N} x_{1k}^2 \quad , \ldots, & \sum_{k=1}^{N} x_{1k}x_{nk} \\ \cdots\cdots & \cdots\cdots & \cdots\cdots \\ \cdots\cdots & \cdots\cdots & \cdots\cdots \\ \sum_{k=1}^{N} x_{nk}x_{0k}, & \sum_{k=1}^{N} x_{nk}x_{1k}, \ldots, & \sum_{k=1}^{N} x_{nk}^2 \end{bmatrix} \tag{9.19}$$

and

$$X^{\mathrm{T}}Y = \begin{bmatrix} \sum_{k=1}^{N} x_{0k}y_k \\ \sum_{k=1}^{N} x_{1k}y_k \\ \cdots\cdots \\ \cdots\cdots \\ \sum_{k=1}^{N} x_{nk}y_k \end{bmatrix} \tag{9.20}$$

respectively.

In the case of an active experiment, the experiment design matrix X can be so chosen as to yield a diagonal matrix $X^{\mathrm{T}}X$, as shown below:

$$X^{\mathrm{T}}X = \begin{bmatrix} \sum_{k=1}^{N} x_{0k}^2, & 0, & 0, \ldots, & 0 \\ 0, & \sum_{k=1}^{N} x_{1k}^2, & 0, \ldots, & 0 \\ 0, & 0, & \sum_{k=1}^{N} x_{2k}^2, \ldots, & 0 \\ \cdots & \cdots & \cdots & \cdots \\ \cdots & \cdots & \cdots & \cdots \\ 0, & 0, & 0, \ldots, & \sum_{k=1}^{N} x_{nk}^2 \end{bmatrix} \tag{9.21}$$

An experiment with a diagonal matrix like (9.21) is called an **orthogonal factorial experiment**. For such an experiment, the polynomial coefficients of (9.10) are calculated as follows:

$$b_i = \frac{1}{\sum_{k=1}^{N} x_{ik}^2} \sum_{k=1}^{N} x_{ik}y_k \tag{9.22}$$

where $i = 0, 1, 2, \ldots, n$ and $k = 1, 2, \ldots, N$.

The following comments can be made concerning the utilization of the orthogonal factorial experiment in the study of the response in analytical chemistry.

(a) only a small effort of calculation is required for the evaluation of the polynomial coefficients and the statistical analysis of the experimental results;

(b) there is no interdependence between the coefficients obtained by an orthogonal factorial experiment and consequently, when the mathematical model is corrected the coefficients already calculated remain valid;

(c) successful investigations of the response can be made by orthogonal factorial experiments when the values of the input variables can be freely altered, and when the fluctuations of the response are small. This last requirement is satisfied, in general, in analytical processes. Consequently, the orthogonal factorial experiment is an efficient technique for the investigation of the response function, with the purpose of optimizing analytical processes.

9.1.2.1 *The complete factorial experiment at two levels – the 2^n experimental design*

In a very widespread method of factorial experimental design, two values are given to each variable – at both sides of a basic value. In principle, a larger number of levels could be given to each variable in the experiment design, but in such cases the organization and accomplishment of the experiment, and also the treatment of data would become much more tedious. For this reason, a larger number of factor levels is used only in cases of stringent necessity. However, two- or three-level factorial experiments are usually all that is required to ascertain the optimal conditions.

In a factorial experiment with two levels for each factor, the number of combinations of the values of the n factors is $N = 2^n$. Such an experiment is called a complete factorial experiment at two levels, or a 2^n experimental design.

The response function can be written in general as

$$\hat{y} = f(x_1, x_2, \ldots, x_n) \tag{9.23}$$

For each variable, let us write the lower and upper levels as $(x_i)_1$ and $(x_i)_2$, respectively, where $i = 1, 2, \ldots, n$. For each variable we introduce the following quantities:

$$x_i^0 = \frac{(x_i)_1 + (x_i)_2}{2}, \text{ the basic level} \tag{9.24}$$

$$\Delta x_i = \frac{(x_i)_2 - (x_i)_1}{2}, \text{ the unit of the interval} \tag{9.25}$$

$$x_i' = \frac{x_i - x_i^0}{\Delta x_i}, \text{ the coded value} \tag{9.26}$$

In this way, by centring and normalization for each factor, the coded values will be +1 for the upper level and −1 for the lower level, while the central co-ordinate will be zero.

For the particular case of two factors, that is $n = 2$, and $N = 2^2 = 4$, the possible combinations of the factor levels are given in Fig. 9.6. The corresponding design matrix of the complete factorial experiment at two levels is given in Table 9.1.

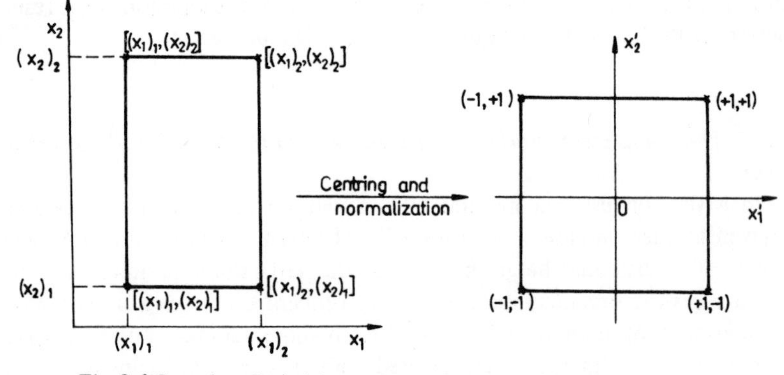

Fig. 9.6 Complete factorial experiment for two factors at two levels.

Table 9.1 Factorial experiment design 2^2

Experiment number	Factors			Response
	x_0'	x_1'	x_2'	
1	+1	+1	+1	y_1
2	+1	−1	+1	y_2
3	+1	+1	−1	y_3
4	+1	−1	−1	y_4

In the case of three factors, with two levels for each factor, the number of possible combinations of the factor levels is $N = 2^3 = 8$. This is shown in Fig. 9.7. The design matrix for a complete experiment for three factors at two levels is given in Table 9.2.

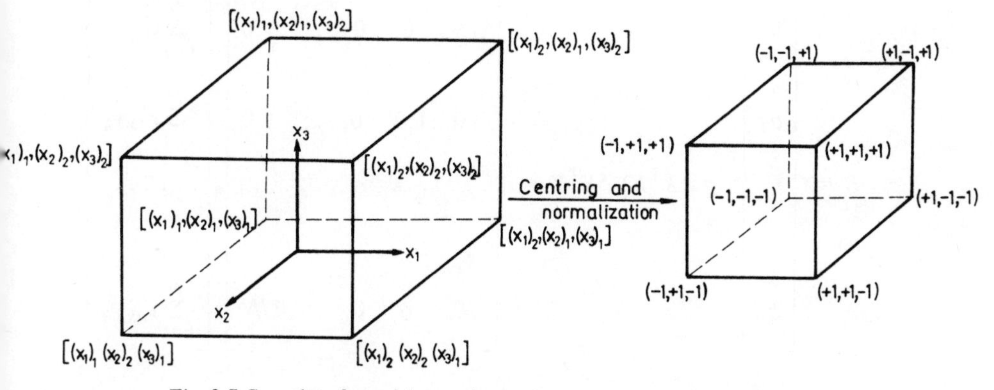

Fig. 9.7 Complete factorial experiment for three factors at two levels.

Table 9.2 Factorial experiment design 2^3

Experiment number	Factors				Response
	x'_0	x'_1	x'_2	x'_3	
1	+1	+1	+1	+1	y_1
2	+1	−1	+1	+1	y_2
3	+1	+1	−1	+1	y_3
4	+1	−1	−1	+1	y_4
5	+1	+1	+1	−1	y_5
6	+1	−1	+1	−1	y_6
7	+1	+1	−1	−1	y_7
8	+1	−1	−1	−1	y_8

A complete factorial experiment at two levels has orthogonal properties. By taking these into account, we find from (9.17) that the estimation values $b_0, b_1, \ldots, b_n$ can be calculated as follows:

$$B = \begin{bmatrix} b_0 \\ b_1 \\ \cdot \\ \cdot \\ \cdot \\ b_n \end{bmatrix} = (X^T X)^{-1} X^T Y = \begin{bmatrix} 1/N, & 0, & 0, \ldots, & 0 \\ 0, & 1/N, & 0, \ldots, & 0 \\ \cdots & \cdots & \cdots & \cdots \\ \cdots & \cdots & \cdots & \cdots \\ \cdots & \cdots & \cdots & \cdots \\ 0, & 0, & 0, \ldots, & 1/N \end{bmatrix} \bullet \begin{bmatrix} \sum_{k=1}^{N} x'_{0k} y_k \\ \sum_{k=1}^{N} x'_{1k} y_k \\ \cdots \\ \cdots \\ \cdots \\ \sum_{k=1}^{N} x'_{nk} y_k \end{bmatrix} \tag{9.27}$$

so

$$b_i = \frac{1}{N} \sum_{k=1}^{N} x'_{ik} y_k ; (i = 1, 2, \ldots, n) \tag{9.28}$$

For example, in the case of a three-factor experiment at two levels (Table 9.2), the coefficient b_1 is calculated as follows:

$$b_1 = \begin{bmatrix} +1 \\ -1 \\ +1 \\ -1 \\ +1 \\ -1 \\ +1 \\ -1 \end{bmatrix} \bullet \begin{bmatrix} y_1 \\ y_2 \\ y_3 \\ y_4 \\ y_5 \\ y_6 \\ y_7 \\ y_8 \end{bmatrix} = \frac{(y_1 + y_3 + y_5 + y_7) - (y_2 + y_4 + y_6 + y_8)}{8} \tag{9.29}$$

The other coefficients of the linear model, b_0, b_2 and b_3, are calculated similarly.

The data for a complete factorial experiment of the type $N = 2^n$ are also sufficient for estimating the coefficients associated with the factor interactions. We shall exemplify this in the case of three factors, and consider the following model which takes into account the factor interactions:

$$\hat{y} = \beta_0 + \beta_1 x_1 + \beta_2 x_2 + \beta_3 x_3 + \beta_{12} x_1 x_2 + \beta_{13} x_1 x_3 + \beta_{23} x_2 x_3 + \beta_{123} x_1 x_2 x_3 \tag{9.30}$$

Supplementary experiments are not required for the estimation of the coefficients β_{12}, β_{13}, and β_{23} (two-factor interaction effects) and β_{123} (three-factor interaction effect). We have only to complete with additional columns the matrix of Table 9.2 (one column for each interaction) shown in Table 9.3.

Table 9.3 Complete experiment design 2^3 with columns for factor interactions

Experiment number	Factors								Response
	x'_0	x'_1	x'_2	x'_3	$x'_1x'_2$	$x'_1x'_3$	$x'_2x'_3$	$x'_1x'_2x'_3$	
1	+1	+1	+1	+1	+1	+1	+1	+1	y_1
2	+1	−1	+1	+1	−1	−1	+1	−1	y_2
3	+1	+1	−1	+1	−1	+1	−1	−1	y_3
4	+1	−1	−1	+1	+1	−1	−1	+1	y_4
5	+1	+1	+1	−1	+1	−1	−1	−1	y_5
6	+1	−1	+1	−1	−1	+1	−1	+1	y_6
7	+1	+1	−1	−1	−1	−1	+1	+1	y_7
8	+1	−1	−1	−1	+1	+1	+1	−1	y_8

The estimation values for the interaction coefficients, b_{12}, b_{13}, b_{23}, and b_{123}, are obtained by the same procedure as for the linear coefficients. For instance, the b_{12} coefficient is calculated as follows:

$$b_{12} = \begin{bmatrix} +1 \\ -1 \\ -1 \\ +1 \\ +1 \\ -1 \\ -1 \\ +1 \end{bmatrix} \bullet \begin{bmatrix} y_1 \\ y_2 \\ y_3 \\ y_4 \\ y_5 \\ y_6 \\ y_7 \\ y_8 \end{bmatrix} = \frac{(y_1 + y_4 + y_5 + y_8) - (y_2 + y_3 + y_6 + y_7)}{8} \tag{9.31}$$

The coefficients b_{13}, b_{23}, and b_{123} are calculated similarly.

The response surface method also includes the statistical analysis of results. Usually, this should establish the significance of the evaluated polynomial coefficients (i.e. the significance of the factor effects) and the agreement between the evaluated polynomial and the true (but unknown) value of the response function.

For the statistical analysis of a factorial experiment, we first verify that the variance of the response is constant, that is, it is independent of the factor values. In order to verify this hypothesis, an experiment is repeated for each row of the design matrix, and for each one of them, the corresponding variance is estimated from the derived results, as follows:

$$s_k^2 = \frac{\sum_{l=1}^{d} (y_{kl} - \bar{y}_k)^2}{d - 1}; \; (k = 1, 2, \ldots, N; \; l = 1, 2, \ldots, d) \tag{9.32}$$

where d is the number of repeated measurements corresponding to a certain line of the design matrix, y_{kl} is the result from an experiment, $\bar{y}_k$ is the mean value of the results pertaining to a line of the design matrix.

Hence, if the design matrix has n rows, there will be N estimation variances: $s_1^2, s_2^2, \ldots, s_N^2$. The homogeneity of this set of estimation variances (i.e. the hypothesis that the response variance is constant with respect to the position of the experiment in the factor space) is then checked by using Bartlett's test (see Section 2.6.3.3).

In the case of an orthogonal experiment, the coefficients $b_1, b_2, \ldots, b_n$ are independently determined, and the variances of the determination errors associated with these coefficients are all equal, that is:

$$s_{b_1}^2 = s_{b_2}^2 = \ldots = s_{b_n}^2 \tag{9.33}$$

The estimation value of these variances is:

$$s_{b_i}^2 = \frac{s_{\text{rep}}^2}{N}; \; (i = 1, 2, \ldots, n) \tag{9.34}$$

where s_{rep}^2 is the estimation value for the variance of the response, as evaluated from repeated measurements.

The significance of the estimated coefficients $b_1, b_2, \ldots, b_n$, that is, the significance of the linear effect of the factors $x_1, x_2, \ldots, x_n$ on the response of the system, is established by using Student's test. For this, we calculate, for each coefficient, the ratio:

$$t_i = \frac{b_i}{s_{b_i}} \tag{9.35}$$

At a significance level α, the hypothesis that the b_i coefficient is significant will be accepted if $|t_i| > |t_{(\alpha;\nu)}|$, where $t_{(\alpha;\nu)}$ is the value of Student's t corresponding to the significance level α and ν degrees of freedom.

The agreement between the evaluated polynomial and the true (but unknown) value of the response function is evaluated by the F-test. For this, we calculate the ratio:

$$F = \frac{s_{\text{res}}^2}{s_{\text{rep}}^2} \tag{9.36}$$

The hypothesis that the evaluated polynomial is significantly different from the true (but unknown) response function (i.e. that there is no agreement between the two functions) will be accepted if:

$$F > F_{(\alpha;\nu_1,\nu_2)} \tag{9.37}$$

Here s_{res}^2 stands for the variance of the residual term, evaluated as:

$$s_{\text{res}}^2 = \frac{\sum_{k=1}^{N} (y_k - \bar{y}_k)^2}{N - (n+1)} \tag{9.38}$$

where y_k is the response value obtained experimentally under the conditions associated with the kth row of the design matrix, and $\bar{y}_k$ is the calculated response value obtained from the evaluated polynomial under the same conditions (i.e. the same factor values). Further, N is the total number of rows in the matrix, $n + 1$ the total number of degrees of freedom, i.e. the number of polynomial coefficients, and $F_{(\alpha;\nu_1,\nu_2)}$ is the F-test parameter for a significance level α and for ν_1 and ν_2 degrees of freedom.

9.1.2.2 *The partial factorial experiment*

When the response surface is sufficiently approximated by a linear polynomial (i.e. by a hyperplane), we do not require all the $N = 2^n$ measurements of a total factorial experiment. For example, in the Box and Wilson steepest ascent method of optimization [7], a linear polynomial (hyperplane) gives a sufficient approximation of the response surface, at positions far from the optimum of the response function. Only in the neighbourhood of the optimum do we resort to a better approximation of the response surface, represented by a polynomial of a higher degree.

When a linear polynomial is a good approximation (i.e. when interactions are absent) it is necessary to complete only a part (half, a quarter, etc.) of a total factorial experiment, but in such a way as to preserve the property of orthogonality. For example, in the case of three factors without (or with negligible) interactions, the estimation of the linear coefficients will require a design matrix containing 2^{3-1} combinations, as given in Table 9.4.

Table 9.4 Partial experiment design 2^{3-1}

Experiment number	Factors				Response
	x'_0	x'_1	x'_2	x'_3	
1	+1	+1	+1	+1	y_1
2	+1	−1	+1	−1	y_2
3	+1	+1	−1	−1	y_3
4	+1	−1	−1	+1	y_4

As seen in this table, the products $x'_1x'_2$, $x'_1x'_3$ and $x'_2x'_3$ take the same values as x'_3, x'_2 and x'_1. Hence, when interactions among factors are absent (i.e. they are equal to zero), the coefficients of the polynomial

$$y = b'_0 + b'_1x'_1 + b'_2x'_2 + b'_3x'_3 \tag{9.39}$$

when evaluated on a 2^{3-1} partial experiment, will appear as a mixed estimation, that is:

$$\begin{aligned} b'_1 &\rightarrow \beta_1 + \beta_{23} \\ b'_2 &\rightarrow \beta_2 + \beta_{13} \\ b'_3 &\rightarrow \beta_3 + \beta_{12} \end{aligned} \tag{9.40}$$

We shall now consider the case of five factors, x_1, x_2, x_3, x_4 and x_5. Assuming, for instance, that we know that β_{123} and β_{12} are zero (that is the interactions $x_1x_2x_3$ and x_1x_2 are zero), the evaluation of the polynomial of linear approximation does not require a complete factorial experiment with $N = 2^5$, but can be obtained from an experiment with $N/4 = 2^{n-2} = 2^3$. The design matrix for such an experiment can be obtained from Table 9.3 (which corresponds to a 2^3 experiment) by making the replacements $x'_4 = x'_1x'_2x'_3$ and $x'_5 = x'_1x'_2$. The resulting design matrix for the partial experiment 2^{5-2} is given in Table 9.5.

Table 9.5 Partial experiment design 2^{5-2}

Experiment number	Factors						Response
	x'_0	x'_1	x'_2	x'_3	x'_4	x'_5	
1	+1	+1	+1	+1	+1	+1	y_1
2	+1	−1	+1	+1	−1	−1	y_2
3	+1	+1	−1	+1	−1	−1	y_3
4	+1	−1	−1	+1	+1	+1	y_4
5	+1	+1	+1	−1	−1	+1	y_5
6	+1	−1	+1	−1	+1	−1	y_6
7	+1	+1	−1	−1	+1	−1	y_7
8	+1	−1	−1	−1	−1	+1	y_8

Obviously, the efficiency of a partial factorial experiment is much diminished as the factor interdependences are more numerous. An estimation of the linear coefficients on the basis of a partial experiment will be correct only if factor interactions are lacking. When factor interactions are absent, the efficiency of a partial experiment will be the same as for the complete experiment.

9.1.2.3 *Factorial experiment at three levels*

When a certain region of a surface has to be evaluated with high accuracy, the model for the surface must be taken as a polynomial of higher degree. A region for which a more correct evaluation is needed is the region around the optimum of the response function and it is usually approximated by a polynomial of second degree. In order to estimate the coefficients of a polynomial of second degree, the experiment must be so designed as to contain three levels for each independent variable (factor).

There are several systems of experimental design at three levels, by which we can obtain an adequate mathematical description of the optimum region of the response surface. Among them we mention the **compositional experiment** [7] and the **rotatable experiment** [8].

The compositional experiment. A complete factorial experiment at three levels, $N = 3^n$, requires a large number of measurements. For example, in the case of five factors, the number of measurements will be $N = 3^5 = 243$. It is possible to diminish the number of measurements by using a compositional experiment design. Such a design is also called a **sequential experiment design,** and this name comes from the fact that the experiment is done in steps.

The compositional experiment design consists of:

(a) a complete factorial experiment at two levels 2^n for $n < 5$, and a partial factorial experiment 2^{n-1} for $n \geqslant 5$;

(b) $2n$ experimental points $(\pm\alpha, 0, 0, \ldots, 0)$, $(0, \pm\alpha, 0, \ldots, 0)$, ..., $(0, 0, 0, \ldots, \pm\alpha)$. The points are situated at a distance α from the centre (the fundamental level);

(c) k_0 experimental points in the centre of the design plane.

Hence, in the case of a compositional experiment, the number of measurements will be:

$$N = 2^n + 2n + k_0 \quad \text{for } n < 5$$

and

$$N = 2^{n-1} + 2n + k_0 \quad \text{for } n \geqslant 5.$$

The structure of a compositional experiment design for two factors is given in Table 9.6, and also in Fig. 9.8.

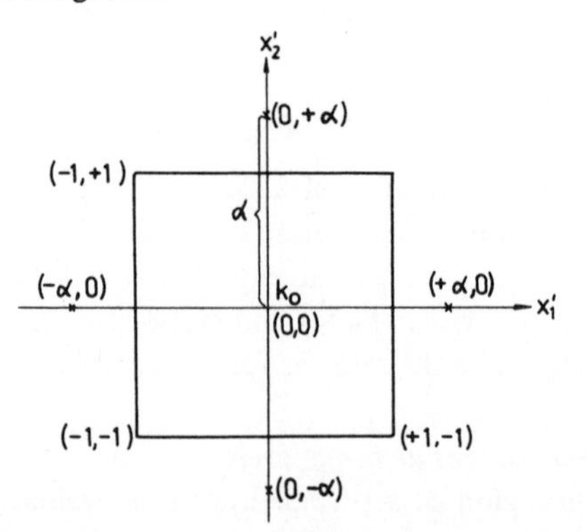

Fig. 9.8 Two-factor compositional experiment design.

Table 9.6 Two-factor compositional experiment

Steps	Experiment number	Factors						Response
		x'_0	x'_1	x'_2	$x'_1x'_2$	$(x'_1)^2$	$(x'_2)^2$	
Factorial experiment 2^2	1	+1	+1	+1	+1	+1	+1	y_1
	2	+1	−1	+1	−1	+1	+1	y_2
	3	+1	+1	−1	−1	+1	+1	y_3
	4	+1	−1	−1	+1	+1	+1	y_4
Supplementary points on axes	5	+1	$+\alpha$	0	0	$+\alpha^2$	0	y_5
	6	+1	$-\alpha$	0	0	$+\alpha^2$	0	y_6
	7	+1	0	$+\alpha$	0	0	$+\alpha^2$	y_7
	8	+1	0	$-\alpha$	0	0	$+\alpha^2$	y_8
Points at centre	9	+1	0	0	0	0	0	y_9

A compositional experiment design for three factors is illustrated in Fig. 9.9. As can be seen, for a design of this kind we first perform a partial 2^{3-1} factorial experiment, by making measurements at the points of co-ordinates (1, 1, 1), (1, 1, −1), (−1, 1, 1), and (−1, −1, −1). From these measurements, a linear polynomial approximation of the response surface is obtained. The agreement between this approximation and the real (but unknown) response function is then verified by using the *F*-test. If the test shows a significant difference between the evaluated polynomial and the real (but unknown) response function, the experiment design will be completed up to 2^3, with six measurements at the points $(\pm\alpha, 0, 0)$, $(0, \pm\alpha, 0)$, and $(0, 0, \pm\alpha)$ and one measurement in the centre (0, 0, 0).

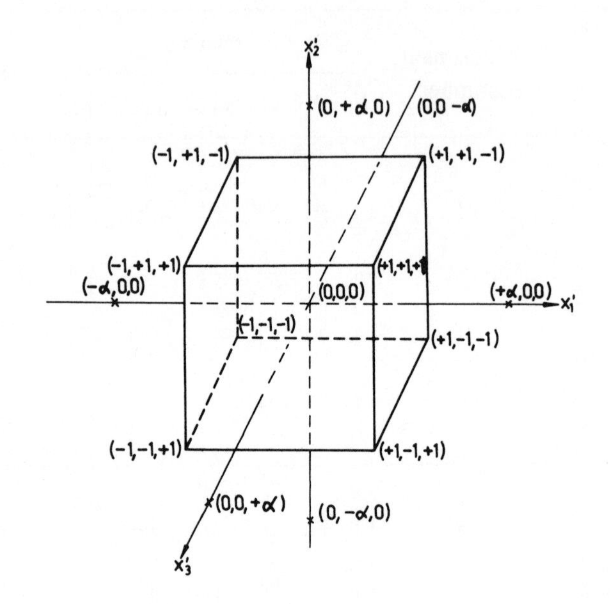

Fig. 9.9 Three-factor compositional experiment design.

In a compositional experiment the values of the parameter α are chosen so as to ensure the orthogonality of the experimental results. In Table 9.7 we give the values of α for several values of the number of factors n.

Table 9.7 Values of the parameter α for several values of n

Number of factors (n)	2	3	4	5
Basic design	2^2	2^3	2^4	2^5
α	1.00	1.215	1.414	1.547

In Table 9.8 we give the general design by which the design matrix of a compositional experiment is completed.

Table 9.8 General compositional experiment design

Steps	Experiment number	Factors					Response
		x'_0	x'_1	x'_2	...	x'_n	
Factorial experiment 2^n	1	+1	+1	+1	...	+1	y_1
	2	+1	−1	+1	...	+1	y_2
	3	+1	+1	−1	...	+1	y_3
	⋮	⋮	⋮	⋮	⋮	⋮	⋮
	$m=2^n$	+1	−1	+1	...	−1	y_m
Supplementary points on axes	$m+1$	+1	$+\alpha$	0	...	0	y_{m+1}
	$m+2$	+1	$-\alpha$	0	...	0	y_{m+2}
	⋮	⋮	⋮	⋮	⋮	⋮	⋮
	$m+2n$	+1	0	0	...	$-\alpha$	y_{m+2n}
Points at centre	$N=m+2n+1$	+1	0	0	...	0	y_N

To simplify the calculation of the coefficients of the polynomial terms of second degree, the columns associated with these terms in the matrix are transformed as follows:

$$x''_k = (x'_k)^2 - (\bar{x}'_i)^2 \tag{9.41}$$

where $(\bar{x}'_i)^2$ is the mean value of the data in the column of squared terms, and $(x'_k)^2$ is one element of the column. For example, by this transformation the matrix from Table 9.6 is brought into the form given in Table 9.9, from which the polynomial coefficients are obtained in the form:

$$\bar{y} = b'_0 + \sum_{i=1}^{n} b_i x'_i + \sum_{\substack{i,j=1 \\ i \neq j}}^{n} b_{ij} x'_i x'_j + \sum_{i=1}^{n} b_{ii}[(x'_i)^2 - (\bar{x}'_i)^2] \tag{9.42}$$

In order to change this into the form:

$$\bar{y} = b_0 + \sum_{i=1}^{n} b_i x_i' + \sum_{\substack{i,j=1 \\ i \neq j}}^{n} b_{ij} x_i' x_j' + \sum_{i=1}^{n} b_{ii}(x_i')^2 \tag{9.43}$$

it is only necessary to calculate the coefficient b_0, and this is obtained from the equation:

$$b_0 = b_0' - \sum_{i=1}^{n} b_{ii}(\overline{x}_i')^2 \tag{9.44}$$

Table 9.9 Two-factor compositional experiment design

Steps	Experiment number	Factors						Response
		x_0'	x_1'	x_2'	$x_1'x_2'$	x_1''	x_2''	
Factorial experiment 2^2	1	+1	+1	+1	+1	$+\frac{1}{3}$	$+\frac{1}{3}$	y_1
	2	+1	−1	+1	−1	$+\frac{1}{3}$	$+\frac{1}{3}$	y_2
	3	+1	+1	−1	−1	$+\frac{1}{3}$	$+\frac{1}{3}$	y_3
	4	+1	−1	−1	+1	$+\frac{1}{3}$	$+\frac{1}{3}$	y_4
Supplementary points on axes	5	+1	+1	0	0	$+\frac{1}{3}$	$-\frac{2}{3}$	y_5
	6	+1	−1	0	0	$+\frac{1}{3}$	$-\frac{2}{3}$	y_6
	7	+1	0	+1	0	$-\frac{2}{3}$	$+\frac{1}{3}$	y_7
	8	+1	0	−1	0	$-\frac{2}{3}$	$+\frac{1}{3}$	y_8
Points at centre	9	+1	0	0	0	$-\frac{2}{3}$	$-\frac{2}{3}$	y_9

The rotatable experiment. A rotatable compositional experiment design has been proposed by Box and Hunter [18] for the evaluation of the coefficients in a model of second order. In order to make a compositional experiment rotatable, we choose the value of the parameter α so as to satisfy the equation:

$$\alpha = 2^{r/4} \tag{9.45}$$

where r is the exponent which characterizes the type of factorial experiment at two levels. For example, for a partial factorial experiment at two levels of the type 2^{n-1}, we have:

$$\alpha = 2^{(n-1)/4} \tag{9.46}$$

The values of the α parameter and also the number of experiments in the design plane centre, k_0, are given in Table 9.10 for a rotatable compositional experiment.

Table 9.10 Values for α and k_0 in a rotatable experiment

Factors number	2	3	4	5	6	6	6	7	7
Type of design	2^2	2^3	2^4	2^5	2^{5-1}	2^6	2^{6-1}	2^7	2^{7-1}
α	1.414	1.682	2.000	2.378	2.000	2.828	2.378	3.333	2.828
k_0	5	6	7	10	6	15	9	21	14

The design matrix for a rotatable experiment with two factors is given in Table 9.11.

For a rotatable experiment, the design matrix is not orthogonal. Therefore, the calculation of the squared term polynomial coefficients is more tedious. Several auxiliary quantities [2,3] have to be evaluated before the calculation of squared term coefficients.

When the model is a second degree polynomial, the statistical analysis of results poses similar problems as for a first degree model. In this case we again have the two fundamental problems: to establish the significance of the evaluated polynomial coefficients, and to establish the significance of the agreement between the evaluated polynomial and the true, but unknown, response function. The statistical methodology by which these two hypotheses can be verified is similar to the one already given in Section 9.1.2.1 for the linear model.

Table 9.11 Two-factor rotatable experiment design

Steps	Experiment number	Factors						Response
		x'_0	x'_1	x'_2	$x'_1x'_2$	$(x'_1)^2$	$(x'_2)^2$	
Factorial experiment 2^2	1	+1	+1	+1	+1	+1	+1	y_1
	2	+1	−1	+1	−1	+1	+1	y_2
	3	+1	+1	−1	−1	+1	+1	y_3
	4	+1	−1	−1	+1	+1	+1	y_4
Supplementary points on axes	5	+1	$+\sqrt{2}$	0	0	+2	0	y_5
	6	+1	$-\sqrt{2}$	0	0	+2	0	y_6
	7	+1	0	$+\sqrt{2}$	0	0	+2	y_7
	8	+1	0	$-\sqrt{2}$	0	0	+2	y_8
Points at centre	9	+1	0	0	0	0	0	y_9
	10	+1	0	0	0	0	0	y_{10}
	11	+1	0	0	0	0	0	y_{11}
	12	+1	0	0	0	0	0	y_{12}

9.1.3 The Box and Wilson method

This method of optimization is also known under the following names: the steepest ascent method, the gradient method, and the maximum slope method.

In this section, we shall start by giving a short description of the classical methodology of optimization of the response in analytical chemistry, and then present the Box and Wilson statistical experimental method [7].

Let us consider the hypothetical situation of Fig. 9.10, which shows the curves $f(X)$ = constant for the case of two factors to be optimized. The response may be the absorbance of a solution, the intensity of a frequency line, the area or height of a chromatographic peak, etc., and the factors may be the pH, temperature, time, pressure, etc.

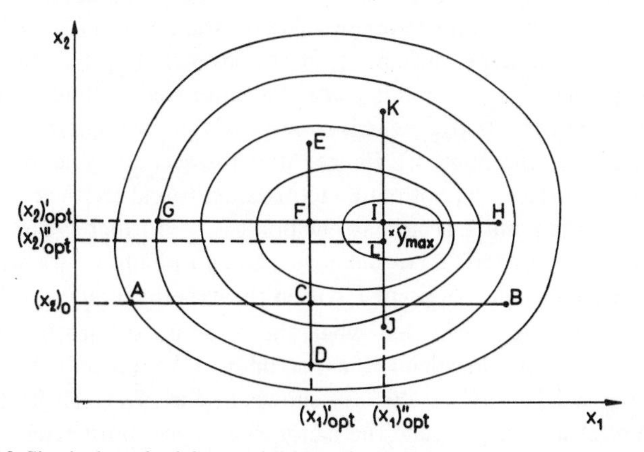

Fig. 9.10 Classical method for establishing the optimum values for the variables.

For the hypothetical situation given in Fig. 9.10, the response optimization consists in finding the co-ordinates corresponding to the maximum response $\hat{y}_{max}$, under such circumstances as entailed from the fact that the response function is unknown.

The classical methodology consists in modifying the factors successively one at a time, while the others are kept unchanged. In order to establish the optimum factor values we use either a cycle during which all the factors are varied one after the other, or several successive cycles. This principle is shown in Fig. 9.10. To establish the optimum values of the two variables (factors), the factor x_2 is kept at a conveniently chosen level $(x_2)_0$ while the factor x_1 is varied in a region AB and we evaluate the corresponding responses. After completing this first stage, we choose as optimum value for the factor x_1 the value $(x_1)'_{opt}$, which gives the maximum response value for this stage (the point C in Fig. 9.10). Then, in order to obtain the optimum value for the factor x_2, the factor x_1 is kept unchanged at the level $(x_1)'_{opt}$, while the factor x_2 is varied in the region

DE, and the corresponding responses are evaluated. For this stage, we take the optimum value of factor x_2 at the value $(x_2)'_{opt}$, which leads to the maximum response on the segment DE (the point F). If a single cycle is prescribed for the response surface investigation, the optimum values of the two factors will be given as $(x_1)'_{opt}$ and $(x_2)'_{opt}$ (the co-ordinates of the point F).

As seen in Fig. 9.10, it is obvious that a response surface investigation by a single cycle in order to establish the optimum values for two variables is insufficient. The point F with co-ordinates $(x_1)'_{opt}$ and $(x_2)'_{opt}$, which has been indicated as optimum after one cycle of investigation, is actually at a large distance from the point $\hat{y}_{max}$. In order to improve the result, the response surface has to be explored further, by several more cycles. For this, we maintain the factor x_2 at the level $(x_2)'_{opt}$, and vary the factor x_1 over a region GH. The improved optimum value for this last factor will be found at the value $(x_1)''_{opt}$, which leads to the maximum response value on the line GH (the point I). After this, the factor x_1 is kept unchanged, at the level $(x_1)''_{opt}$ and the factor x_2 is varied over a range JK. After this stage, the improved optimum value for the factor x_2 is found to be $(x_2)''_{opt}$, which corresponds to the highest response in the region investigated (the point L). Hence, after the second cycle of investigation of the response surface, the optimum factor values found are the co-ordinates of the point L. As the point L is close to the point $\hat{y}_{max}$, further cycles of investigation of the response surface would give new optimum values which would reproduce the co-ordinates of point L within the experimental error limits.

In Fig. 9.11 it is shown that when the response surface has a ridge, the classical procedure for localizing the maximum (the procedure given above) does not give satisfactory results. As shown in Fig. 9.11, if the factor x_2 is kept constant at a level $(x_2)_0$ and the factor x_1 changed in the range AB, we find the value $(x_1)'_{opt}$ as the optimum value for the factor x_1. Then, if the factor x_1 is kept at the level $(x_1)'_{opt}$, and the factor x_2 varied in the range DE, the optimum value for this factor will be found to be $(x_2)'_{opt}$. Hence, after this cycle of investigation, the optimum factor values found are $(x_1)'_{opt}$ and $(x_2)'_{opt}$, i.e. the co-ordinates of the point F. Owing to the presence of the ridge on the surface, if the investigation of the surface is repeated for an arbitrary number of cycles the optimum values of the two factors will reproduce (within the experimental error limits) the co-ordinates of the point F, found before. However, the point F is far from $\hat{y}_{max}$, so that such a determination of the optimum conditions would not lead to a correct result in this case.

A much more correct determination of the optimum values of the variables is given by the Box and Wilson method. This method was first given by Box and Wilson [7] and subsequently investigated and improved by Box [20], Brook and Mickey [19], and other authors.

In this method of optimization, the response surface is investigated along the gradient by a factorial experiment in several stages, i.e. in the direction of steepest ascent (higher slope) and the response surface is approximated successively by hyperplanes. After localizing the region which contains the optimum point, we proceed to a detailed investigation of this region and to the determination of the optimum values. In the region close to the optimum, the response surface is modelled by a polynomial of higher degree.

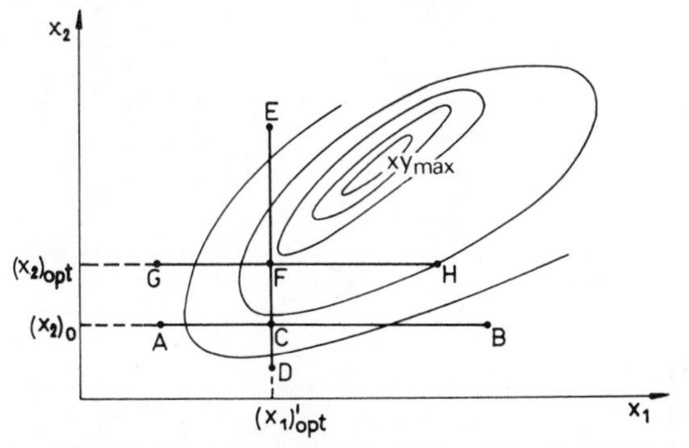

Fig. 9.11 Incorrect localization of the optimum values when the response surface has a ridge.

The procedure, based on the Box and Wilson method, for obtaining the optimum values for two factors is shown in Fig. 9.12. In the same figure we have also represented the classical procedure for establishing the optimum values. In the Box and Wilson method, the optimum is approached by steps on the response surface. At points far from the optimum, the response surface is approximated by a polynomial of first degree, which is evaluated on the basis of a complete or partial factorial experiment at two levels. After each factorial experiment, the direction of the next step on the response surface is determined. Once located, the region around the optimum is studied by a compositional or rotatable compositional factorial experiment. In the case shown in Fig. 9.12, the factorial experiments 1, 2, 3, and 4 are of the type 2^2, while the factorial experiment number 5, which is made in the optimum region, consists of a 2^2 experiment, four experimental points on the axis, and one point in the centre.

For a response surface described by the general form:

$$\hat{y} = f(x_1, x_2, \ldots, x_n) \tag{9.47}$$

the gradient is given by:

$$\nabla\hat{y} = \frac{\partial\hat{y}}{\partial x_1}\delta_1 + \frac{\partial\hat{y}}{\partial x_2}\delta_2 + \ldots + \frac{\partial\hat{y}}{\partial x_n}\delta_n \tag{9.48}$$

where $\delta_1, \delta_2, \ldots, \delta_n$ are the unit vectors of the co-ordinate axis. By normalizing the gradient with respect to its norm $||\nabla\hat{y}||$, we shall find the unit vector, the components of which indicate the relative dimension of the step on each co-ordinate axis x_i, necessary in order to obtain the steepest ascent.

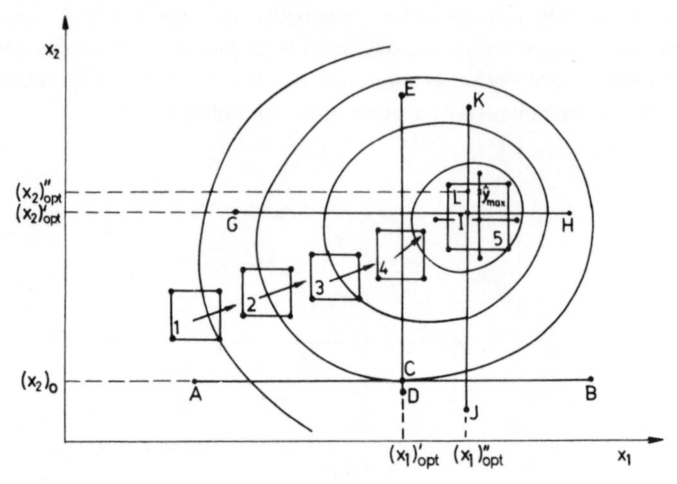

Fig. 9.12 Evaluation of a maximum by the Box and Wilson method.

For a first degree polynomial

$$\hat{y} = \beta_0 + \beta_1 x_1 + \beta_2 x_2 \tag{9.49}$$

the normalization of the gradient $\nabla\hat{y}$ with the norm $||\nabla\hat{y}||$ gives the unit vector:

$$\frac{\nabla\hat{y}}{||\nabla\hat{y}||} = \frac{\beta_1\delta_1 + \beta_2\delta_2}{\sqrt{\beta_1^2 + \beta_2^2}} \tag{9.50}$$

and for a second degree polynomial:

$$\hat{y} = \beta_0 + \beta_1 x_1 + \beta_2 x_2 + \beta_{11} x_1^2 + \beta_{22} x_2^2 + \beta_{12} x_1 x_2 \tag{9.51}$$

the unit vector is:

$$\frac{\nabla\hat{y}}{||\nabla\hat{y}||} = \frac{(\beta_1 + 2\beta_{11}x_1 + \beta_{12}x_2)\delta_1 + (\beta_2 + 2\beta_{22}x_2 + \beta_{12}x_1)\delta_2}{[(\beta_1 + 2\beta_{11}x_1 + \beta_{12}x_2)^2 + (\beta_2 + 2\beta_{22}x_2 + \beta_{12}x_1)^2]^{1/2}} \tag{9.52}$$

As the unit vector (9.52) depends on the variables x_1 and x_2, it results that, in order to obtain the steepest ascent towards the optimum region, the displacement should not be made in a single stage; instead we must select here and there new experimental regions on the response surface. After each factorial experiment, a new experimental region is selected at a certain distance along the steepest ascent vector. Step by step, this procedure leads to higher and higher points on the response surface, which makes it possible to identify in this way the region of the maximum (optimum).

The Box and Wilson method of optimization given above is often applied to response optimization in analytical chemistry. Detailed comments can be found in the works of Nalimov [21], and Alimarin *et al.* [22]. We mention here the following more important applications: the optimization of the conditions for determination of arginine, glutamic acid, and lysine [23]; for polarographic determination of cadmium [24], tellurium [25], and molybdenum [26]; photometric determination of antipyrene [27] and phenol [28]; optimization of the sensitivity of arc spectrochemical analysis of solutions [29]; optimization of conditions for gel-chromatography and arc spectrochemical analysis [30]; titration of ferric ion with ascorbic acid [31]; determination of chalcosine [32]; X-ray fluorescence analysis of soda-potash [33]; extraction of copper from minerals [34]; extraction of lead from anglesite, cerussite and galena [35]; optimization of spectrochemical analysis of graphite powder for trace impurities [36].

It is to be expected that in future the Box and Wilson method will find new applications in optimization of the response of analytical processes. In trace analysis this technique can find many applications in optimization of the detection and determination capacity of analytical processes, i.e. finding the conditions under which an analytical system is able to detect and determine the smallest possible amounts.

9.1.4 The simplex method

Another technique widely used for response optimization in analytical chemistry is the **simplex** method. In order to avoid misunderstanding we emphasize the fact that the simplex method of interest here has nothing in common with the linear programming simplex method.

Like the Box and Wilson method, the simplex method is used for investigation of an unknown response surface, with the purpose of establishing the maximum or minimum response. It was developed by Spendley *et al.* [11] and modified by Nelder and Mead [12] to make it faster.

By a regular simplex we mean a geometrical figure which, in n-dimensional factor space, has $n + 1$ equidistant vertices. In one-dimensional space, i.e. when there is a single factor, a simplex is a line segment. When there are two factors, i.e. in two-dimensional space, a simplex is an equilateral triangle. In the case of three factors, i.e. in three-dimensional space, a simplex is a tetrahedron.

The displacement towards the maximum (optimum) in the Box and Wilson method and in the simplex method for the case of two factors is shown in Fig. 9.13.

To investigate a response surface by the simplex method, we start with a basic simplex, for example ABC. The response is evaluated at the three vertices of the simplex, and the vertex with the smallest response is rejected (in our case, the rejected vertex is B). The vertex B is replaced with a new point D, which is the reflexion image of the point B on the hypersurface (or surface) defined by the other vertices. In this way, the new simplex, ACD, is constructed. The procedure is continued until the maximum region is attained. In Fig. 9.13, the point I in the neighbourhood of the maximum is attained in six such steps.

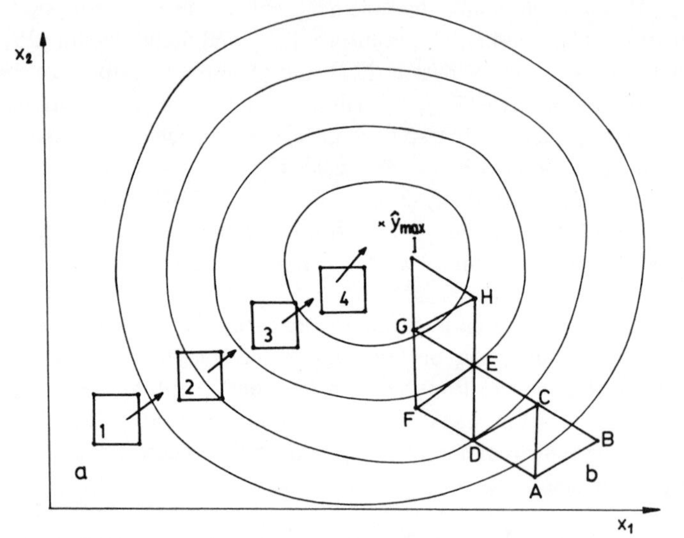

Fig. 9.13 Shifting towards a maximum (optimum) in the Box and Wilson method (a); in the simplex method (b).

In the general n-dimensional case (when there are n factors) the simplex vertices are represented by the co-ordinates of the n-dimensional vectors: $W_1, W_2, \ldots, W_i, \ldots, W_n, W_{n+1}$. The vertex where the response is the smallest has to be eliminated: let us say that this is the vertex W_i. Its removal leaves the hypersurface determined by the vectors $W_1, W_2, \ldots, W_{i-1}, W_{i+1}, \ldots, W_n, W_{n+1}$, with the centroid

$$\overline{W} = 1/n(W_1 + W_2 + \ldots + W_{i-1} + W_{i+1} + \ldots + W_n + W_{n+1}) \quad (9.53)$$

The new simplex will be defined by the hypersurface $W_1, W_2, \ldots, W_{i-1}, W_{i+1}, \ldots, W_n, W_{n+1}$ and the new vertex W_i^* at the point:

$$W_i^* = \overline{W} + (\overline{W} - W_i) \quad (9.54)$$

If the simplex investigation shows that the smallest response of the new simplex is found at the new vertex (9.54), further continuation of the simplex procedure will come back to the previous simplex from which the vertex (9.54) was obtained. This is shown in Fig. 9.14, where we have illustrated the simplex procedure for the case when the response surface has a ridge. The simplex ABC is chosen as the basic simplex. The smallest response is found at the vertex B, so this vertex is replaced the vertex D′, and in the simplex ACD′ this vertex would itself give the smallest response. Consequently, we come back to the original simplex ABC, and pick up the vertex A which is the smaller of the vertices which remain if vertex B, which has already been considered, is disregarded. The vertex A is replaced by the vertex D, thus giving the new simplex BCD. After this, the displacement towards the maximum passes through the simplexes CDE, EFG, and FGH.

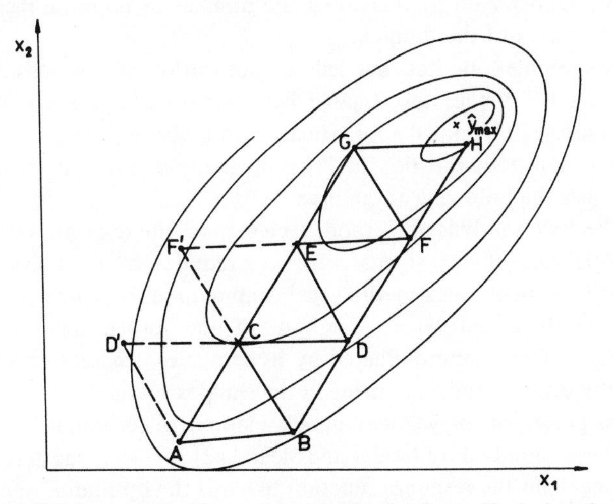

Fig. 9.14 Shifting towards the maximum in the simplex method when the response surface has a ridge.

When the maximum of the response function has been attained by the simplex investigation, the simplex starts to rotate, as shown in Fig. 9.15.

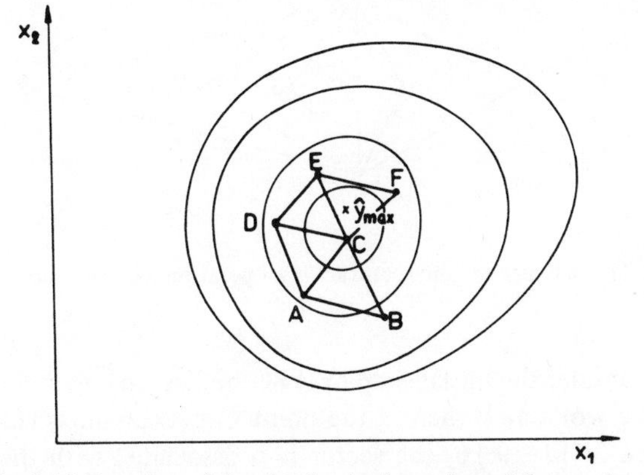

Fig. 9.15 Rotation of the simplex in the region of the maximum.

In order to accelerate the investigation of the surface by the simplex method, a simplex of larger dimension may be used in the initial stages. This will insure a fast movement towards the optimum, and also an approximate localization of the optimum. After this, a much better localization of the optimum will be obtained by using simplexes of smaller dimension. In the final stage of investigation, the simplex dimension can be reduced to a limit determined by the precision of the process response investigated. When response fluctuations are large, the reliability of the displacement on the response surface towards the optimum can be increased by increasing the number of response measurements taken at every vertex of the simplex.

When the simplex method has led to localization of the optimum region, this region can be further investigated by a factorial experiment based on a model which uses a polynomial approximation of higher degree.

In comparison with the Box and Wilson method, the simplex method of optimization has the following advantages.

(a) In the Box and Wilson method, each step on the response surface results from a factorial experiment, so that each step requires several individual experiments, whereas in the simplex method each step requires only one experiment.

(b) In the Box and Wilson method, in the displacement towards the optimum, the surface is approximated by hyperplanes, whereas no assumptions concerning the surface model are made in the simplex method.

We have presented above the simplex method in its initial form, but this method has been modified by Nelder and Mead [12] in such a way as to accelerate the displacement on the response function towards the optimum. For this, these authors have supplemented the reflection operation of the original simplex procedure with two other operations: expansion and contraction. The procedure supplemented with these operations is explained in the example given in Fig. 9.16.

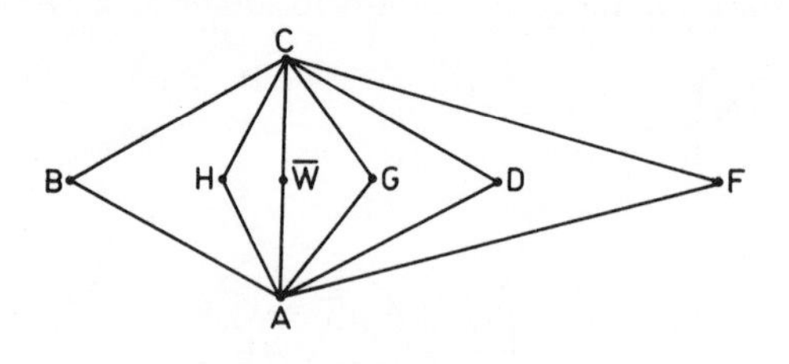

Fig. 9.16 The expansion and contraction operations in the modified simplex method.

We shall consider the initial simplex ABC, and assume that the best response is at A, and the worst at B (hence, the point C gives an intermediate response). As can be seen in Fig. 9.16, the vector $\overline{W}$ is associated with the centre of the surface CA, and W_A, W_B, and W_C are the position vectors for the three vertices.

The image of the vertex B in the mirror CA is the point D with the position vector given by:

$$W_D = \overline{W} + (\overline{W} - W_B) \quad (9.55)$$

With respect to the response values at A, B, and C, the response at D can be one of the following.

(a) Better than at A. This recommends an operation of expansion, i.e. a further displacement along the direction BD. The segment BD is extended to the point F with co-ordinates

$$W_F = \overline{W} + \gamma(\overline{W} - W_B) \quad (9.56)$$

where γ is the expansion coefficient ($\gamma > 1$). If the response at F is higher than at B, the investigation is further developed on the simplex ACF. If the response at F is smaller than at B, the investigation will be further developed on the simplex ACD.

(b) Intermediate between the responses at A and C. Neither an expansion nor a contraction is indicated in this case, so we further develop the investigation on the simplex ACD.

(c) Worse than at C. This recommends a contraction operation, and one of the following vertices can be generated:

(i) if the response at D is smaller than at C, but higher than at B, we generate the point G with co-ordinates

$$W_G = \overline{W} + \beta(\overline{W} - W_B) \quad (9.57)$$

where β is the contraction coefficient ($\beta < 1$) and the response surface is further investigated with the new simplex, ACG;

(ii) if the response at D is smaller than at B, we generate the point H with co-ordinates:

$$W_H = \overline{W} - \beta(\overline{W} - W_B) \quad (9.58)$$

and the response surface is further investigated starting with the simplex ACH.

The process of investigation of the response surface by the modified simplex method is stopped when the replacement of one point with another results in a response variation smaller than a certain previously set value, or when the response variation is comparable with the amplitude of the response fluctuation.

Details of applications of the simplex method to problems of response optimization in analytical chemistry can be found in a paper by Long [37], where the author gives a matrix design for establishing the vertices of an initial simplex containing up to ten factors, and also the method for advancing towards the maximum. The technique is illustrated by its application to the optimization of the colorimetric determination of sulphur dioxide with pararosaniline and of phosphorus by the molybdenum blue method.

The analytical applications of the simplex method have been reviewed by Deming and Morgan [38]; in another paper [39] these authors give an interesting example of optimization by the simplex method. They prove the capacity of the simplex method to optimize systems with many factors, using the example of the optimization of the determination of blood cholesterol.

For the case of a single factor, the simplex optimization technique, under the name of the **uniplex method**, is given by King and Deming [40].

The simplex method has been applied to pattern recognition problems by Ritter *et al.* [41] and by Kaberline and Wilkins [42].

Other reviews of applications have been given by Deming *et al.* [43–45]. An improved simplex method is treated by Routh *et al.* [46].

We may also mention the following applications: optimization of magnetic field homogeneity in NMR spectroscopy [47], sensitivity optimization in atomic absorption determination of calcium [48], optimization of separation in gas–liquid chromatography [49], optimization of the determination of zinc titration with potassium ferrocyanide [50], optimization of spectrophotometric multicomponent analysis [51]; optimization of a solvent extraction process [52].

9.2 SIGNAL MEASURING TECHNIQUES FOR INCREASING THE SIGNAL-TO-NOISE RATIO

The dependence of the signal amplitude S, and signal-to-noise ratio S/N, on the values of the technical characteristics of analytical instruments can be found in specialized works. The functions which describe the dependence are usually called the 'signal equation' and 'signal-to-noise ratio equation'. These equations, theoretically derived, give suggestions concerning the steps to be taken in order to improve analytical instruments technically, such as increasing the S/N ratio.

This section is devoted to the improvement of the S/N ratio by using suitable measurement techniques. A brief examination will be made of the main sources of noise in analytical measurement, the properties of such sources, and the technique for increasing the S/N ratio by signal filtration and modulation.

9.2.1 Noise sources and their properties

The noise sources acting in analytical measurement instruments are the sources common to technical measurement instruments in general.

(a) *The noise in electrical devices.* There are two fundamental sources of such noise. One of them originates in the thermal agitation of the material particles (atoms, molecules, ions, or electrons), and the other is due to the fact that light and electrical charge are quantized.

In connection with these two fundamental sources of noise, there are usually three kinds of noise distinguishable in electrical devices [53, 54].

(i) The *Johnson* or *resistance noise.* This noise appears in electrical resistances. It is due to the random motion of the electrons, so it has a thermal origin. The standard deviation for current fluctuations in a resistor, which is the measure of the Johnson noise, is evaluated as:

$$\sigma_i = \left(\frac{4\,kT\Delta f}{R}\right)^{1/2} \tag{9.59}$$

where σ_i is the standard deviation of the current intensity, R is the value of the resistance in which the noise is produced, T is the absolute temperature, k the Boltzmann constant, and Δf the frequency bandwidth of the measuring instrument.

In terms of the noise power spectra, the Johnson noise is a white noise. By noise power spectrum we mean the relation between noise power and frequency. For a white noise, as shown in Fig. 9.17, the power spectrum is flat, so white noise is a mixture containing all the frequencies, in which the components have random phases and amplitudes.

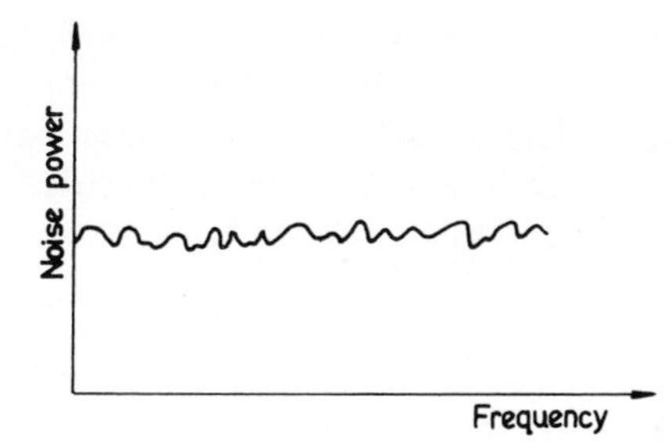

Fig. 9.17 The power spectrum of a white noise.

(ii) The *shot noise.* This kind of noise appears in diodes, triodes, photomultipliers and transistors, and is due to the random emission of electrons from the cathode or junctions. Like Johnson noise, the shot noise is white, i.e. it has a flat power spectrum.

(iii) The *flicker noise.* Unlike the other two kinds of noise, the flicker noise has a power spectrum in which the noise power is approximately proportional to $1/f^m$, where f is the frequency and m is a constant which has a value close to unity (Fig. 9.18). For this reason, such a noise is also called '$1/f$ noise'.

The flicker noise is generally present in all electrical instruments, and is especially associated with current amplifiers.

(b) The *environmental noise*. Under real conditions, perturbations of the most varied nature, coming from the environment, act on analytical measurement instruments. Among them, we mention fluctuations of power supply, temperature, humidity, illumination, degree of purity of the environment, quality of the materials employed in the analysis process, various external vibrations transmitted to the instruments, electromagnetic noise coming from various electrical apparatus (motors, radio and television, etc.).

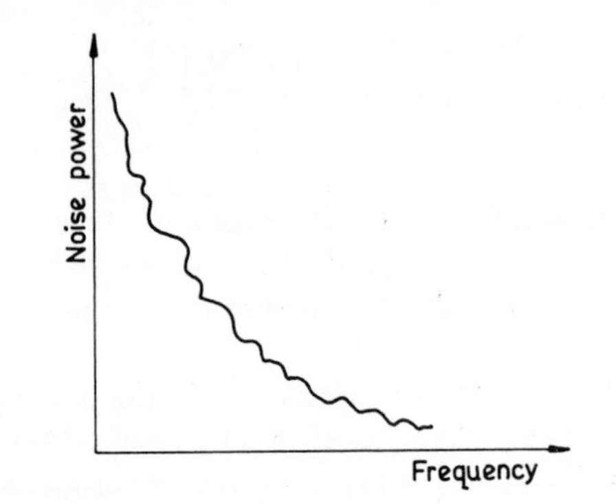

Fig. 9.18 The power spectrum of a low-frequency noise.

The overall noise due to the environment (i.e. the noise due to fluctuations of power supply, temperature, humidity, illumination, purity of environment, quality of materials, and vibrations) has a $1/f^m$ power spectrum [53].

The interference noise due to electrical motors, electrical cables, radio and television, has a power spectrum as illustrated in Fig. 9.19.

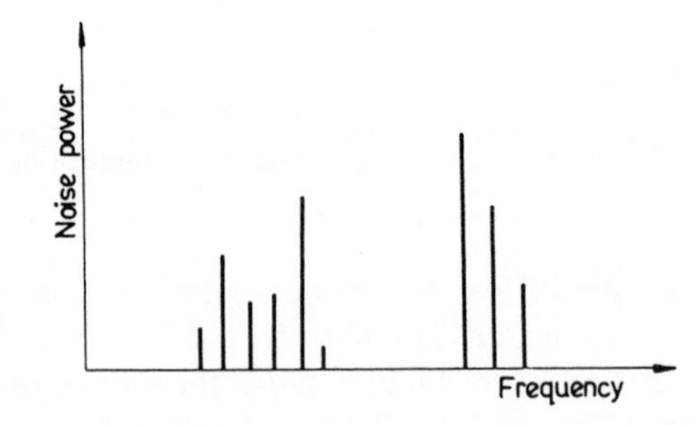

Fig. 9.19 The power spectrum for an interference noise.

(c) The *noise due to quantization of measurements.* In a process of analytical determination several measurements are usually taken. In a measurement, the quantity that is being measured appears to be quantized. That is, the measured result does not give the true value of the measured quantity, but represents the mark on the instrument scale which is the closest to the true value which is being measured. This is shown in Fig. 9.20, where we have represented the scale of a measuring instrument, with a distance Δy between two successive marks.

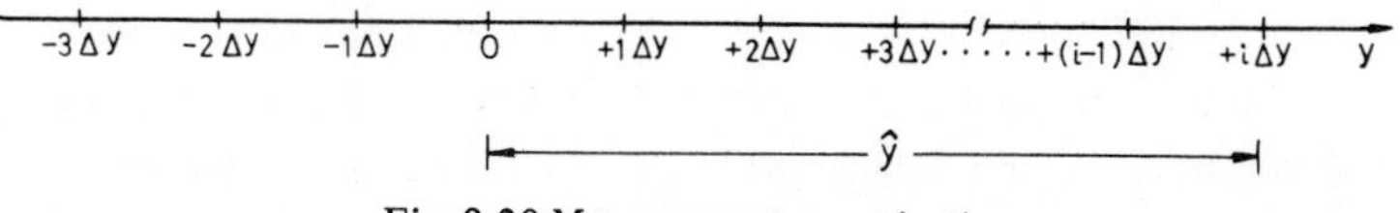

Fig. 9.20 Measurement quantization.

In the measurement, the true value $\hat{y}$ of the quantity being measured is replaced by the closest scale value $i\Delta y$. Hence, the quantized measurement result y_q will always include some uncertainty, i.e. the quantization noise q:

$$y_\mathrm{q} = \hat{y} + q \tag{9.60}$$

The process of quantization is illustrated in Fig. 9.21. Quantization also appears when analytical measurement results are processed, and generally in data processing.

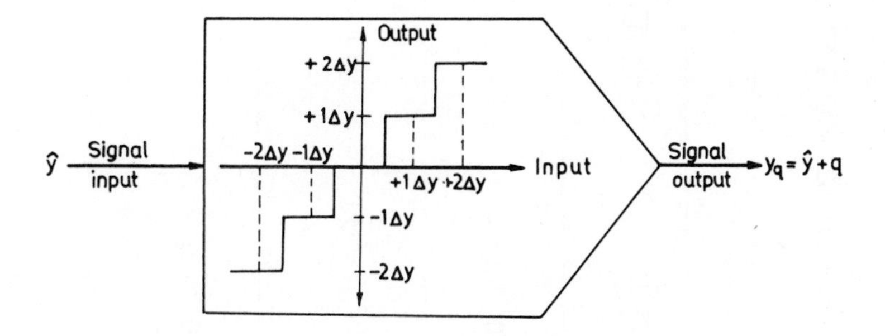

Fig. 9.21 The process of quantization for one quantity.

In Fig. 9.22 we represent the process of quantization of a normally distributed random quantity, with a probability density $f(y)$ and a mean value equal to zero. The output function of the quantizer, $f(y_\mathrm{q})$, is a discrete random variable.

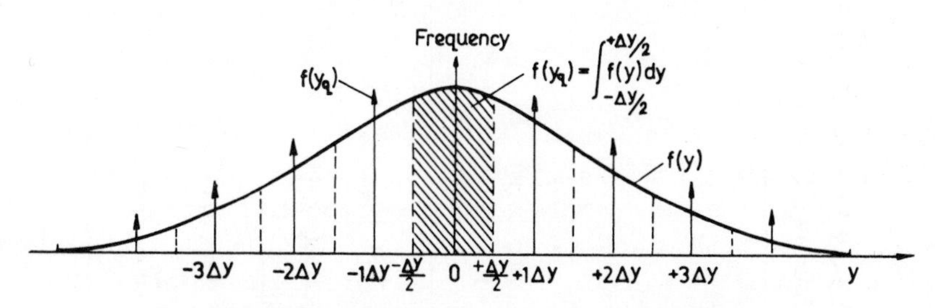
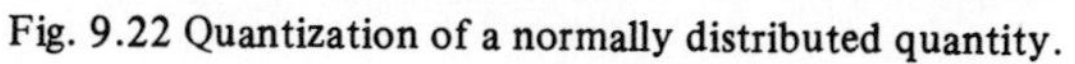

Fig. 9.22 Quantization of a normally distributed quantity.

Table 9.12 Errors from the application of Eqs. (9.61), (9.62) and (9.63) to Gaussian variables [55] (By permission of the McGraw-Hill Book Co.)

	Δy		
	$\sqrt{\mathrm{Var}\,(y)}$	$2\sqrt{\mathrm{Var}\,(y)}$	$3\sqrt{\mathrm{Var}\,(y)}$
$\max \lvert E(y) - E(y_q) \rvert$, if $E(y) \neq 0$	$8.3 \times 10^{-10}\,\Delta y$	$2.3 \times 10^{-3}\,\Delta y$	$3.5 \times 10^{-2}\,\Delta y$
$\mathrm{Var.}\,(y_q) - \mathrm{Var}\,(y) - \frac{(\Delta y)^2}{12}$, if $E(y) = 0$	$-1.1 \times 10^{-8}\,\mathrm{Var}\,(y)$	$-3.1 \times 10^{-2}\,\mathrm{Var}\,(y)$	$-0.54\,\mathrm{Var}\,(y)$
$\mathrm{Var.}\,(q) - \frac{(\Delta y)^2}{12}$, if $E(y) = 0$	$-2.6 \times 10^{-10}\,(\Delta y)^2$	$-7.2 \times 10^{-4}\,(\Delta y)^2$	$-1.1 \times 10^{-2}\,(\Delta y)^2$

In the theory of quantization of data it is proved that if a non-quantized random variable is normally distributed with a variance Var(y), then, in agreement with (9.60), the quantized variable will have a variance Var(y_q) sufficiently approximated by:

$$\mathrm{Var}(y_q) = \mathrm{Var}(y) + \frac{1}{12}(\Delta y)^2 \qquad (9.61)$$

Hence the variance of the quantization noise Var(q) is approximated by:

$$\mathrm{Var}(q) = \frac{1}{12}(\Delta y)^2 \qquad (9.62)$$

The expected value for the non-quantized variable, $E(y)$, is well approximated by the expected value for the quantized variable $E(y_q)$, that is:

$$E(y_q) = E(y) \qquad (9.63)$$

In Table 9.12 we give the errors made by using (9.61), (9.62), and (9.63) as Gaussian variables, for three values of the interval of quantization Δy [55].

9.2.2 The noise figure

An analytical system can be treated as a channel of information transmission (see Section 3.2) resolvable into a succession of subsystems and elements. At the input of the channel we find the material to be analysed (the source of information), and at the output the information derived by decoding the analytical measurement. In Fig. 9.23 we exemplify such a representation for an emission spectroscopy system. The system consists of a succession of subsystems, and each subsystem can be further decomposed into several elements.

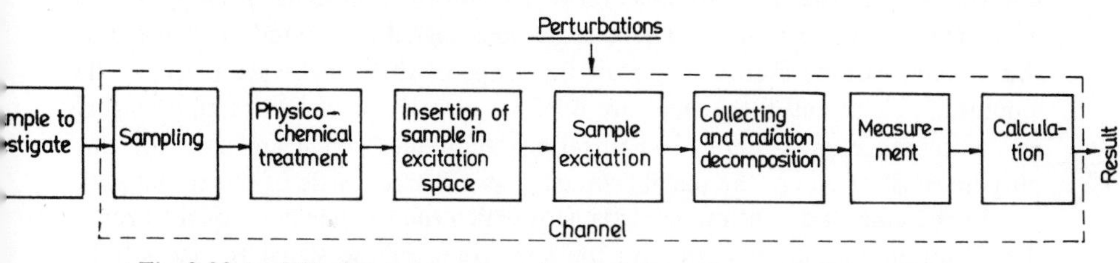

Fig. 9.23 Analytical system represented as an information transmission channel.

The material to be analysed (the source of information) is unavoidably characterized by a certain degree of non-homogeneity, i.e. by a variability of the amount of component to be determined. Hence, the material to be analysed will itself induce a certain value $(S/N)_0$ in the signal-to-noise ratio.

We would have an ideal analytical system if the signal-to-noise ratio were not at all degraded by performing the process of analysis. For such a system, the signal-to-noise ratio associated with the analytical determination results would be precisely the same as the input value of the signal-to-noise ratio. Under real conditions, channel and environment noise sources always result in a certain degradation of the signal-to-noise ratio, so the ratio associated with the analytical results, $(S/N)_r$, is always smaller than the input signal-to-noise ratio, $(S/N)_0$.

It is very useful to express quantitatively the extent to which the information is degraded by transmission through a system, subsystem or element. Such a measurement is given by the noise figure, which is defined as follows:

$$\mathrm{NF} = 10 \log \frac{(S/N)_0}{(S/N)_r} \tag{9.64}$$

where $(S/N)_0$ and $(S/N)_r$ are the signal-to-noise ratios evaluated at the input and output of the system, subsystem, or element. For example, if information transmission degrades the signal-to-noise ratio by a factor of two, then the noise figure for this will be NF = 10 log 2 = 3 dB.

Hence, the noise figure can be used as a characteristic for the quality of the overall system, or of a subsystem or element of the system.

9.2.3 Signal modulation and filtration

Environmental noise sources can be diminished by various technical procedures, for example using stabilized supply sources, screening electrical measurement instruments, placing the measurement systems in controlled-climate rooms, vibration damping, etc. However, no matter how carefully applied, such technical precautions can only diminish the noise, and not suppress it completely.

The noise due to quantization of measurement can be reduced by using instruments with expanded scales.

The technique of signal filtration and modulation must be used for further improvement of the signal-to-noise ratio. The information signal usually consists of a wave of a certain shape, whereas the noise signal has a random shape. For this reason, various characteristics of the shape of the signal wave (such as the amplitude, phase and frequency) are used for identifying and extracting a signal in the presence of a noise. The filtration techniques which are used for this purpose most often use the signal frequency as a characteristic of signal shape.

In the case of a continuous signal (for which the frequency is equal to zero) the filtration, or, more correctly, the smoothing of the signal is successfully accomplished by using recorders with a high time-constant (high damping). Except for the frequency of a continuous signal (zero), all the frequencies are attenuated.

As can be seen in Fig. 9.18, continuous information signals, or generally, information signals of low frequency, are strongly affected by $1/f$-type noises. If we wish to remove $1/f$ noises, which are situated in the low-frequency range, the information signal has to have a frequency far from zero (i.e. far from the region where a continuous signal is strongly affected by a $1/f$ noise).

In order to increase the *S/N* ratio, the information signal can be transferred, by modulation techniques, from a frequency range where there is a high noise power, to a frequency range where the noise power is low. The modulation consists of imprinting an information signal on a carrier wave; the frequency of this wave should be chosen so as to have, at that frequency, a low noise power.

Signal filtration and modulation techniques are applied with very good results to the improvement of the *S/N* ratio in such systems as radio and television equipment, in complex systems for signal transmission, measurement and processing, as well as in instruments for measuring analytical signals.

In the process of modulation, the information signal is usually shaped (imprinted) on a high-frequency wave (the carrier) by modifying one or several properties of the carrier wave according to the information signal properties. The carrier wave properties which are modified are usually the amplitude, phase and frequency.

The process of recovering the information signal from the modulated signal is called demodulation. Increasing the *S/N* ratio also requires a suitable demodulation technique.

The signal modulation technique has found many applications to the improvement of the *S/N* ratio in analytical measurement systems. The most important applications are found in spectroscopic methods of analysis [56-58]. The general block-diagram for the application of the modulation technique in atomic absorption spectroscopy is given in Fig. 9.24. The signal from a hollow-cathode lamp is chopped, either mechanically [56] or electronically [57]; the electronic method is shown in Fig. 9.24. By using electronic chopping methods we can obtain several kinds of wave-form for the light-radiation from the hollow-cathode lamp, for example sine-wave, square-wave, pulse, pseudo-random pulse, and so on [58]. The chopped radiation then goes through the flame, where the information signal is imprinted on it. The modulated signal is received by a detector (photomultiplier) and after this, is demodulated by a suitable device.

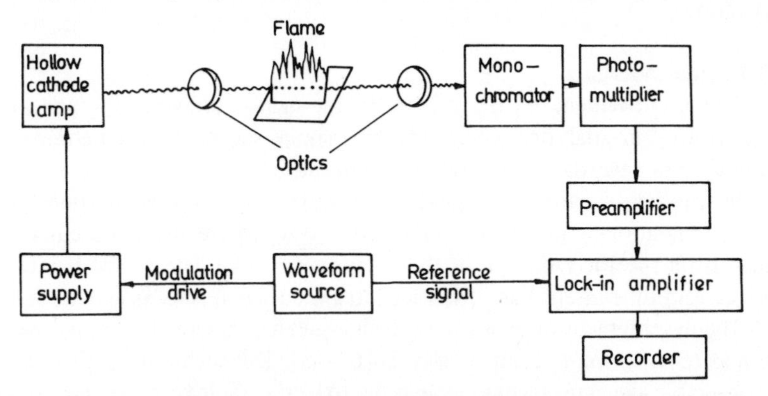

Fig. 9.24 Block diagram in atomic absorption spectroscopy with electronic modulation and phase-sensitive amplifier.

In order to obtain a high *S/N* ratio in atomic absorption spectroscopy the modulated signal is measured with a frequency-selective (or tuned) amplifier [54], or a phase-selective (or lock-in) amplifier [54, 59-61].

The frequency-selective amplifier responds only within a narrow frequency band. For the amplifier to be able to make, receive and amplify the modulated information signal, the frequency band of this signal must coincide with the band of the amplifier. Hence, both the amplifier and the carrier wave generator should be as stable as possible, in order to have the best possible correspondence between the two frequency bands. A frequency-selective amplifier gives an *S/N* ratio which is proportional to the reciprocal of the square of the amplifier bandwidth (the width of the frequency band). Usually, with conventional frequency-selective amplifiers we cannot obtain a bandwidth smaller than 1 Hz [54]. A further decrease of the bandwidth for this type of amplifier can be obtained by frequency-stabilization with quartz crystals.

The phase-sensitive amplifier has a narrower bandwidth than the tuned amplifier. In such an amplifier, the noise is removed not only from the frequency characteristic, but also from the phase characteristic. The application of the phase-sensitive amplifier to signal demodulation in atomic absorption spectroscopy can be seen in Fig. 9.24. The reference and the modulated signal have the same phase, and, consequently, only the signals which have this phase will be amplified. With phase-sensitive amplifiers we can obtain an amplifier bandwidth of 0.01 Hz [54], hence the efficiency for increasing the *S/N* ratio is much better with these amplifiers than with tuned amplifiers.

9.3 IMPROVEMENT OF THE *S/N* RATIO BY SIGNAL PROCESSING

The procedures for increasing the *S/N* ratio by processing the signal have been developed in close relation with the advance in the electronic techniques of storage and processing of data. In this section we shall present briefly the most important procedures of signal processing with the purpose of increasing the *S/N* ratio.

9.3.1 Signal averaging

The improvement of the *S/N* ratio by signal averaging is applied in the simplest way in analytical chemistry, by making several determinations on the material to be investigated and taking the mean.

In another variant, the signal is integrated over a certain time interval. Signal integration is often used in spectroscopy, where the photocurrent produced by a specific radiation of the component to be determined is integrated for a certain time interval and then the integrated signal is measured.

The development of data storage techniques has extended the signal averaging method to much more complex signals [62-65]. This technique is also known as the ensemble averaging technique. In order to be separable from the accompanying noise by an averaging technique, a signal should be repeatable. The recorded signal, for example the spectrum or chromatogram of a material, is sampled at certain intervals in the range of measurement, and the results are digitized and stored in a multi-channel memory system. These operations – the signal recording, sampling, digitization, and storage to memory at the same locations – are repeated several times. This method for increasing the *S/N* ratio assumes that the signals obtained by repeated measurement are coherent, while the noise is random and has an average equal to zero. The principle of ensemble averaging is shown in Fig. 9.25. Here we have represented n hypothetical recordings, $y_1, y_2, \ldots, y_n$, of a complex signal (a spectrum or chromatogram) over the range x', x''.

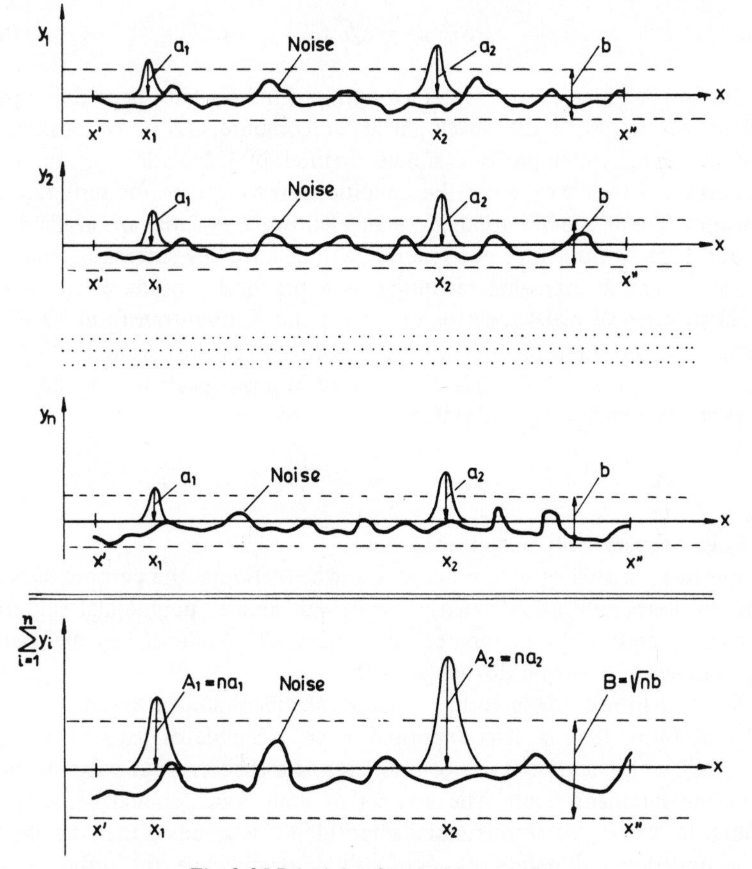

Fig. 9.25 Principle of ensemble averaging.

In order to clarify the method of improving the *S/N* ratio by averaging, the result of each measurement is represented as decomposed into a hypothetical signal which has two maxima, of amplitudes a_1 and a_2, and a random noise which has an average equal to zero and a width of fluctuation equal to b. In the lower part of Fig. 9.25 we represent separately the summed signal. Because the condition of coherence is satisfied, the pure signal sum (the signal without noise) is proportionally increasing with the number of accumulated recordings, while the random noise is increasing only in proportion to the square root of the number of accumulated recordings. Hence, the pure signal sum will have two maxima of amplitudes $A_1 = na_1$ and $A_2 = na_2$ at the points x_1 and x_2, respectively. The summed noise will have a width of fluctuation $B = \sqrt{n}\, b$. Hence, under such conditions of summation, if the signal-to-noise ratio for an individual measurement is $(S/N)_1$, the signal-to-noise ratio for summed measurements or averages of n individual measurements will be:

$$(S/N)_n = \sqrt{n}\,(S/N)_1 \qquad (9.65)$$

The applications of the signal averaging technique have been developed in close connection with the development of computers. Strictly speaking, the signal averaging technique is a simple method of calculation (summing and averaging) in which we ensure the condition of coherence for repeated signal recordings. For ensemble averaging many hardware systems are available [62, 63], but it can be done on any computer with an adequate software system.

The ensemble averaging technique is a practical procedure, and is often applied in classical NMR spectroscopy and pulse Fourier-transform NMR spectroscopy for the purpose of increasing sensitivity, i.e. increasing the S/N ratio [66, 67], etc. This technique is also applied in mass spectroscopy [68, 69], as well as in other analytical methods.

9.3.2 Least-squares polynomial smoothing

Another method of data processing – which also has the purpose of making easier the extraction of information – is least-squares polynomial smoothing. The great interest in this method of data processing is evidenced by the relatively large number of papers on this subject [70–77].

In order to use this smoothing method, the measurement result is expressed in digital form, that is, it is expressed as an ensemble of pairs of data x_1y_1, x_2y_2, . . . , which represent the co-ordinates of a succession of points belonging to the measurement result. The number of such points should be as large as possible. In least-squares polynomial smoothing it is assumed that the abscissae of the experimental points are rigorously known, while the ordinates y are affected by Gaussian random errors. The first $2m + 1$ pairs of consecutive values from the set of primary experimental data are used in the least-squares method to evaluate the coefficients for a polynomial of degree smaller than $2m + 1$. The first smoothed value is obtained by interpolating in this polynomial the mean value of the abscissæ of the $2m + 1$ data. In order to obtain the next smoothed value, we take another ensemble of $2m + 1$ primary data by leaving out the first pair $(x_1 y_1)$ and including a new one (x_{2m+2}, y_{2m+2}). As before, the new polynomial approximation is evaluated, and the new smoothed value derived. The procedure is repeated until all the primary data are exhausted. The smoothing procedure may be repeated on the set of smoothed data obtained in the first stage.

The smoothing method given above has been investigated in detail by Savitzky and Golay [75] and by Steiner *et al.* [76]. These authors have given the weighting coefficients c_i which can be used in convolution with the primary experimental data in order to derive the smoothed data. The evaluation of the smoothed values is obtained in this case from the equation:

$$\bar{y}_j = \sum_{i=-m}^{i=m} \frac{c_i y_{j+i}}{M} \tag{9.66}$$

where y are the primary data, $\bar{y}$ the smoothed data, c_i the weighting coefficients, M a normalization coefficient, and $y_{j-m}, \ldots, y_{j+m}$ the set of primary data from which the smoothed value $\bar{y}_j$ is evaluated.

The method of Savitzky and Golay [75] can be applied in the case when the abscissa values are equidistant. When we take the weighting coefficients $c_i = 1$ and $M = 2m + 1$, we have the so-called moving average method for smoothing data.

A practical application of the Savitzky and Golay method is exemplified by the hypothetical ensemble of data given in Fig. 9.26a. In Fig. 9.26b we give the smoothed data obtained for a cubic polynomial and $m = 2$, in which case the smoothed data are calculated from the equation:

$$\bar{y}_j = \frac{1}{35}(-3y_{j-2} + 12y_{j-1} + 17y_j + 12y_{j+1} - 3y_{j+2}) \tag{9.67}$$

For example, if $j = 3$, from the first group of $2m + 1 = 5$ data, we obtain:

$$\bar{y}_3 = \frac{1}{35}(-3 \times 2.9 + 12 \times 0.7 + 17 \times 10.4 + 12 \times 9.5 - 3 \times 0.7) = 8.7.$$

The least-squares smoothing method has been reviewed and examined in detail by Enke and Nieman [77]. In connection with the noise problem, these authors have investigated the distortion of the smoothed signal and the increase of the S/N ratio when the smoothing is done by fitting the polynomial by the least-squares method. This method belongs to the class of digital filters [65].

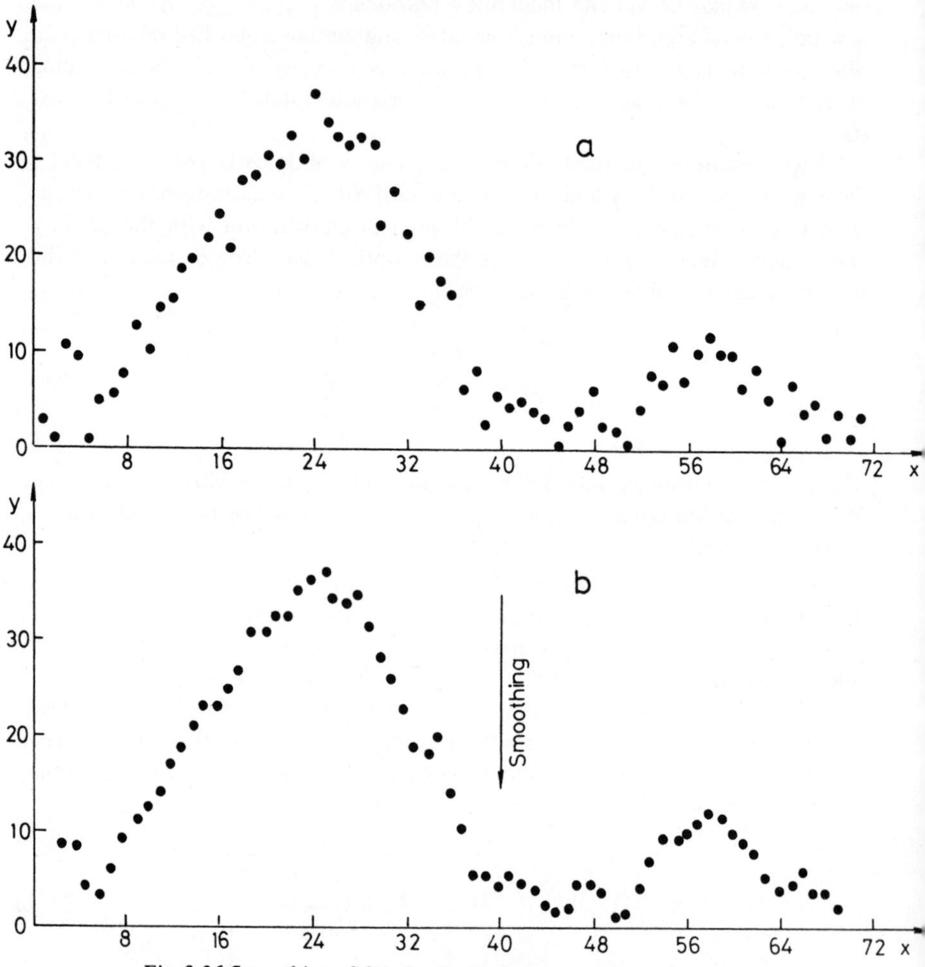

Fig. 9.26 Smoothing of data by the least-squares polynomial method.

9.3.3 The derivative of the signal

In analytical chemistry (in spectroscopy, chromatography, polarography, thermogravimetry, etc.) the signal derivative method is applied in order to simplify the extraction of the information contained in a measurement. The most significant applications of this procedure are in spectroscopy. Known as derivative spectroscopy, this technique has been applied for more than 20 years [78, 79], and the remarkable interest in this spectroscopic technique is shown by the large number of works on this subject [78–89].

The main advantage of derivative spectroscopy comes from the fact that it improves the detection of superimposed spectral bands. Usually, in the derivative spectrum the details are clearer and more numerous than in the normal spectrum. This can be seen in Fig. 9.27 where we give the first and second order derivatives for a Gaussian signal a and a sum signal b which is obtained by summing two Gaussian signals. The figure clearly shows that by differentiation the superimposition of signals is shown up more clearly, i.e. the resolution is improved. Hence, generally, in the derivative signal the systematic errors of interference are smaller. Obviously, the random errors (due to random noises) have higher values in derivative spectroscopy. Hence, derivative spectroscopy will be the recommended method of analysis when the most important errors are those due to the mutual interference of spectral bands.

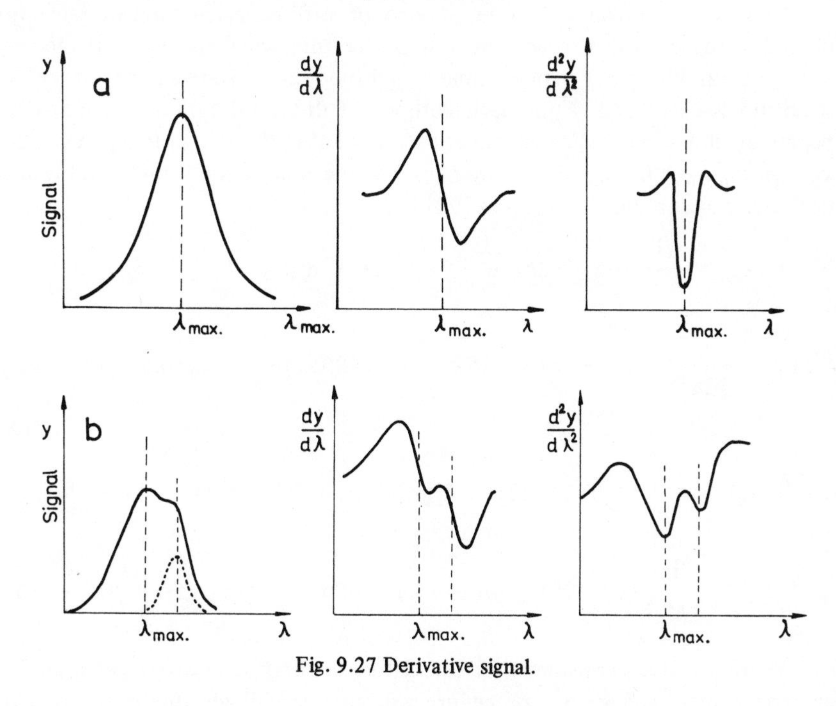

Fig. 9.27 Derivative signal.

In derivative spectroscopy, quantitative analysis is performed by the same procedure as in normal spectroscopy. Instead of measurements of intensity of absorption we make use of the derivative measurements. In derivative spectroscopy, the transfer (calibration) functions usually have the same form as in normal spectroscopy.

The methods for obtaining the derivative spectrum are divided into two classes. One includes the methods in which the derivative is obtained by operating on the output signal of the spectrometer, for instance in the case of the methods of mechanical, electronic, or numerical differentiation. The other contains the methods in which we operate on the light-beam, i.e. within the optical part of the spectrometer.

By the development of electronic computers, numerical methods in signal differentiation have been much extended. Once digitized, the derivative spectrum can be obtained by several numerical methods. Savitzky and Golay [75] obtain the derivative spectrum by the least-squares method. At any point of the spectrum, the derivative is calculated by using (9.66). The experimentally obtained spectrum is sampled at equal intervals on the abscissa, and the derivative is obtained from the convolution of these experimental data with convolution coefficients corresponding to the order of differentiation and also to the degree of the polynomial approximation of the data.

Generally, we can use either on-line or off-line calculation systems for obtaining the derivative spectrum. For a spectrum which is much affected by noise, a smoothing must first be applied, and then the derivative of the smoothed spectrum is calculated. Numerical methods of differentiation are described in papers on this subject. For example, by considering the five points at $x_0 - 2h$, $x_0 - h$, x_0, $x_0 + h$, $x_0 + 2h$, on the abscissa the first, second, third, and fourth derivatives are obtained as follows [90]:

$$f'(x_0) = \frac{1}{12h}[f(x_0-2h) - 8f(x_0-h) + 8f(x_0+h) - f(x_0+2h)]$$

$$f''(x_0) = \frac{1}{12h^2}[-f(x_0-2h) + 16f(x_0-h) - 30f(x_0) + 16f(x_0+h) - f(x_0+2h)]$$

$$\tag{9.68}$$

$$f'''(x_0) = \frac{1}{2h^3}[-f(x_0-2h) + 2f(x_0-h) - 2f(x_0+h) + f(x_0+2h)]$$

$$f''''(x_0) = -\frac{1}{h^4}[f(x_0-2h) - 4f(x_0-h) + 6f(x_0) - 4f(x_0+h) + f(x_0+2h)]$$

There are also expressions for calculating the derivative with more than five abscissa values, and such a procedure will correspondingly diminish errors of estimation.

9.3.4 Fourier-transform data processing

The Fourier transform is a mathematical procedure by which two functions $F(\nu)$ and $F(t)$ can be generated, according to the equations:

$$F(t) = \int_{-\infty}^{\infty} F(\nu) e^{2\pi i \nu t}\, d\nu \tag{9.69}$$

$$F(\nu) = \int_{-\infty}^{\infty} F(t)e^{-2\pi i\nu t}\, dt \tag{9.70}$$

Here the function $F(t)$ is the Fourier transform of the function $F(\nu)$, and $F(\nu)$ is the Fourier transform of $F(t)$.

The exponential function in (9.70) is equal to $\cos(2\pi\nu t) - i\sin(2\pi\nu t)$. Hence, if $F(t)$ is an even function, (9.70) becomes:

$$F(\nu) = 2\int_0^{\infty} F(t)\cos 2\pi\nu t\, dt, \tag{9.71}$$

which is known as the cosine Fourier transform. If $F(t)$ is an odd function, we find an equation which is similar to (9.71), but with the cosine replaced by the sine, that is:

$$F(\nu) = 2\int_0^{\infty} F(t)\sin 2\pi\nu t\, dt \tag{9.71'}$$

The Fourier transform procedure is qualitatively illustrated in Fig. 9.28. In the left upper side of this figure is represented a cosine wave of the general form $F(t) = \cos 2\pi\nu_0 t$, where ν_0 is the wave frequency, and t the time. The Fourier transform of this function is shown in the right upper side of Fig. 9.28.

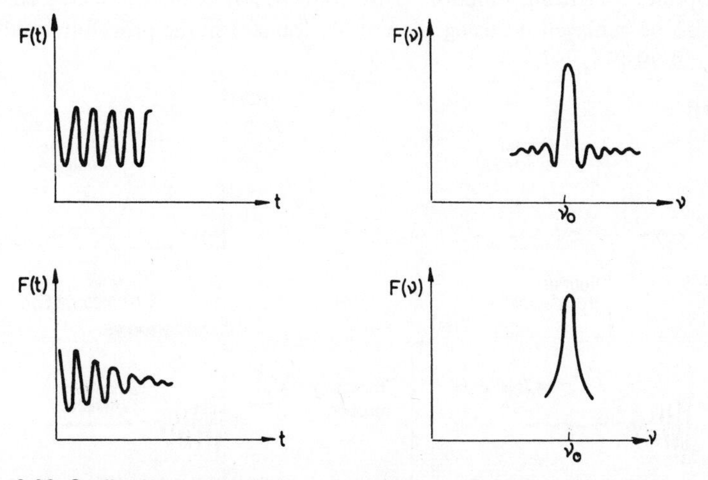

Fig. 9.28 Qualitative representation of the Fourier transform: (a) cosine wave undamped; (b) cosine wave with Gaussian damping.

In the lower part of the figure we have considered qualitatively the Fourier transform of a cosine wave with Gaussian damping. The Fourier transform of the cosine wave has a maximum corresponding to the frequency of the wave ν_0, the width of the transform decreasing with increasing wave life-time, and the shape of the transform is dependent on the truncation to which the wave is subjected.

With few exceptions, the integrals (9.72), (9.73), and (9.74) cannot be easily calculated, but correct and fast evaluations for them can be obtained on digital computers. The method of evaluation of the Fourier transform by using computers is known as the fast Fourier transform [91, 92].

The Fourier transform is fundamental in all spectral determinations. In order to obtain the spectrum of an electromagnetic radiation with a certain wave-shape, a frequency analysis of the wave must be performed. This analysis means effecting the Fourier transform of the electromagnetic wave.

Such a Fourier transform can be obtained indirectly in the optical region by using prisms or diffraction gratings, and in the X-ray region by using crystals. Another method for obtaining the spectrum is through the interferogram of the radiation, obtained by using the Michelson interferometer. The spectrum is obtained by Fourier transform of the interferogram by use of a digital computer. This technique is called Fourier transform spectroscopy [93, 94].

The Fourier transform is also applied to data processing with good results – the object being to make easier the process of extracting the information [95–101]. The operations to which spectral data can be subjected by Fourier transform, with the object of making the information more accessible, i.e. to improve the resolution, are smoothing, deconvolution and differentiation.

Smoothing. The smoothing of data has been discussed in Section 9.2. One of the most applied smoothing methods is the convolution of data with an appropriate weighting function. This convolution (which is called smoothing) can also be achieved by using the Fourier transform; the procedure is illustrated in Fig. 9.29 [95, 99].

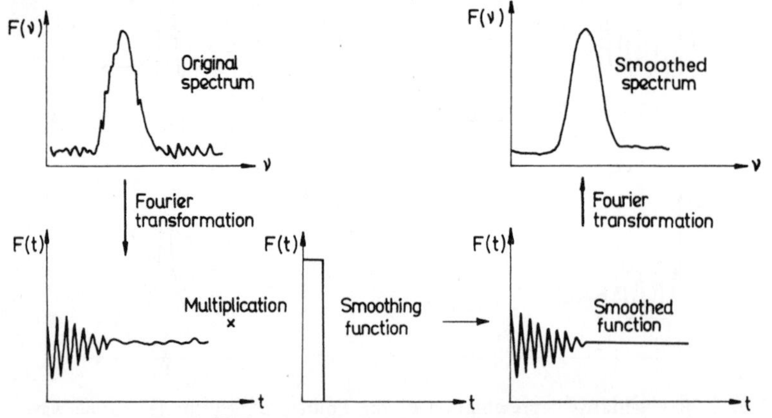

Fig. 9.29 Fourier-transform smoothing for a spectrum.

The original spectrum $F(\nu)$, which includes a noise, is Fourier-transformed to give the function $F(t)$, which gives the wave-shape in the presence of noise. This function $F(t)$, which also includes a noise, is multiplied with an adequate smoothing function to give the smoothed wave-shape function. Taking the Fourier transform of this yields the smoothed spectrum.

Deconvolution. We shall again consider data given in the form of a spectrum. As is well known, there are various processes which result in broadening and distortion of spectral lines, e.g. Doppler and collisional broadening of atomic lines. The finite resolving power of a monochromator causes a distortion of the atomic and molecular spectral lines, and electronic filters lead to corresponding truncations. All these interactions are described mathematically by the convolution integral:

$$F(\nu) = \int_{-\infty}^{\infty} f(\nu)B(\nu)\mathrm{d}\nu \qquad (9.72)$$

where $f(\nu)$ is the line profile free from distortions, $B(\nu)$ the distorting function, and $F(\nu)$ the recorded line profile.

The procedure by which we can calculate the non-distorted profile $f(\nu)$ from $F(\nu)$ and $B(\nu)$ is known as deconvolution. Such a calculation can be done by two methods. In one of them [101], we convolute the observed profile $F(\nu)$ with the function $B(\nu)$, which results in the convolution profile $C(\nu)$. The true line profile $f(\nu)$ is then obtained from:

$$f(\nu) = F^2(\nu)/C(\nu) \qquad (9.73)$$

The other method of deconvolution is based on the Fourier transform [97, 98]. The observed profile $F(\nu)$ and the distorting function $B(\nu)$ are Fourier-transformed, thus giving the functions $F(t)$ and $B(t)$. Their ratio:

$$f(t) = F(t)/B(t) \qquad (9.74)$$

is the Fourier transform of the true profile. By taking the Fourier transform of the function $f(t)$ we obtain the true profile $f(\nu)$.

Differentiation. This method was discussed in Section 9.3.3. This method of data processing can also be done by using Fourier transforms [95, 98]. If the imaginary part of the Fourier transform from a function is multiplied by a function with linear slope, the product is the derivative of the initial function.

9.3.5 Correlation method for increasing the *S*/*N* ratio

Mathematically, the correlation of two signals $y_1(x)$ and $y_2(x)$ is evaluated by the function:

$$C_{y_1y_2}(\tau) = \lim_{X\to\infty} \frac{1}{2X}\int_{-X}^{X} y_1(x)y_2(x \pm \tau)\mathrm{d}x \qquad (9.75)$$

where $C_{y_1y_2}(\tau)$ is the correlation function of the two signals, τ is the relative displacement of the two signals, and x is the variable on which the signals are dependent (time, wavelength, frequency, voltage, etc.).

If $y_1(x)$ and $y_2(x)$ are identical, Eq. (9.75) will give the auto-correlation function. This function gives indications of the possible coherence of the two signals.

If the signals are different, Eq. (9.75) will give the cross-correlation function of the two signals. This function gives an evaluation of the degree of resemblance of the two signals.

The auto-correlation function is evaluated as shown in Fig. 9.30.

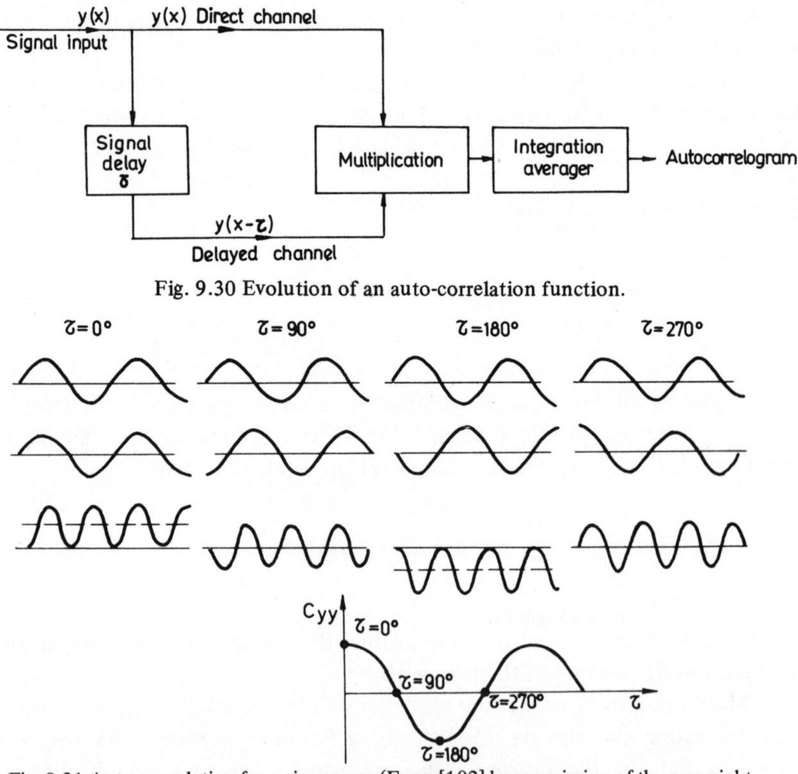

Fig. 9.30 Evolution of an auto-correlation function.

Fig. 9.31 Auto-correlation for a sine wave. (From [102] by permission of the copyright holder, the American Chemical Society)

The auto-correlation function for a sine-wave [102] is shown in Fig. 9.31, at the following four values of the signal displacement: $\tau = 0°, 90°, 180°, 270°$. The figure shows that a sine-wave is converted into a cosine-wave by the auto-correlation operation.

Figure 9.32 illustrates a noise auto-correlation function, C_N (non-periodic signal with random wave profile). The ideal auto-correlation shown in Fig. 9.32a corresponds to an ideal white noise (noise which contains all the frequencies, i.e. with infinite bandwidth). Under real conditions the noise bandwidth is finite, and the auto-correlation function has the form shown in Fig. 9.32b.

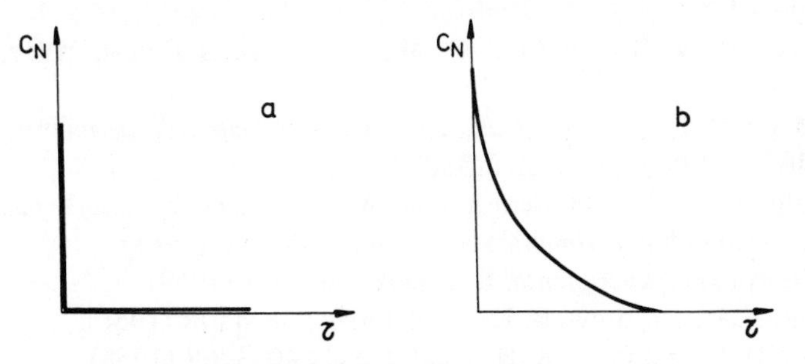

Fig. 9.32 Noise auto-correlation.

As can be seen in Figs. 9.31 and 9.32, by comparing the auto-correlation functions we can distinguish between a periodic signal and a noise.

The auto-correlation function C for a sine-wave in the presence of a noise which has a finite bandwidth is shown in Fig. 9.33.

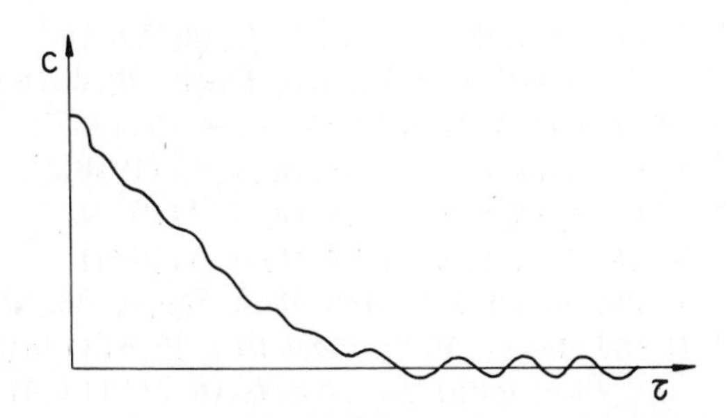

Fig. 9.33 Auto-correlation of a noise-perturbed sine wave.

The correlation function for two different signals can be obtained as shown in Fig. 9.30. In this case, the direct channel is connected to one signal source and the delay channel is connected to the other signal source.

The correlation function is obtained either on specialized hardware systems (correlators) or on any computer by an appropriate software system.

The correlation technique has many applications in the domain of communications [103–105]. In analytical chemistry, this procedure of signal processing already has many valuable applications [62, 102]. Hieftje and co-workers [106] have used this technique as a demodulator in flame spectrometry, with a view to increasing the *S/N* ratio. Horlick and Codding [107] have applied it as a smoothing method and to resolution improvement and differentiation of spectra. The signal correlation technique is also applied as a method for extracting information from various spectra [108–110].

REFERENCES

[1] Cochran, W. G. and Cox, G. M., *Experimental Design,* Wiley, London (1957).
[2] Davies, O. L., *The Design and Analysis of Industrial Experiments,* 2nd Ed., Oliver and Boyd, London (1960).
[3] Nalimov, V. V. and Cernova, N. A., *Statisticeskie metodi planirovania extremalnikh experimentov,* Izd. Nauka, Moscow (1965).
[4] *Planirovanie experimenta,* Izd. Nauka, Moscow (1969).
[5] Hill, H. M. and Brow, R. H., *Anal. Chem.,* **38**, 440R (1966).
[6] Hill, H. M. and Brow, R. H., *Anal. Chem.,* **40**, 376R (1968).
[7] Box, G. E. P. and Wilson, K. B., *J. Roy. Statist. Soc., Ser. B,* **13**, 1 (1951).
[8] Box, G. E. P., *Biometrics,* **10**, 16 (1954).
[9] Box, G. E. P., *Appl. Statist.* **6**, 81 (1957).
[10] Currie, L. A., Filliben, J. J. and DeVoe, J. R., *Anal. Chem.,* **44**, 497R (1972).
[11] Spendley, W., Hext, G. R. and Himsworth, F. R., *Technometrics,* **4**, 441 (1962).
[12] Nelder, J. A. and Mead, R., *Comput. J.,* **7**, 308 (1965).
[13] Shoenfeld, P. S. and DeVoe, J. R., *Anal. Chem.,* **48**, 403R (1976).
[14] Bradley, R. A., *Ind. Qual. Control,* **15**, No. 4, 16 (1958).
[15] Hunter, J. S., *Ind. Qual. Control,* **15**, No. 6, 16 (1958).
[16] Hunter, J. S., *Ind. Qual. Control,* **15**, No. 7, 7 (1959).
[17] Hunter, J. S., *Ind. Qual. Control,* **15**, No. 8, 6 (1959).
[18] Box, G. E. P. and Hunter, J. S., *Ann. Math. Statist.,* **28**, 195 (1957).
[19] Brooks, S. H. and Mickey, M. R., *Biometrics,* **17**, 81 (1961).
[20] Box, G. E. P., *J. Roy. Statist. Soc., Ser. B,* **16**, 211 (1954).
[21] Nalimov, V. V., *Wiss. Z. Techn. Hochschule für Chemie, Leuna Merseburg,* **11**, 16 (1969).
[22] Alimarin, I. P., Petrukhin, O. M. and Malofeeva, G. I., *Zh. Analit. Khim.,* **26**, 2019 (1971).
[23] Rubin, I. B., Mitchell, T. J. and Goldstein, G., *Anal. Chem.,* **43**, 717 (1971).
[24] Kamenev, A. I., Granovskii, Yu. V. and Vinogradova, E. N., *Zavodsk. Lab.,* **34**, 1041 (1968).

[25] Kamenev, A. I., Luk'yanov, V. B., Figurovskaya, V. N. and Vinogradova, E. N., *Zh. Analit. Khim.*, **21**, 793 (1966).
[26] Bardina, S. M. and Chikrizova, E. G., *Zh. Analit. Khim.*, **31**, 2101 (1976).
[27] Belikov, V. G., Kokovkin-Shcherbak, N. I. and Mutsueva, S. Kh., *Zavodsk. Lab.*, **33**, 1049 (1967).
[28] Barskii, V. D. and Noskov, V. V., *Zavodsk. Lab.*, **31**, 349 (1965).
[29] Silakova, V. G., Makulov, N. A., Manova, T. G. and Bozhevol'nov, E. A., *Zh. Analit. Khim.*, **29**, 1683 (1974).
[30] Arpadjan, S., Doerffel, K., Holland-Letz, K., Much, H. and Pannach, M., *Z. Anal. Chem.*, **270**, 257 (1974).
[31] Davies, L., *Talanta* 22, 371 (1975).
[32] Ospanov, Kh. K., Oksengoit, E. A., Brook, B. S. and Akhnazarova, S. L., *Zh. Analit. Khim.*, **33**, 453 (1978).
[33] Bazykina, E. N., Smagunova, A. N. and Molchanova, A. P., *Zh. Analit. Khim.*, **32**, 2003 (1977).
[34] Oksengoit, E. A., Brook, B. S. and Futerman, Yu. S., *Zh. Analit. Khim.*, **34**, 251 (1979).
[35] Oksengoit, E. A., Brook, B. S., Ospanov, Kh. K. and Akhnazarova, S. L., *Zh. Analit. Khim.*, **34**, 728 (1979).
[36] Sokolowska, W., *Chem. Anal. (Warsaw)*, **24**, 577 (1979).
[37] Long, D. E., *Anal. Chim. Acta*, **46**, 193 (1969).
[38] Deming, S. N. and Morgan, S. L., *Anal. Chem.*, **45**, 278A (1973).
[39] Morgan, S. L. and Deming, S. N., *Anal. Chem.*, **46**, 1170 (1974).
[40] King, P. G. and Deming, S. N., *Anal. Chem.*, **46**, 1476 (1974).
[41] Ritter, G. L., Lowry, S. R., Wilkins, C. L. and Isenhour, T. L., *Anal. Chem.*, **47**, 1951 (1975).
[42] Kaberline, S. L. and Wilkins, C. L., *Anal. Chim. Acta*, **103**, 417 (1978).
[43] Deming, S. N., Morgan, S. L. and Willcott, M. R., *Int. Lab.*, Nov./Dec., 11 (1976).
[44] Deming, S. N., Morgan, S. L. and Willcott, M. R., *Am. Lab.* **8**, No. 10, 13 (1976).
[45] Deming, S. N. and Parker, L. R., *CRC Crit. Rev. Anal. Chem.*, **7**, 187 (1978).
[46] Routh, M. W., Swartz, P. A. and Denton, M. B., *Anal. Chem.*, **49**, 1422 (1977).
[47] Ernst, R. R., *Rev. Sci. Instrum.* **39**, 998 (1969).
[48] Parker, L. R., Morgan, S. L. and Deming, S. N., *Appl. Spectry.*, **29**, 429 (1975).
[49] Morgan, S. L. and Deming, S. N., *J. Chromatog.*, **112**, 267 (1975).
[50] Meus, M., Parczewski, A. and Rokosz, A., *Chem. Anal. (Warsaw)*, **20**, 247 (1975).
[51] Parczewski, A., *Chem. Anal. (Warsaw)*, **22**, 27 (1977).
[52] Turoff, M. L. H. and Deming, S. N., *Talanta*, **24**, 567 (1977).
[53] Coor, T., *J. Chem. Educ.* **45**, A533 (1968).

[54] Hieftje, G. M., *Anal. Chem.*, **44**, (6), 81A (1972).
[55] Korn, G. A., *Random-Process Simulation and Measurements,* McGraw-Hill, New York (1966).
[56] Russell, B. J., Shelton, J. P. and Walsh, A., *Spectrochim. Acta,* **8**, 317 (1957).
[57] Box, G. F. and Walsh, A., *Spectrochim. Acta,* **16**, 255 (1960).
[58] Hieftje, G. M., Holder, B. E., Maddux, A. S., Jr. and Lim, R., *Anal. Chem.*, **45**, 238 (1973).
[59] Coor, T., *J. Chem. Educ.,* **45**, A583 (1968).
[60] O'Haver, T. C., *J. Chem. Educ.,* **49**, A131 (1972).
[61] O'Haver, T. C., *J. Chem. Educ.,* **49**, A211 (1972).
[62] Hieftje, G. M., *Anal. Chem.*, **44**, (7), 69A (1972).
[63] Rowell, R. L., *J. Chem. Educ.,* **51**, A71 (1974).
[64] Kraus, J. B., *Industrial Research,* June, 62 (1975).
[65] Dulaney, G., *Anal. Chem.*, **47**, 24A (1975).
[66] Ernst, R. R., *Rev. Sci. Instrum.*, **36**, 1689 (1965).
[67] *Jeol News,* **9a**, 16 (1971).
[68] Biros, F. J., *Anal. Chem.*, **42**, 537 (1970).
[69] Harrison, W. W. and Mattson, W. A., *Anal. Chem.*, **46**, 1979 (1974).
[70] Yule, H. P., *Anal. Chem.*, **38**, 103 (1966).
[71] Jones, R. N., Venkataraghavan, R. and Hopkins, J. W., *Spectrochim. Acta,* **23A**, 941 (1967).
[72] Yule, H. P., *Anal. Chem.*, **44**, 1245 (1972).
[73] Nieman, T. A. and Enke, C. G., *Anal. Chem.*, **48**, 619 (1976).
[74] Overton, M. W., Alber, L. L. and Smith, D. E., *Anal. Chem.*, **47**, 363A (1975).
[75] Savitzky, A. and Golay, M.J.E., *Anal. Chem.*, **36**, 1627 (1964).
[76] Steinier, J., Termonia, Y. and Deltour, J., *Anal. Chem.*, **44**, 1906 (1972).
[77] Enke, C. G. and Nieman, T. A., *Anal. Chem.*, **48**, 705A (1976).
[78] Hammond, V. J. and Price, W. C., *J. Opt. Soc. Amer.*, **43**, 924 (1953).
[79] Morrison, J. D., *J. Chem. Phys.*, **21**, 1767 (1953).
[80] Martin, A. E., *Spectrochim. Acta,* **14**, 97 (1959).
[81] Collier, G. L. and Panting, A. C. M., *Spectrochim. Acta,* **14**, 104 (1959).
[82] Porro, T. J., *Anal. Chem.*, **44**, (4), 93A (1972).
[83] McWilliam, I. G., *Anal. Chem.*, **41**, 674 (1969).
[84] Elser, R. C. and Winefordner, J. D., *Anal. Chem.*, **44**, 698 (1972).
[85] Fowler, W., Knapp, D. and Winfordner, J. D., *Anal. Chem.*, **46**, 4 (1974).
[86] Hager, R. N., Jr., *Anal. Chem.*, **45**, 1131A (1973).
[87] Green, G. L. and O'Haver, T. C., *Anal. Chem.*, **46**, 2191 (1974).
[88] O'Haver, T. C. and Green, G. L., *Int. Lab.*, May/June, 11 (1975).
[89] O'Haver, T. C., *Anal. Chem.*, **51**, 91A (1979).
[90] Darn, W. S. and McCracken, D. D., *Numerical Methods with Fortran IV, Case Studies,* Wiley, New York (1972).

[91] Cooley, J. W. and Tukey, J. W., *Math. Comp.*, **19**, 297 (1965).
[92] Mertz, L., *Appl. Opt.*, **10**, 386 (1971).
[93] Horlick, G., *Anal. Chem.*, **43**, (8), 61A (1971).
[94] Horlick, G. and Yuen, W. K., *Anal. Chem.*, **47**, 775A (1975).
[95] Horlick, G., *Anal. Chem.*, **44**, 943 (1972).
[96] Kirmse, D. W. and Westerberg, A. W., *Anal. Chem.*, **43**, 1035 (1971).
[97] den Harder, A. and de Galan, L., *Anal. Chem.*, **46**, 1464 (1974).
[98] Kelly, P. C. and Horlick, G., *Anal. Chem.*, **46**, 2130 (1974).
[99] Bush, C. A., *Anal. Chem.*, **46**, 890 (1974).
[100] Jones, R. N., Venkataraghavan, R. and Hopkins, J. W., *Spectrochim. Acta,* **23A**, 925 (1967).
[101] Hayes, J. W., Glover, D. E., Smith, D. E. and Overton, M. W., *Anal. Chem.*, **45**, 277 (1973).
[102] Betty, K. R. and Horlick, G., *Anal. Chem.*, **48**, 1899 (1976).
[103] Lee, Y. W., Cheatom, T. P., Jr., and Wiesner, J. B., *Proc. IRE,* **38**, 1165 (1950).
[104] Lee, Y. W., *Statistical Theory of Communication,* Wiley, New York (1960).
[105] Lange, F. H., *Correlation Techniques,* Van Nostrand, Princeton (1967).
[106] Hieftje, G. M., Bystroff, R. I. and Lim, R., *Anal. Chem.*, **45**, 253 (1973).
[107] Horlick, G. and Codding, E. G., *Anal. Chem.*, **45**, 1749 (1973).
[108] Horlick, G., *Anal. Chem.*, **45**, 319 (1973).
[109] MacDonald, R., Robertson, A., Kennett, T. J. and Prestwich, W. V., *J. Radioanal. Chem.*, **23**, 123 (1973).
[110] Powell, L. A. and Hieftje, G. M., *Anal. Chim. Acta,* **100**, 313 (1978).

Appendix

Table I

Values of the Laplace function $\Phi(z) = \frac{1}{\sqrt{2\pi}} \int_0^z \exp\left[-\frac{z^2}{2}\right] dz$

z	0	1	2	3	4	5	6	7	8	9
0.0	0.0000	0.0040	0.0080	0.0120	0.0160	0.0199	0.0239	0.0279	0.0319	0.0359
0.1	0.0398	0.0438	0.0478	0.0517	0.0557	0.0596	0.0636	0.0675	0.0714	0.0753
0.2	0.0793	0.0832	0.0871	0.0910	0.0948	0.0987	0.1026	0.1064	0.1103	0.1141
0.3	0.1179	0.1217	0.1255	0.1293	0.1331	0.1368	0.1406	0.1443	0.1480	0.1517
0.4	0.1554	0.1591	0.1628	0.1664	0.1700	0.1736	0.1772	0.1808	0.1844	0.1879
0.5	0.1915	0.1950	0.1985	0.2019	0.2054	0.2088	0.2123	0.2157	0.2190	0.2224
0.6	0.2257	0.2291	0.2324	0.2357	0.2389	0.2422	0.2454	0.2486	0.2517	0.2549
0.7	0.2580	0.2611	0.2642	0.2673	0.2704	0.2734	0.2764	0.2794	0.2823	0.2852
0.8	0.2881	0.2910	0.2939	0.2967	0.2995	0.3023	0.3051	0.3078	0.3106	0.3133
0.9	0.3159	0.3186	0.3212	0.3238	0.3264	0.3289	0.3315	0.3340	0.3365	0.3389
1.0	0.3413	0.3438	0.3461	0.3485	0.3508	0.3531	0.3554	0.3577	0.3599	0.3621
1.1	0.3643	0.3665	0.3686	0.3708	0.3729	0.3749	0.3770	0.3790	0.3810	0.3830
1.2	0.3849	0.3869	0.3888	0.3907	0.3925	0.3944	0.3962	0.3980	0.3997	0.4015
1.3	0.4032	0.4049	0.4066	0.4082	0.4099	0.4115	0.4131	0.4147	0.4162	0.4177
1.4	0.4192	0.4207	0.4222	0.4236	0.4251	0.4265	0.4279	0.4292	0.4306	0.4319
1.5	0.4332	0.4345	0.4357	0.4370	0.4382	0.4394	0.4406	0.4418	0.4429	0.4441
1.6	0.4452	0.4463	0.4474	0.4484	0.4495	0.4505	0.4515	0.4525	0.4535	0.4545
1.7	0.4554	0.4564	0.4573	0.4582	0.4591	0.4599	0.4608	0.4616	0.4625	0.4633
1.8	0.4641	0.4649	0.4656	0.4664	0.4671	0.4678	0.4686	0.4693	0.4699	0.4706
1.9	0.4713	0.4719	0.4726	0.4732	0.4738	0.4744	0.4750	0.4756	0.4761	0.4767
2.0	0.4772	0.4778	0.4783	0.4788	0.4793	0.4798	0.4803	0.4808	0.4812	0.4817
2.1	0.4821	0.4826	0.4830	0.4834	0.4838	0.4842	0.4846	0.4850	0.4854	0.4857
2.2	0.4861	0.4864	0.4868	0.4871	0.4875	0.4878	0.4881	0.4884	0.4887	0.4890
2.3	0.4893	0.4896	0.4898	0.4901	0.4904	0.4906	0.4909	0.4911	0.4913	0.4916
2.4	0.4918	0.4920	0.4922	0.4925	0.4927	0.4929	0.4931	0.4932	0.4934	0.4936
2.5	0.4938	0.4940	0.4941	0.4943	0.4945	0.4946	0.4948	0.4949	0.4951	0.4952
2.6	0.4953	0.4955	0.4956	0.4957	0.4959	0.4960	0.4961	0.4962	0.4963	0.4964
2.7	0.4965	0.4966	0.4967	0.4968	0.4969	0.4970	0.4971	0.4972	0.4973	0.4974
2.8	0.4974	0.4975	0.4976	0.4977	0.4977	0.4978	0.4979	0.4979	0.4980	0.4981
2.9	0.4981	0.4982	0.4983	0.4983	0.4984	0.4984	0.4985	0.4985	0.4986	0.4986
3.0	0.4987	0.4987	0.4987	0.4988	0.4988	0.4989	0.4989	0.4989	0.4990	0.4990
3.1	0.4990	0.4991	0.4991	0.4991	0.4992	0.4992	0.4992	0.4992	0.4993	0.4993

Table II

Values of the function $F(x) = 0.5 + \Phi(z) = \frac{1}{\sqrt{2\pi}} \int_{-\infty}^{z} \exp\left[-\frac{z^2}{2}\right] dz;\ z > 0$

z	0	1	2	3	4	5	6	7	8	9
0.0	0.5000	0.5040	0.5080	0.5120	0.5160	0.5199	0.5239	0.5279	0.5319	0.5359
0.1	0.5398	0.5438	0.5478	0.5517	0.5557	0.5596	0.5636	0.5675	0.5714	0.5753
0.2	0.5793	0.5832	0.5871	0.5910	0.5948	0.5987	0.6026	0.6064	0.6103	0.6141
0.3	0.6179	0.6217	0.6255	0.6293	0.6331	0.6368	0.6406	0.6443	0.6480	0.6517
0.4	0.6554	0.6591	0.6628	0.6664	0.6700	0.6736	0.6772	0.6808	0.6844	0.6879
0.5	0.6915	0.6950	0.6985	0.7019	0.7054	0.7088	0.7123	0.7157	0.7190	0.7224
0.6	0.7257	0.7291	0.7324	0.7357	0.7389	0.7422	0.7454	0.7486	0.7517	0.7549
0.7	0.7580	0.7611	0.7642	0.7673	0.7704	0.7734	0.7764	0.7794	0.7823	0.7852
0.8	0.7881	0.7910	0.7939	0.7967	0.7995	0.8023	0.8051	0.8078	0.8106	0.8133
0.9	0.8159	0.8186	0.8212	0.8238	0.8264	0.8289	0.8315	0.8340	0.8365	0.8389
1.0	0.8413	0.8438	0.8461	0.8485	0.8508	0.8531	0.8554	0.8577	0.8599	0.8621
1.1	0.8643	0.8665	0.8686	0.8708	0.8729	0.8749	0.8770	0.8790	0.8810	0.8830
1.2	0.8849	0.8869	0.8888	0.8907	0.8925	0.8944	0.8962	0.8980	0.8997	0.9015
1.3	0.9032	0.9049	0.9066	0.9082	0.9099	0.9115	0.9131	0.9147	0.9162	0.9177
1.4	0.9192	0.9207	0.9222	0.9236	0.9251	0.9265	0.9279	0.9292	0.9306	0.9319
1.5	0.9332	0.9345	0.9357	0.9370	0.9382	0.9394	0.9406	0.9418	0.9429	0.9441
1.6	0.9452	0.9463	0.9474	0.9484	0.9495	0.9505	0.9515	0.9525	0.9535	0.9545
1.7	0.9554	0.9564	0.9573	0.9582	0.9591	0.9599	0.9608	0.9616	0.9625	0.9633
1.8	0.9641	0.9649	0.9656	0.9664	0.9671	0.9678	0.9686	0.9693	0.9699	0.9706
1.9	0.9713	0.9719	0.9726	0.9732	0.9738	0.9744	0.9750	0.9756	0.9761	0.9767
2.0	0.9772	0.9778	0.9783	0.9788	0.9793	0.9798	0.9803	0.9808	0.9812	0.9817
2.1	0.9821	0.9826	0.9830	0.9834	0.9838	0.9842	0.9846	0.9850	0.9854	0.9857
2.2	0.9861	0.9864	0.9868	0.9871	0.9875	0.9878	0.9881	0.9884	0.9887	0.9890
2.3	0.9893	0.9896	0.9898	0.9901	0.9904	0.9906	0.9909	0.9911	0.9913	0.9916
2.4	0.9918	0.9920	0.9922	0.9925	0.9927	0.9929	0.9931	0.9932	0.9934	0.9936
2.5	0.9938	0.9940	0.9941	0.9943	0.9945	0.9946	0.9948	0.9949	0.9951	0.9952
2.6	0.9953	0.9955	0.9956	0.9957	0.9959	0.9960	0.9961	0.9962	0.9963	0.9964
2.7	0.9965	0.9966	0.9967	0.9968	0.9969	0.9970	0.9971	0.9972	0.9973	0.9974
2.8	0.9974	0.9975	0.9976	0.9977	0.9977	0.9978	0.9979	0.9979	0.9980	0.9981
2.9	0.9981	0.9982	0.9983	0.9983	0.9984	0.9984	0.9985	0.9985	0.9986	0.9986
3.0	0.9987	0.9987	0.9987	0.9988	0.9988	0.9989	0.9989	0.9989	0.9990	0.9990
3.1	0.9990	0.9991	0.9991	0.9991	0.9992	0.9992	0.9992	0.9992	0.9993	0.9993

Table III

Values of the variable χ^2 for $P = F(\chi^2) = \int_0^{\chi^2} f(\chi^2; \nu)\, d\chi^2$

$\nu \backslash P$	0.01	0.025	0.05	0.10	0.90	0.95	0.975	0.99	0.995
1	0.00016	0.00098	0.0039	0.0158	2.71	3.84	5.02	6.63	7.88
2	0.0201	0.0506	0.1026	0.2107	4.61	5.99	7.38	9.21	10.60
3	0.115	0.216	0.352	0.584	6.25	7.81	9.35	11.34	12.84
4	0.297	0.484	0.711	1.064	7.78	9.49	11.14	13.28	14.86
5	0.554	0.831	1.15	1.61	9.24	11.07	12.83	15.09	16.75
6	0.872	1.24	1.64	2.20	10.64	12.59	14.45	16.81	18.55
7	1.24	1.69	2.17	2.83	12.02	14.07	16.01	18.48	20.28
8	1.65	2.18	2.73	3.49	13.36	15.51	17.53	20.09	21.96
9	2.09	2.70	3.33	4.17	14.68	16.92	19.02	21.67	23.59
10	2.56	3.25	3.94	4.87	15.99	18.31	20.48	23.21	25.19
11	3.05	3.82	4.57	5.58	17.28	19.68	21.92	24.73	26.76
12	3.57	4.40	5.23	6.30	18.55	21.03	23.34	26.22	28.30
13	4.11	5.01	5.89	7.04	19.81	22.36	24.74	27.69	29.82
14	4.66	5.63	6.57	7.79	21.06	23.68	26.12	29.14	31.32
15	5.23	6.26	7.26	8.55	22.31	25.00	27.49	30.58	32.80
16	5.81	6.91	7.96	9.31	23.54	26.30	28.85	32.00	34.27
18	7.01	8.23	9.39	10.86	25.99	28.87	31.53	34.81	37.16
20	8.26	9.59	10.85	12.44	28.41	31.41	34.17	37.57	40.00
24	10.86	12.40	13.85	15.66	33.20	36.42	39.36	42.98	45.56
30	14.95	16.79	18.49	20.60	40.26	43.77	46.98	50.89	53.67
40	22.16	24.43	26.51	29.05	51.81	55.76	59.34	63.69	66.77
60	37.48	40.48	43.19	46.46	74.40	79.08	83.30	88.38	91.95
120	86.92	91.58	95.70	100.62	140.23	146.57	152.21	158.95	163.64

Table IV

Values of Student's variable t for $P = F(t) = \int_{-\infty}^{t} f(t;\nu)\mathrm{d}t$

ν \ P	0.70	0.80	0.90	0.95	0.99	0.999
1	1.963	3.078	6.314	12.706	63.657	636.619
2	1.386	1.886	2.920	4.303	9.925	31.598
3	1.250	1.638	2.353	3.182	5.841	12.941
4	1.190	1.533	2.132	2.776	4.604	8.610
5	1.156	1.476	2.015	2.571	4.032	6.859
6	1.134	1.440	1.943	2.447	3.707	5.959
7	1.119	1.415	1.895	2.365	3.499	5.405
8	1.108	1.397	1.860	2.306	3.355	5.041
9	1.100	1.383	1.833	2.262	3.250	4.781
10	1.093	1.372	1.812	2.228	3.169	4.587
11	1.088	1.363	1.796	2.201	3.106	4.437
12	1.083	1.356	1.782	2.179	3.055	4.318
13	1.079	1.350	1.771	2.160	3.012	4.221
14	1.076	1.345	1.761	2.145	2.977	4.140
15	1.074	1.341	1.753	2.131	2.947	4.073
16	1.071	1.337	1.746	2.120	2.921	4.015
17	1.069	1.333	1.740	2.110	2.898	3.965
18	1.067	1.330	1.734	2.101	2.878	3.922
19	1.066	1.328	1.729	2.093	2.861	3.883
20	1.064	1.325	1.725	2.086	2.845	3.850
21	1.063	1.323	1.721	2.080	2.831	3.819
22	1.061	1.321	1.717	2.074	2.819	3.792
23	1.060	1.319	1.714	2.069	2.807	3.767
24	1.059	1.318	1.711	2.064	2.797	3.745
25	1.058	1.316	1.708	2.060	2.787	3.725
26	1.058	1.315	1.706	2.056	2.779	3.707
27	1.057	1.314	1.703	2.052	2.771	3.690
28	1.056	1.313	1.701	2.048	2.763	3.674
29	1.055	1.311	1.699	2.045	2.756	3.659
30	1.055	1.310	1.697	2.042	2.750	3.646
40	1.050	1.303	1.684	2.021	2.704	3.551
60	1.046	1.296	1.671	2.000	2.660	3.460
120	1.041	1.289	1.658	1.980	2.617	3.373
∞	1.036	1.282	1.645	1.960	2.576	3.291

Table V
Values of the variable F for $P = F(F) = \int_0^{\mp} f(F; \nu_1, \nu_2)\mathrm{d}F = 0.95$

ν_2 \ ν_1	1	2	3	4	5	6	7	8	9	10	11	12	13	14	15	16	17
1	161	200	216	225	230	234	237	239	241	242	243	244	245	245	246	246	247
2	18.5	19.0	19.2	19.2	19.3	19.3	19.4	19.4	19.4	19.4	19.4	19.4	19.4	19.4	19.4	19.4	19.4
3	10.1	9.55	9.28	9.12	9.01	8.94	8.89	8.85	8.81	8.79	8.76	8.74	8.73	8.71	8.70	8.69	8.68
4	7.71	6.94	6.59	6.39	6.26	6.16	6.09	6.04	6.00	5.96	5.94	5.91	5.89	5.87	5.86	5.84	5.83
5	6.61	5.79	5.41	5.19	5.05	4.95	4.88	4.82	4.77	4.74	4.70	4.68	4.66	4.64	4.62	4.60	4.59
6	5.99	5.14	4.76	4.53	4.39	4.28	4.21	4.15	4.10	4.06	4.03	4.00	3.98	3.96	3.94	3.92	3.91
7	5.59	4.74	4.35	4.12	3.97	3.87	3.79	3.73	3.68	3.64	3.60	3.57	3.55	3.53	3.51	3.49	3.48
8	5.32	4.46	4.07	3.84	3.69	3.58	3.50	3.44	3.39	3.35	3.31	3.28	3.26	3.24	3.22	3.20	3.19
9	5.12	4.26	3.86	3.63	3.48	3.37	3.29	3.23	3.18	3.14	3.10	3.07	3.05	3.03	3.01	2.99	2.97
10	4.96	4.10	3.71	3.48	3.33	3.22	3.14	3.07	3.02	2.98	2.94	2.91	2.89	2.86	2.85	2.83	2.81
11	4.84	3.98	3.59	3.36	3.20	3.09	3.01	2.95	2.90	2.85	2.82	2.79	2.76	2.74	2.72	2.70	2.69
12	4.75	3.88	3.49	3.26	3.11	3.00	2.91	2.85	2.80	2.75	2.72	2.69	2.66	2.64	2.62	2.60	2.58
13	4.67	3.81	3.41	3.18	3.03	2.92	2.83	2.77	2.71	2.67	2.63	2.60	2.58	2.55	2.53	2.51	2.50
14	4.60	3.74	3.34	3.11	2.96	2.85	2.76	2.70	2.65	2.60	2.57	2.53	2.51	2.48	2.46	2.44	2.43
15	4.54	3.68	3.29	3.06	2.90	2.79	2.71	2.64	2.59	2.54	2.51	2.48	2.45	2.42	2.40	2.38	2.37
16	4.49	3.63	3.24	3.01	2.85	2.74	2.66	2.59	2.54	2.49	2.46	2.42	2.40	2.37	2.35	2.33	2.32
17	4.45	3.59	3.20	2.96	2.81	2.70	2.61	2.55	2.49	2.45	2.41	2.38	2.35	2.33	2.31	2.29	2.27
18	4.41	3.55	3.16	2.93	2.77	2.66	2.58	2.51	2.46	2.41	2.37	2.34	2.31	2.29	2.27	2.25	2.23
19	4.38	3.52	3.13	2.90	2.74	2.63	2.54	2.48	2.42	2.38	2.34	2.31	2.28	2.26	2.23	2.21	2.20
20	4.35	3.49	3.10	2.87	2.71	2.60	2.51	2.45	2.39	2.35	2.31	2.28	2.25	2.22	2.20	2.18	2.17
21	4.32	3.47	3.07	2.84	2.68	2.57	2.49	2.42	2.37	2.32	2.28	2.25	2.22	2.20	2.18	2.16	2.14
22	4.30	3.44	3.05	2.82	2.66	2.55	2.46	2.40	2.34	2.30	2.26	2.23	2.20	2.17	2.15	2.13	2.11
23	4.28	3.42	3.03	2.80	2.64	2.53	2.44	2.37	2.32	2.27	2.23	2.20	2.18	2.15	2.13	2.11	2.09
24	4.26	3.40	3.01	2.78	2.62	2.51	2.42	2.36	2.30	2.25	2.21	2.18	2.15	2.13	2.11	2.09	2.07
25	4.24	3.39	2.99	2.76	2.60	2.49	2.40	2.34	2.28	2.24	2.20	2.16	2.14	2.11	2.09	2.07	2.05
26	4.23	3.37	2.98	2.74	2.59	2.47	2.39	2.32	2.27	2.22	2.18	2.15	2.12	2.09	2.07	2.05	2.03
27	4.21	3.35	2.96	2.73	2.57	2.46	2.37	2.31	2.25	2.20	2.17	2.13	2.10	2.08	2.06	2.04	2.02
28	4.20	3.34	2.95	2.71	2.56	2.45	2.36	2.29	2.24	2.19	2.15	2.12	2.09	2.06	2.04	2.02	2.00
29	4.18	3.33	2.93	2.70	2.55	2.43	2.35	2.28	2.22	2.18	2.14	2.10	2.08	2.05	2.03	2.01	1.99

Table VI

Values of the variable F for $P = F(F) = \int_0^F f(F; \nu_1, \nu_2)\mathrm{d}F = 0.99$

ν_2 \ ν_1	1	2	3	4	5	6	7	8	9	10	11	12	13	14	15	16	17
1	4052	4999	5403	5625	5764	5859	5928	5981	6022	6056	6002	6106	6130	6142	6157	6169	6180
2	98.5	99.0	99.2	99.2	99.3	99.3	99.4	99.4	99.4	99.4	99.4	99.4	99.4	99.4	99.4	99.4	99.4
3	34.1	30.8	29.5	28.7	28.2	27.9	27.7	27.5	27.3	27.1	27.1	27.1	27.0	26.9	26.9	26.8	26.8
4	21.2	18.0	16.7	16.0	15.5	15.2	15.0	14.8	14.7	14.5	14.4	14.4	14.3	14.2	14.2	14.2	14.1
5	16.3	13.3	12.1	11.4	11.0	10.7	10.5	10.3	10.2	10.1	9.96	9.89	9.82	9.77	9.72	9.68	9.64
6	13.7	10.9	9.78	9.15	8.75	8.47	8.26	8.10	7.98	7.98	7.79	7.72	7.66	7.60	7.56	7.52	7.48
7	12.2	9.55	8.45	7.85	7.46	7.19	6.99	6.84	6.72	6.62	6.54	6.47	6.41	6.36	6.31	6.27	6.24
8	11.3	8.65	7.59	7.01	6.63	6.37	6.18	6.03	5.91	5.81	5.73	5.67	5.61	5.56	5.52	5.48	5.44
9	10.6	8.02	6.99	6.42	6.06	5.80	5.61	5.47	5.35	5.26	5.18	5.11	5.05	5.00	4.96	4.92	4.89
10	10.0	7.56	6.55	5.99	5.64	5.39	5.20	5.06	4.94	4.85	4.77	4.71	4.65	4.60	4.56	4.52	4.49
11	9.65	7.21	6.22	5.67	5.32	5.07	4.89	4.74	4.63	4.54	4.46	4.40	4.34	4.29	4.25	4.21	4.18
12	9.33	6.93	5.95	5.41	5.06	4.82	4.64	4.50	4.39	4.30	4.22	4.16	4.10	4.05	4.01	3.97	3.94
13	9.07	6.70	5.74	5.21	4.86	4.62	4.44	4.30	4.19	4.10	4.02	3.96	3.91	3.86	3.82	3.78	3.75
14	8.86	6.51	5.56	5.04	4.70	4.46	4.28	4.14	4.03	3.94	3.86	3.80	3.75	3.70	3.66	3.62	3.59
15	8.68	6.36	5.42	4.89	4.56	4.32	4.14	4.00	3.89	3.80	3.73	3.67	3.61	3.56	3.52	3.49	3.45
16	8.53	6.23	5.29	4.77	4.44	4.20	4.03	3.89	3.78	3.69	3.62	3.55	3.50	3.45	3.41	3.37	3.34
17	8.40	6.11	5.18	4.67	4.34	4.10	3.93	3.79	3.68	3.59	3.52	3.46	3.40	3.35	3.31	3.27	3.24
18	8.29	6.01	5.09	4.58	4.25	4.01	3.84	3.71	3.60	3.51	3.43	3.37	3.32	3.27	3.23	3.19	3.16
19	8.18	5.93	5.01	4.50	4.17	3.94	3.77	3.63	3.52	3.43	3.36	3.30	3.24	3.19	3.15	3.12	3.08
20	8.10	5.85	4.94	4.43	4.10	3.87	3.70	3.56	3.46	3.37	3.29	3.23	3.18	3.13	3.09	3.05	3.02
21	8.02	5.78	4.87	4.37	4.04	3.81	3.64	3.51	3.40	3.31	3.24	3.17	3.12	3.07	3.03	2.99	2.96
22	7.95	5.72	4.82	4.31	3.99	3.76	3.59	3.45	3.35	3.26	3.18	3.12	3.07	3.02	2.98	2.94	2.91
23	7.88	5.66	4.76	4.26	3.94	3.71	3.54	3.41	3.30	3.21	3.14	3.07	3.02	2.97	2.93	2.89	2.86
24	7.82	5.61	4.72	4.22	3.90	3.67	3.50	3.36	3.26	3.17	3.09	3.03	2.98	2.93	2.89	2.85	2.82
25	7.77	5.57	4.68	4.18	3.86	3.63	3.46	3.32	3.22	3.13	3.06	2.99	2.94	2.89	2.85	2.81	2.78
26	7.72	5.53	4.64	4.14	3.82	3.59	3.42	3.29	3.18	3.09	3.02	2.96	2.90	2.86	2.82	2.78	2.74
27	7.68	5.49	4.60	4.11	3.78	3.56	3.39	3.26	3.15	3.06	2.99	2.93	2.87	2.82	2.78	2.75	2.71
28	7.64	5.45	4.57	4.07	3.75	3.53	3.36	3.23	3.12	3.03	2.96	2.90	2.84	2.79	2.75	2.72	2.68
29	7.60	5.42	4.54	4.04	3.73	3.50	3.33	3.20	3.09	3.00	2.93	2.87	2.81	2.77	2.73	2.69	2.66
30	7.56	5.39	4.51	4.02	3.70	3.47	3.30	3.17	3.07	2.98	2.91	2.84	2.79	2.74	2.70	2.66	2.63

Table VII Values of the variable R

n \ P	0.95	0.99
3	123	31.4
4	7.17	16.27
5	5.08	9.00
6	4.34	6.85
7	3.98	5.88
8	3.77	5.33
9	3.63	4.98
10	3.54	4.75
11	3.48	4.58
12	3.42	4.45
14	3.36	4.28
16	3.32	4.17
18	3.30	4.08
20	3.28	4.02
25	3.26	3.94

Table VIII Values of the variable Q

n \ P	0.95	0.98	0.99	0.995
3	0.941	0.976	0.988	0.994
4	0.765	0.846	0.889	0.926
5	0.642	0.729	0.780	0.821
6	0.560	0.644	0.698	0.740
7	0.507	0.586	0.637	0.680
8	0.468	0.543	0.590	0.634
9	0.437	0.510	0.555	0.598
10	0.412	0.483	0.527	0.568
11	0.392	0.460	0.502	0.542
12	0.376	0.441	0.482	0.522
13	0.361	0.425	0.465	0.503
14	0.349	0.411	0.450	0.488
15	0.338	0.399	0.438	0.475
16	0.329	0.388	0.426	0.463
18	0.313	0.370	0.407	0.442
20	0.300	0.356	0.391	0.425
25	0.277	0.329	0.362	0.393
30	0.260	0.309	0.341	0.372

Table IX Values of the Kolmogorov function $K(\lambda) = \sum_{\nu=-\infty}^{+\infty} (-1)^{\nu} \exp(-2\nu^2\lambda^2)$

λ	K(λ)	λ	K(λ)	λ	K(λ)	λ	K(λ)
0.34	0.0002	0.71	0.3054	1.08	0.8061	1.45	0.9701
0.35	0.0003	0.72	0.3222	1.09	0.8143	1.46	0.9718
0.36	0.0005	0.73	0.3391	1.10	0.8222	1.47	0.9734
0.37	0.0008	0.74	0.3559	1.11	0.8299	1.48	0.9749
0.38	0.0012	0.75	0.3728	1.12	0.8373	1.49	0.9764
0.39	0.0019	0.76	0.3896	1.13	0.8445	1.50	0.9777
0.40	0.0028	0.77	0.4063	1.14	0.8513	1.51	0.9790
0.41	0.0039	0.78	0.4230	1.15	0.8580	1.52	0.9803
0.42	0.0054	0.79	0.4395	1.16	0.8644	1.53	0.9814
0.43	0.0073	0.80	0.4558	1.17	0.8706	1.54	0.9825
0.44	0.0097	0.81	0.4720	1.18	0.8765	1.55	0.9836
0.45	0.0125	0.82	0.4880	1.19	0.8822	1.56	0.9846
0.46	0.0160	0.83	0.5038	1.20	0.8877	1.57	0.9855
0.47	0.0200	0.84	0.5193	1.21	0.8930	1.58	0.9864
0.48	0.0246	0.85	0.5346	1.22	0.8981	1.59	0.9872
0.49	0.0300	0.86	0.5497	1.23	0.9029	1.60	0.9880
0.50	0.0360	0.87	0.5645	1.24	0.9076	1.61	0.9887
0.51	0.0428	0.88	0.5790	1.25	0.9121	1.62	0.9894
0.52	0.0503	0.89	0.5933	1.26	0.9164	1.63	0.9901
0.53	0.0585	0.90	0.6072	1.27	0.9205	1.64	0.9907
0.54	0.0674	0.91	0.6209	1.28	0.9245	1.65	0.9913
0.55	0.0771	0.92	0.6342	1.29	0.9282	1.66	0.9919
0.56	0.0875	0.93	0.6473	1.30	0.9319	1.67	0.9924
0.57	0.0986	0.94	0.6600	1.31	0.9353	1.68	0.9929
0.58	0.1103	0.95	0.6725	1.32	0.9386	1.69	0.9933
0.59	0.1227	0.96	0.6846	1.33	0.9418	1.70	0.9938
0.60	0.1357	0.97	0.6964	1.34	0.9448	1.71	0.9942
0.61	0.1492	0.98	0.7079	1.35	0.9477	1.72	0.9946
0.62	0.1632	0.99	0.7191	1.36	0.9505	1.73	0.9949
0.63	0.1777	1.00	0.7300	1.37	0.9531	1.74	0.9953
0.64	0.1926	1.01	0.7405	1.38	0.9556	1.75	0.9956
0.65	0.2079	1.02	0.7508	1.39	0.9580	1.76	0.9959
0.66	0.2236	1.03	0.7607	1.40	0.9603	1.77	0.9962
0.67	0.2395	1.04	0.7704	1.41	0.9624	1.78	0.9964
0.68	0.2557	1.05	0.7797	1.42	0.9645	1.79	0.9967
0.69	0.2721	1.06	0.7888	1.43	0.9665	1.80	0.9969
0.70	0.2887	1.07	0.7976	1.44	0.9683	1.81	0.9971

Table X Values of the Wilcoxon test: $100\,P(u \leqslant u_\alpha) = 100\alpha$

u	n_1 and n_2													
	2:5	2:6	2:7	2:8	2:9	2:10	3:3	3:4	3:5	3:6	3:7	3:8	3:9	3:10
0	4.76	3.57	2.78	2.22	1.82	1.52	5.00	2.86	1.79	1.19	0.83	0.61	0.45	0.35
1				4.44	3.64	3.03			3.57	2.38	1.67	1.21	0.91	0.70
2										4.76	3.33	2.42	1.82	1.40
3												4.24	3.18	2.45
4													5.00	3.85

u	n_1 and n_2													
	4:4	4:5	4:6	4:7	4:8	4:9	4:10	5:5	5:6	5:7	5:8	5:9	5:10	6:6
0	1.43	0.79	0.48	0.30	0.20	0.14	0.10	0.40	0.22	0.13	0.08	0.05	0.03	0.11
1	2.86	1.59	0.95	0.61	0.40	0.28	0.20	0.79	0.43	0.25	0.16	0.10	0.07	0.22
2		3.17	1.90	1.21	0.81	0.56	0.40	1.59	0.87	0.51	0.31	0.20	0.13	0.43
3			3.33	2.12	1.41	0.98	0.70	2.78	1.52	0.88	0.54	0.35	0.23	0.76
4				3.64	2.42	1.68	1.20	4.76	2.60	1.52	0.93	0.60	0.40	1.30
5					3.64	2.52	1.80		4.11	2.40	1.48	0.95	0.63	2.06
6						3.78	2.70			3.66	2.25	1.45	0.97	3.25
7							3.80				3.26	2.10	1.40	4.65
8											4.66	3.00	2.00	
9												4.15	2.76	
10													3.76	
11													4.96	

u	n_1 and n_2													
	6:7	6:8	6:9	6:10	7:7	7:8	7:9	7:10	8:8	8:9	8:10	9:9	9:10	10:10
0	0.06	0.03	0.02	0.01	0.03	0.02	0.01	0.01	0.01	0.00				
1	0.12	0.07	0.04	0.02	0.06	0.03	0.02	0.01	0.02	0.01	0.00	0.00		
2	0.23	0.13	0.08	0.05	0.12	0.06	0.03	0.02	0.03	0.02	0.01	0.01	0.00	
3	0.41	0.23	0.14	0.09	0.20	0.11	0.06	0.04	0.05	0.03	0.02	0.01	0.01	0.00
4	0.70	0.40	0.24	0.15	0.35	0.19	0.10	0.06	0.09	0.05	0.03	0.02	0.01	0.01
5	1.11	0.63	0.38	0.24	0.55	0.30	0.17	0.10	0.15	0.08	0.04	0.04	0.02	0.01
6	1.75	1.00	0.60	0.37	0.87	0.47	0.26	0.15	0.23	0.12	0.07	0.06	0.03	0.02
7	2.56	1.47	0.88	0.55	1.31	0.70	0.39	0.23	0.35	0.19	0.10	0.09	0.05	0.02
8	3.67	2.13	1.28	0.80	1.89	1.03	0.58	0.34	0.52	0.28	0.15	0.14	0.07	0.04
9		2.96	1.80	1.12	2.65	1.45	0.82	0.48	0.74	0.39	0.22	0.20	0.10	0.05
10		4.06	2.48	1.56	3.64	2.00	1.15	0.68	1.03	0.56	0.31	0.28	0.15	0.08
11			3.32	2.10	4.87	2.70	1.56	0.93	1.41	0.76	0.43	0.39	0.21	0.10
12			4.40	2.80		3.61	2.09	1.25	1.90	1.03	0.58	0.53	0.28	0.14
13				3.63		4.69	2.74	1.65	2.49	1.37	0.78	0.71	0.38	0.19
14				4.67			3.56	2.15	3.25	1.80	1.03	0.94	0.51	0.26
15							4.54	2.77	4.15	2.32	1.33	1.22	0.66	0.34
16								3.51		2.96	1.71	1.57	0.86	0.45
17								4.39		3.72	2.17	2.00	1.10	0.57
18										4.64	2.73	2.52	1.40	0.73
19											3.38	3.13	1.75	0.93
20											4.16	3.85	2.17	1.16

Table XI

Values of the variable X (van der Waerden)

n	$P=0.95$			$P=0.98$			$P=0.99$		
	n_1-n_2 0 and 1	n_1-n_2 2 and 3	n_1-n_2 4 and 5	n_1-n_2 0 and 1	n_1-n_2 2 and 3	n_1-n_2 4 and 5	n_1-n_2 0 and 1	n_1-n_2 2 and 3	n_1-n_2 4 and 5
6	∞	∞	∞	∞	∞	∞	∞	∞	∞
7	∞	∞	∞	∞	∞	∞	∞	∞	∞
8	2.40	2.30	∞	∞	∞	∞	∞	∞	∞
9	2.38	2.20	∞	2.80	∞	∞	∞	∞	∞
10	2.60	2.49	2.30	3.00	2.90	2.80	3.20	3.10	∞
11	2.72	2.58	2.40	3.20	3.00	2.90	3.40	3.40	∞
12	2.86	2.79	2.68	3.29	3.30	3.20	3.60	3.58	3.40
13	2.96	2.91	2.78	3.50	3.36	3.18	3.71	3.68	3.50
14	3.11	3.06	3.00	3.62	3.55	3.46	3.94	3.88	3.76
15	3.24	3.19	3.06	3.74	3.68	3.57	4.07	4.05	3.88
16	3.39	3.36	3.28	3.92	3.90	3.80	4.26	4.25	4.12
17	3.49	3.44	3.36	4.06	4.01	3.90	4.44	4.37	4.23
18	3.63	3.60	3.53	4.23	4.21	4.14	4.60	4.58	4.50
19	3.73	3.69	3.61	4.37	4.32	4.23	4.77	4.71	4.62
20	3.86	3.84	3.78	5.52	4.50	4.44	4.94	4.92	4.85
21	3.96	3.92	3.85	4.66	4.62	4.53	5.10	5.05	4.96
22	4.08	4.06	4.01	4.80	4.78	4.72	5.26	5.24	5.17
23	4.18	4.15	4.08	4.92	4.89	4.81	5.40	5.36	5.27
24	4.29	4.27	4.23	5.06	5.04	4.99	5.55	5.53	5.48
25	4.39	4.36	4.30	5.18	5.14	5.08	5.68	5.65	5.58
26	4.50	4.48	4.44	5.30	5.29	5.24	5.83	5.81	5.76
27	4.59	4.56	4.51	5.42	5.39	5.33	5.95	5.92	5.85
28	4.69	4.68	4.64	5.54	5.52	5.48	6.09	6.07	6.03
29	4.78	4.76	4.72	5.65	5.62	5.57	6.22	6.19	6.13
30	4.88	4.87	4.84	5.77	5.75	5.72	6.35	6.34	6.30

Table XII

Values of the variable R_1 for the significance levels $\alpha = 0.05$ and 0.01

n	Positive tail		Negative tail	
	$P = 0.025$	$P = 0.005$	$P = 0.975$	$P = 0.995$
5	0.253	0.297	–0.753	–0.798
6	0.345	0.447	–0.708	–0.863
7	0.370	0.510	–0.674	–0.799
8	0.371	0.531	–0.625	–0.764
9	0.366	0.533	–0.593	–0.737
10	0.360	0.525	–0.564	–0.705
11	0.353	0.515	–0.539	–0.679
12	0.348	0.505	–0.515	–0.655
13	0.341	0.495	–0.497	–0.634
14	0.335	0.485	–0.479	–0.615
15	0.328	0.475	–0.462	–0.597
20	0.299	0.432	–0.399	–0.524
25	0.276	0.398	–0.356	–0.473
30	0.257	0.370	–0.325	–0.433
(35)	0.242	0.347	–0.300	–0.401
(40)	0.229	0.329	–0.279	–0.376
45	0.218	0.314	–0.262	–0.356
(50)	0.208	0.301	–0.248	–0.339
(55)	0.199	0.289	–0.236	–0.324
(60)	0.191	0.278	–0.225	–0.310
(65)	0.184	0.268	–0.216	–0.298
(70)	0.178	0.259	–0.207	–0.287
75	0.173	0.250	–0.199	–0.276

General References

Acton, F. S., *Analysis of Straight-Line Data,* Wiley, New York (1959).
Bendat, J. S. and Piersol, A. G., *Random Data, Analysis and Measurement Procedures,* Wiley, New York (1971).
Bendat, J. S., *Principles and Application of Random Noise,* Wiley, New York (1958).
Bennett, C. A. and Franklin, N. L., *Statistical Analysis in Chemical and the Chemical Industry,* Wiley, New York (1954).
Beveridge, G. S. G., *Optimization Theory and Practice,* McGraw-Hill, New York (1970).
Birnbaum, Z. W., *Introduction to Probability and Mathematical Statistics,* Harper, New York (1962).
Brillouin, L., *La science et la théorie de l'information,* Masson, Paris (1959).
Brownlee, K., *Statistical Theory and Methodology in Science and Engineering,* Wiley, New York (1960).
Cochran, W. G. and Cox, C. M., *Experimental Design,* Wiley, New York (1957).
Cramér, H., *Mathematical Methods of Statistics,* Princeton Univ. Press (1946).
Cramér, H., *The Elements of Probability Theory and Some of its Applications,* Wiley, New York (1967).
Dixon, W. J. and Massey, F. J., Jr., *Introduction to Statistical Analysis,* McGraw-Hill, New York (1957).
Doerffel, K., *Statistik in der analytischen Chemie,* VEB Deutscher Verlag für Grundstoffindustrie, Leipzig (1966).
Draper, N. R. and Smith, H., *Applied Regression Analysis,* Wiley, New York (1966).
Dunin-Barkovskii, I. V. and Smirnov, I. V., *Teoria veroiatnostei i matematicheskaia statistika v tehnike,* Iz. Moscow (1955).
Eckschlager, K., *Fehler bei de chemischen Analyse,* Akad. Verlag, Leipzig (1965); *Errors and Measurement in Chemical Analysis,* Van Nostrand, London (1969).
Eckschlager, K. and Štěpánek, V., *Information Theory as Applied to*

Chemical Analysis, Wiley-Interscience, New York (1979).
Fraser, D. A. S., *Non-parametric Statistics,* Wiley, New York (1957).
Gibra, I. N., *Probability and Statistical Inference for Scientists,* Prentice-Hall, Englewood Cliffs, N.J. (1972).
Gottschalk, G., *Statistik in der quantitativen chemischen Analyse,* F. Enke Verlag, Stuttgart (1962).
Hald, A., *Statistical Theory with Engineering Applications,* Wiley, New York (1952).
Hancock, J. C. and Wintz, P. A., *Signal Detection Theory,* McGraw-Hill, New York (1966).
Helstrom, C. W., *Statistical Theory of Signal Detection,* Pergamon, London (1960).
Himmelblau, D. M., *Process Analysis by Statistical Methods,* Wiley, New York (1970).
Jurs, P. and Isenhour, T. C., *Chemical Applications of Pattern Recognition,* Wiley-Interscience, New York (1975).
Lange, F. H., *Correlation Techniques,* Van Nostrand, Princeton (1967).
Lark, P. D., Craven, B. R. and Bosworth, R. C., *The Handling of Chemical Data,* Pergamon, Oxford (1960).
Larson, H. J., *Introduction to the Theory of Statistics*, Wiley, New York (1973).
Mandel, J., *The Statistical Analysis of Experimental Data,* Wiley, New York (1964).
Massart, D. L., Dijkstra, A. and Kaufman, L., *Evaluation and Optimization of Laboratory Methods and Analytical Procedures,* Elsevier, Amsterdam (1978).
Nalimov, V. V., *The Application of Mathematical Statistics to Chemical Analysis,* Pergamon, London (1963).
Noether, G. E., *Elements of Non-parametric Statistics,* Wiley, New York (1957).
Wald, A., *Sequential Analysis,* Wiley, New York (1947).
Youden, W. J., *Statistical Methods for Chemists,* Wiley, New York (1962).

Index

A

Absorption spectroscopy methods, 298
acceptability threshold, 342
accuracy, 21, 351
activation analysis, 287
additive interference, 144
amplifier, frequency-sensitive, 410
  phase-selective, 410
  phase-sensitive, 410
analytical conditions, optimization, 365
  detection, 184
  determination, 185
  function, 150
  identification, 186
  measurement, 16
  measuring element, 18
  method, 16
    selection, 330
  performance, 329, 337
  procedure, 16
  regulation, 137
  reliability, 134, 333
  signal, 181
    resolution, 214
  stability, 115
    control, 139
    improvement, 138
    non-stationary state, 117, 122, 134
    response function, 142
    stationary state, 117, 134
anodic stripping voltammetry, 310
atomic absorption, 287, 295
  emission, 295, 296
  fluorescence, 287, 295
  spectroscopy, 300
    detection limits, 295, 296, 301, 302
    phase-sensitive detector, 302
    signal averaging, 302
    signal modulation, 409, 410
    tuned amplification, 302
Auger spectroscopy, 296
auto-correlation, 420
average – *see* mean

B

Background correction, 163, 175
  signal, 147
Bartlett's test, 51, 157, 272
Bayes criterion, 193
  test, 193
bias, 36
binomial distribution, 35, 40
  and decision formulation, 229
bit, 93
'blocked' determination, 146
Box and Wilson method, 371, 393
  applications, 397

C

Calculation function, multivariable, 175
  non-linear, 172
calibration, 141
  error, 151
  function, 150
    linear, 151, 161
    multivariable, 175
    non-linear, 172
  matrix, 146
central limit theorem, 31
channel, 89, 90
chelate gas chromatography, 287
chi-squared distribution, 31
  function, 54
  test, 54, 157, 274
chromatography, 310
  detection limits, 313
  detectors, 311
  dilution effect, 311
  informational capacity, 108

circular serial correlation, 130
coder, 89, 90
coefficient of quality, 156
  variation, 329
colour reactions in solution, 287
compositional experiment, 386
  design, general, 388
    three-factor, 386
    two-factor, 387
concentration, relation to signal, 141
confidence interval, 36
  of binomial distribution, 40
  of calibration function, 160
  of decision threshold, 276
  of detection limit, 269, 276
  of least squares, 79, 153
  of linear regression, 78
  of mean, 37
  of regression line, 79
  of retention time, 314
  of variance, 38
control chart, 140
  of stability of analytical system, 139
convolution, 416
correlation, 83, 420
  applications, 422
  coefficient, serial, 129
    table, 437
  serial, circular, 130
  set, method of pattern recognition, 251
correlators, 205, 421
coulometry, 310
criterion of ideal observer, 197
  of least-squares mean error, 205
  of maximal likelihood, 197

D

Decision formulation, 220, 227
  rule, 26
  surface, 195
  threshold, 189, 190, 221, 263, 276
  informational definition, 201
decoder, 89, 90
deconvolution, 419
derivative signals, 414
  spectroscopy, 414
detection, analytical, 184
  capacity, *see* detection limit
  characteristic, 208, 263
  decision, 23, 192, 220
  in chromatography, 314
  in presence of noise, 187
  limit, 255, 263
    and determination limit, 348, 351
    confidence limits, 269, 276
detection, analytical – *cont.*
  limit – *cont.*
    estimation, 266, 268
    frequentometric method, 281
    from standard deviation, 278, 280
    in atomic absorption spectroscopy, 295, 296, 301, 302
    in chromatography, 313
    in electron spin resonance, 304
    in flame spectrometry, 296
    in mass spectroscopy, 302
    in neutron activation, 305
    in nuclear magnetic resonance, 304
    in photon activation, 305
    in radiometric methods, 304
    informational definition, 261
    signal-to-noise rato, 258, 265
  non-parametric, 205
  sequential, 202
  threshold, 189
  t-test, 223
  two-step model, 190
detector, coulometric, 312
  flame ionization, 312
  flame photometric, 312
  microcoulometric, 312
  microwave plasma, 312
  phase-sensitive, 302
  refractive index, 312
  thermal conductivity, 311
  ultraviolet absorption, 312
deterioration of analytical systems, 137
determination, analytical, 185
  capacity, *see* determination limit
  characteristic, 337, 346, 351
  limit, 341, 351
    and detection limit, 348, 351
    and entropy, 344
    and performance parameter, 344
    and signal-to-noise ratio, 343
    improvement, 350
differentiation of signals, 419
dilution effect in chromatography, 311
discriminant function method of pattern recognition, 251
dispersion band, 155
distribution of random variables, 27
  binomial, 35
    confidence interval, 40
  chi-squared, 31
    table, 429
  F, 34
    table 431, 432
  function, test of type, 54
  Gauss, 29
    confidence interval, 37
  log-normal, 31

distribution of random variables – *cont.*
nomenclature (use of ⇒), 77
non-normal, 31
normal, 29
confidence interval, 37
Student t, 33
t, 33
table 430
dit, 346
Dixon's Q criterion, 44
table, 433

E

Ebel and Ebel equation, 297
electrical noise, 403
electrochemical methods, 309
electron microprobe analysis, 295
electron spin resonance, 304
emission spectroscopy, 287, 288
ensemble averaging, 412
entropy, 92
and determination limit, 261
and information, 96
characteristic of determination, 339
of errors, 100
residual, 102
environmental noise, 403
errors, 19
absolute, 20, 327
classification, 21
control chart, 140
entropy value, 100
first kind, 20, 189, 190, 202
gross, 20, 328
limits, and reliability of a method, 333
random, 20, 328
distribution, 339
relative, 20, 327
second kind, 20, 189, 190, 202
systematic, 20, 328, 330
tests for, 70, 331
total, 332
estimation, 36
calibration function parameters, 159
variance, 159
detection limit, 266, 268
evolutionary operation method, 371
expectance, 36
experimental design, Box and Wilson, 393
applications, 397
compositional, 386
factorial, 374
rotatable, 391
extreme values, testing, 43, 44

F

Factor analysis method of pattern recognition, 251
factorial experiments, 374
orthogonal, 374
three-level, 385
two-level, complete, 377
partial, 383
F-distribution, 34
table, 341, 342
file search method of pattern recognition, 233
filtration of signal, 408
Fisher transformation, 85
flame photometry, 292
flicker noise, 403
frequency-selective amplifier, 410
frequentometric method for detection limit, 281
fluorescence reactions, 287
Fourier transform methods, 416
F-test, 50, 272
of variances, 50

G

Gas chromatography, 311
and mass spectrometry, 312
Gauss distribution, 29
Glivenko's theorem, 22
gravimetric analysis, 287
Grubb's R criterion, 43
table, 433

H

Hartley, 346
hypothesis testing, 41
Kolmogorov test, 56
table, 434
of systematic error, 331
of type of distribution, 54
order statistics, 65
sequential tests, 68
Smirnov test, 66
statistical, and linear response function, 156
tests for stability, 124
Van der Waerden test, 67
table, 436
variance analysis, 56
Wilcoxon test, 66
table, 435

I

Ideal observer criterion, 197
identification, 186

information, 95
  amount of, 96, 98
  and entropy, 96
  transmission, 89, 90
informational capacity, 108
  definition of detection limit, 261
  flow, 110
  power of instruments, 106
    of system, 111
  processing, 89
inhomogeneity, evaluation, 356
  of materials, 24, 353
instruments, informational power, 106
interaction of factors, 61
interference, 144
  additive, 144
  limit, 146
  multiplicative, 145
  ratio, 146
internal standards in chromatography, 316
inverse voltammetry, 287
ion-selective electrodes, 310
ionic fluorescence, 295, 296
isotopic dilution, 307, 308

J

Johnson noise, 403

K

Kaiser's definition of detection limit, 256
kinetic measurements, 287
K nearest neighbour method of pattern recognition, 247
Kolmogorov's test, 56, 158
  table 434

L

Lambert–Beer law, 143
Laplace function, 31
  table 427, 428
largest-vote method of pattern recognition, 249
laser spectroscopy, 292
learning machine method of pattern recognition, 240
least-square mean error, 205
least-squares method of pattern recognition, 248
likelihood criteria, 192
  ratio, 202
linear regression, 76
log-normal distribution, 31
luminescence methods, 292

M

Mass spectroscopy, 287
  and gas chromatography, 312
  detection limits, 302
  secondary ion, 303
  signal averaging, 412
matrix effects, 175
maximal likelihood criterion, 197
mean, 22, 29
  confidence interval, 37
  testing, 45
    of two means, 46
minimax criterion, 196
minimum risk criterion, 196
modulation of signals, 408
moments, statistical, 29
multiple hypotheses, 206
  regression, 83
  standard addition, 162
multiplicative interference, 145

N

Nearest neighbour method of pattern recognition, 247
Nernst equation, 143
neutron activation analysis, 305
Neyman–Pearson theory, 41
  criterion, 198, 217, 221
noise 19, 402
  and signal detection, 187
  electrical, 402
  environmental, 403
  figure, 407
  flicker, 403
  Johnson, 403
  quantization, 404
  resistance, 403
  shot, 403
  white, 403
nomenclature (use of ⇒ for distribution), 77
non-homogeneity, 407, *see* also inhomogeneity
non-homogeneous materials, 24, 353
non-linear regression, 82
non-normal distribution, 31
non-parametric detection, 204
  tests, 226
normal distribution, 29
  and decision formulation, 227
nuclear magnetic resonance, 303
  detection capacity (limit), 304

O

Objective function, 366

optimization, 365
order statistics, 65, 66, 67

P

Pattern recognition methods, 231
  correlation set, 251
  discriminant function, 251
  factor analysis, 251
  file search, 233
  largest vote, 249
  learning machine, 240
  least squares, 248
  nearest neighbour, 247
  spectrum sum, 249
performance, 329, 337
  and determining limit, 341, 344, 351
  improvement, 365
phase-selective amplifier, 410
phase-sensitive amplifier, 410
  detector, 302
phosphorimeter, 107
photon activation analysis, 305
polarography, 309
  differential pulse, 310
  pulse, 310
potentiometry, ion-selective electrode, 310
precision, 21, 328
probability density, 28, 91, 92
purity, criterion for, 359
purpose function, 366

Q

Quality coefficient, 156
quantization noise, 404
  of measurement, 404
Q-criterion, 44
  table, 433

R

Radioactive tracers, 307
radioisotope techniques, 286, 307
radiometric methods, limit of detection, 304
  limit of determination, 341
randomness tests, 124
R-criterion, 43
  table, 433
receiver, 89, 90
redundancy, 104, 110
regression analysis, 76
  and stability of systems, 124
regression anlysis – *cont.*
  intersection of two lines, 166
  line, confidence interval, 79, 153
  linear, 76
    and statistical hypotheses, 156
    confidence interval, 78
  non-linear, 82
regulation of analytical systems, 137
relative standard deviation, 329
reliability of analytical systems, 134
  of analytical methods, 333
resolution, 147, 214
  parameter, 219
response function, 25, 142, 366
  confidence intervals, 153
  interference, 144
    additive, 144
    multiplicative, 145
response surface method, 371, 372, 393
retention time, 313
  confidence interval, 314
  standard deviation, 314
rotatable experiment, 391

S

Sample, inhomogeneous, 23, 353
sampling, 16, 353
  constant, 355
  non-homogeneous material, 353
  plan, 24, 353
  probability, tests, 51, 52
Scheibe-Lamakin equation, 288
secondary-ion mass spectroscopy, 303
segregation, 355
sensitivity, 149
separation, degree of, 147
sequential probability ratio test, 68, 204, 227
  binomial distribution, 74
  normal distribution, 69
serial correlation coefficient, 129
shot noise, 403
signal amplitude, as function of time, 182
  averaging, 302, 410
  correlation, 420
    applications, 422
  deconvolution, 419
  derivative, 414, 419
  detection, 23, 181
  filtration, 408
  Fourier transform, 416, 419
  modulation, 408
  processing, 410
  smoothing, 412, 419

signal-to-noise ratio, 190, 365
 and detection limit, 258, 265
 and determination limit, 343
 and precision, 330
 improvement, 402
significance hypotheses, testing, 42
simplex method, 371, 397
 applications, 402
 uniplex method, 402
six-sigma rule, 257, 265
Smirnov's test, 66
smoothing, 412, 419
spectral methods, 106
spectrograph, determination limit, 341
 informational capacity, 108
spectrography, 288
spectrometers, 107
 infrared, 127
spectroscopy, derivative, 414
spectrum sum method, 249
stability of analytical systems, 115, 117
 control, 139
 improvement, 138
 non-stationary, 117
  time-dependent systematic deviation, 122, 134
   variance, 122
 stationary, 117, 118, 134
  constant systematic deviation, 120
  covariant, 119
  intermittent gross error, 121
  random, 120
standard addition, multiple, 162
standard deviation, 22, *also see* variance
 relative, 329
steepest ascent method, 371, 393
Student's *t* distribution, 33
 table, 430
 testing extreme values, 43
substoichiometric method, 309

T

t-distribution, 33
 table, 430
testing hypotheses, 41
 means, 45, 46
 sampling probability, 51
 type of distribution, 54
 variances, 49
time correlation of errors, 328, 329
trace analysis methods, 286
 determination, 327
tracer methods, 307
t-test, and detection, 223
 of means, 45, 47, 48
tuned amplification, 302

U

Uniplex method, 402

V

Van der Waerden's test, 67
 table, 436
variance, 29
 analysis, 56, 131
  and stability, 131
  single-factor, 57, 131
  two-factor, 61
 confidence interval, 38
 testing, 49
  two variances, 50, 51
voltammetry, anodic stripping, 310
volumetric analysis, 287

W

White noise, 403
Wilcoxon's test, 66
 table, 435

X

X-ray fluorescence, 175, 287, 293
 photoelectron spectroscopy, 297

Z

z-test, and means, 45, 46
z-values, table 427, 428